Instructor's Guide

for

The Cosmic Perspective

Fourth Edition

Jeffrey Bennett
University of Colorado at Boulder

Megan Donahue
Michigan State University

Nicholas Schneider
University of Colorado at Boulder

Mark Voit
Michigan State University

San Francisco Boston New York
Capetown Hong Kong London Madrid Mexico City
Montreal Munich Paris Singapore Sydney Tokyo Toronto

ISBN 0-8053-9284-X

3 4 5 6 7 8 9 10—OPM— 08 07
www.aw-bc.com

About the Instructor's Guide

This Instructor's Guide contains resources designed for use with the textbook *The Cosmic Perspective, Fourth Edition*, in either its complete form or in the two available split forms (*The Solar System: The Cosmic Perspective, Volume 1* or *Stars, Galaxies, and Cosmology: The Cosmic Perspective, Volume 2*).

The introductory sections contain material designed to help you prepare for your course. Thebulk of the guide then is organized chapter-by-chapter, with teaching hints and answers toquestions and problems from the text.

This *Instructor's Guide* is also available in the Instructor's Resource area on the text website, www.masteringastronomy.com.

If you have additional questions or comments, you may contact the authors directly by e-mailing the lead author, Jeffrey Bennett—jeffrey.bennett@comcast.net.

Table of Contents

Which Version of *The Cosmic Perspective* Fits Your Course?

The first step in choosing a textbook is making sure that it is suitable for your course. The following information should help you make sure that you have chosen the right version of *The Cosmic Perspective* for your course.

Courses Suited to *The Cosmic Perspective*

The Cosmic Perspective is designed for use in introductory astronomy courses aimed primarily at nonscience majors, with essentially no prerequisites. In particular, this text should fit your course if it is any of the following:

- A college course in general astronomy.
- A college course focusing on the solar system.
- A college course focusing on stars, galaxies, and/or cosmology.
- A high school course in astronomy.

It is not designed for courses intended for physics majors or for upper level college courses, although some professors have had success using it in such courses (often by supplementing it with additional mathematics or physics material).

Alternate Versions of *The Cosmic Perspective*

The Cosmic Perspective is available in four versions tailored to particular course types. From the list below, you may choose the version best suited to the length, depth, and coverage of your course:

- *The Cosmic Perspective, Fourth Edition* (full version), ISBN 0-8053-9283-1. This is the most complete version of the book, containing sufficient material for astronomy courses ranging in length from one quarter to a full year.
- *The Solar System: The Cosmic Perspective, Fourth Edition*, ISBN 0-8053-9296-3. This version contains Chapters 1–14, S1 and 24 of the full version of the book. It is designed for courses that focus on the solar system rather than on stars, galaxies, and cosmology.
- *Stars, Galaxies, and Cosmology: The Cosmic Perspective, Fourth Edition,* ISBN 0-8053-9210-3. This version contains Chapters 1–6, 14–24, and S2–S4 of the full version of the book. It is designed for courses that focus on stars, galaxies, and cosmology.
- *The Essential Cosmic Perspective, Third Edition,* Media Update ISBN 0-8053-8956-3. This condensed and simplified version of the full book is designed primarily for courses that provide a survey of all areas of astronomy in just ones semester. *Note:* The fourth edition of *The Essential Cosmic Perspective* will be available in December 2006.

NOTE: Any of these versions may be ordered bundled with the Addison Wesley Astronomy Tutor Center (see below for information). There is no extra cost to students for the Tutor Center, but it will be included only if you request the special bundle. Contact your local Addison Wesley sales representative if you would like to do this.

Resources and Supplements for *The Cosmic Perspective*

The Cosmic Perspective is much more than just a textbook. It is a complete package of resources designed to help you and your students. In this section, we briefly describe the available resources.

The MasteringAstronomy™ Website (www.masteringastronomy.com)—the Online Resource for You and Your Students

As teachers, we've all had frustrating experiences with websites that claim to be educational but have steep learning curves, or with text websites that claim to be designed as study aids but don't really follow the text pedagogy or have much that is useful for students. As authors, we therefore decided to take matters into our hands with the MasteringAstronomy™ website—the website for *The Cosmic Perspective*—by working closely with other educators, programmers, and our publishing team in every aspect of the site's development. If you haven't already tried it out, we think you'll be very pleasantly surprised. Indeed, in its former incarnation as The Astronomy Place, our content has already proven to be the key study resource for tens of thousands of students using our text—and many of these students have made use of it even when their instructors had not made the site a requirement. For instructors, MasteringAstronomy offers a wealth of course management tools and other resources, as well as the peace of mind that comes from knowing that your students have an outstanding set of study aids available to them.

MasteringAstronomy is the most advanced astronomy tutorial, assessment, and self-study system ever built. It provides the first library of activities and problems pre-tested by students nationally. Sophisticated analysis of student performance (including difficulty, time spent, and most common errors) has allowed every item to be systematically refined for educational effectiveness and assessment accuracy. MasteringAstronomy allows you **to tutor and assess your class** like never before with the following instructor resources:

- **Deliver powerful pre- and post-lecture diagnostic tests.** Assign cutting-edge diagnostic tests before and after your class to gauge your students' understanding. See class stats at a glance, compare results with the national average, identify the most difficult problem for your students, even the most difficult step, and their most-common wrong answers, or zero in on time spent by each student.
- **Provide effective homework and tutoring assignment with automatic grading.** Use pre-built, research-based weekly assignments or quickly build your own choosing from an unprecedented variety of tutorial and testing activities, including ranking and sorting tasks, dynamic visualization activities (using

Interactive Figures and Interactive Photos from the text), vocabulary-in-context, Interactive Tutorials, and end-of-chapter problems (including true/false, multiple choice, and numerical with randomized values).

- **Generate ideal exams and tests.** Take the guesswork out of building mid-terms and finals using unique tools that instantly identify assessment questions of ideal difficulty, duration, and concept coverage for your course.
Whether you assign activities or not, your students also have access to the most widely used and highly rated **self-study** media available, refined over five years of use by more than 100,000 students. The following are among the many resources your students will find on the MasteringAstronomy website:
- **Interactive, educational tutorials:** Our web tutorials focus on topics that traditionally give students the most difficulty and are designed to mimic the type of interaction that typically would occur during office hours. We now have a total of 22 full-length tutorials to date, which together include more than 60 individual tutorial lessons (each focused on a key concept), more than 100 interactive tools, more than 125 animations, and some 500 tutorial questions. Working a full tutorial will typically take students between 15 to 45 minutes, depending on how well they understand the material to begin with. The text includes icons in section headers to point to relevant tutorials, and suggested tutorial activities can also be found in the Media Explorations section at the end of each chapter. The grid in Appendix 3 of this Instructor's Guide summarizes the places where the tutorials correspond to text material. Note: The authors have been heavily involved in tutorial development to ensure that the tutorial pedagogy matches the text pedagogy and emphasizes key text ideas.
- **On-line, multiple-choice chapter quizzes:** The MasteringAstronomy Web site has two quizzes for each chapter in the book. The first quiz focuses on basic definitions and ideas, and the second asks more conceptual questions. The quizzes are designed not only to test but also to teach: feedback is provided for both correct and incorrect answers, helping students consolidate their understanding (feedback with correct answers) and helping students understand where they went wrong (for incorrect answers). Quizzes vary slightly in length, but most have between 15–25 multiple-choice questions. Students can use them for self-study, or you can use the course management resources to have scores sent directly to you as students complete the quizzes. Note: Every one of the roughly 1,000 quiz questions was written by the authors of the book, ensuring that they focus on the truly important points and perfectly match the book pedagogy.
- **Animated Movies:** Ten 5-minute animated and narrated mini-movies that provide engaging summaries of key topics. The grid in Appendix 3 of this Instructor's Guide summarizes the places where the movies correspond to text material.
- **And much more:** Keyed to each chapter in the book, MasteringAstronomy includes student study resources such as chapter summaries, additional information on Common Misconceptions, and chapter-specific links (including links for the end-of-chapter Web Projects). In addition, all Think About It questions and all end-of-chapter Problems are digitized for online assignments, and the entire textbook is available on-line in XML and can be browsed by topic.

All new copies of *The Cosmic Perspective* are shipped with a free personal access kit for MasteringAstronomy. Access can also be purchased on-line with a credit card (e.g., for students who purchase used books).

Additional Student Supplements for *The Cosmic Perspective*

In addition to the MasteringAstronomy website, several other very useful student supplements areavailable; note that the first one below is shipped free with every new copy of *The Cosmic Perspective.*

- ***Voyager SkyGazer,* College Edition (CD included free with all new books):** Based on VoyagerIII, one of the world's most popular planetarium programs, SkyGazer makes it easy for students to learn constellations and explore the wonders of the sky through interactive exercises. Note: This software retails for $50 as a stand-alone package, so it's a great deal that it comes free with the book! Appendix 1 offers general suggestions on how to integrate SkyGazer with your course, and the grid in Appendix 3 summarizes the places where the SkyGazer activities correspond to text material.
- **The Addison Wesley Astronomy Tutor Center:** This center provides one-on-one tutoring by qualified college instructors in any of four ways—phone, fax, email, and the Internet—during evening and weekend hours. Tutor Center instructors will answer questions and provide help with examples and exercises from the text. Students who register to use the Astronomy Tutor Center can receive as much tutoring as they wish for the duration of their course. Registration is free to students if you order a package that comes with the Tutor Center. Alternatively, students can purchase access to the tutoring separately. See www.aw.com/tutorcenter or contact your local Addison Wesley sales representative for more information about bundling the Tutor Center with your book order.
  - **Astronomy In-Class Active Learning Tutorials (ISBN 0-8053-8296-8):** This new workbook provides fifty, 20-minute in-class tutorial activities to choose from. Designed for use in large lecture classes, these activities are also suitable for labs. Each activity targets specific learning objectives, such as understanding Newton's laws, Mars' retrograde motion, tracking stars on the H-R diagram, or comparing the properties of planets.
  - **Themes of the Times on Astronomy (ISBN 0-8053-9028-6):** A collection of fifty astronomy articles from *The New York Times*, Themes of the Times on Astronomy, brings the excitement of astronomical discovery to you and your students. Each article correlates to a chapter in *The Cosmic Perspective* and includes a series of follow-up questions for homework or class discussion.

Instructor Supplements for *The Cosmic Perspective*

The following supplements are available only to instructors:

- **Instructor Resources on MasteringAstronomy:** MasteringAstronomy provides an instructor's area with the most advanced, educationally effective, and easy-to-use assessment system (cutting-edge diagnostic tests, weekly homework and tutoring assignments, and exams and tests). It also includes an online version of this Instructor's Guide and numerous other useful resources.

- **Cosmic Lecture Launcher CDv3.0 (ISBN 0-8053-9297-1):** This CD provides the largest library available of purpose-built in-class presentation materials to help you prepare your course lectures, including: a set of prepared PowerPoint® slides and CRS (clicker) questions for every chapter in the textbook, which you can use as-is or customize; a comprehensive collection of high-resolution figures from the book and other astronomical sources that you can integrate into your lectures; and a library of more than 250 interactive applets and simulations. This CD is available free to qualified instructors; ask your local Addison Wesley sales representative if you have not yet received it.
- **Carl Sagan's *Cosmos* on DVD or Video:** The Best of Cosmos and the complete, revised, enhanced, and updated Cosmos series are available free to qualified adopters of *The Cosmic Perspective*. (Ask your local Addison Wesley sales representative about obtaining the series if your school has not already received it.) See Appendix 2 for suggestions on how to integrate the Cosmos series with the textbook.
- **Computerized Test Bank (ISBN 0-8053-9286-6) and Printed Test Bank (ISBN 0-8053-9287-4):** The test bank contains a broad set of multiple-choice, true/false, and free-response questions for each chapter. The CD version of the test bank allows you to edit questions, export questions to tests, and print the questions in a variety of formats.
- **Transparency Acetates (ISBN 0-8053-9289-0):** For those of you who use overhead projectors in your lectures, this set contains nearly 200 images from the text. (Note: if you use a computer in your lectures, these are unnecessary, as all these images and many more are on the Cosmic Lecture Launcher CD, described above.)
- **Instructor's Guide: (ISBN 0-8053-9284-X):** This is the guide you are now reading, available in print or in the instructor's area of the MasteringAstronomy Web site.

Sample Course Outlines

The Cosmic Perspective, Fourth Edition is a comprehensive introduction to astronomy with enough material for two full semesters of course work. If you happen to have a two-semester sequence, you can cover the entire book (a total of 28 chapters, including the 4 supplementary chapters) at an average pace of one chapter per week. In all other cases, you will need to be selective in your coverage, depending on the length and emphasis of your course. To help you plan your course, the next several pages offer the following nine sample course outlines:

- Sample Outline 1: one-semester (14-week) course focusing on the solar system. (No prerequisite.)
- Sample Outline 2: one-semester (14-week) course focusing on the solar system but with more extensive coverage of the sky than Sample Outline 1. (No prerequisite.)
- Sample Outline 3: one-semester (14-week) course focusing on stars, galaxies, and cosmology. (No prerequisite.)
- Sample Outline 4: one-semester (14-week) course focusing on stars, galaxies, and cosmology with relativity and quantum ideas. (Assumes prerequisite covering earlier chapters in book or an "honors" level course.)
- Sample Outline 5: one-semester (14-week) course covering "everything"—i.e., solar system, stars, galaxies, and cosmology. (No prerequisite.)
- Sample Outline 6: one-semester (14-week) course covering "everything" with the solar system last. (No prerequisite.)
- Sample Outline 7: one-quarter (10-week) course focusing on the solar system. (No prerequisite.)
- Sample Outline 8: one-quarter (10-week) course focusing on stars, galaxies, and cosmology. (No prerequisite.)
- Sample Outline 9: one-quarter (10-week) course focusing on stars, galaxies, and cosmology with relativity and quantum ideas. (Assumes prerequisite covering earlier chapters in book or an "honors" level course.)

Before we get into the specific outlines, however, we address two common questions that may come up when you are deciding how to organize your course.

What if I Want to Teach Stars and Galaxies Before the Solar System?

The astronomical teaching community is somewhat split on the question of whether stars and galaxies should be taught before or after teaching about the solar system. Our opinion is that it does not matter very much which order you choose, as long as you follow good pedagogical practices. As a result, we have designed *The Cosmic Perspective* so that the sections on the solar system (Part III of the book) are essentially independent of the sections on stars, galaxies, and cosmology (Parts V and VI).

You therefore will have no difficulty if you choose to have your students cover the solar system last, and the only impact will be that you will need to ask your students to read the chapters in a different order than they appear in the book. We have included one sample outline (#6) for this ordering.

To Math or Not to Math?

One of the challenges of any astronomy course is deciding how much mathematics to include. We have therefore designed *The Cosmic Perspective* to accommodate a wide range of levels of mathematics. In particular, nearly all the mathematics in our textbook is found in the Mathematical Insight boxes, so you can easily tailor your course to the appropriate mathematical level for your students. Thus, if you want a nonmathematical course, simply skip all the Mathematical Insight boxes. At the other extreme, if you want a course with a substantial amount of algebra-based mathematics, cover all the Mathematical Insight boxes. You can create a course with any intermediate level of mathematics by covering only selected Mathematical Insight boxes. Note that basic mathematical skills (such as scientific notation, unit analysis, and ratios) are covered in Appendix C of the textbook.

When it comes to assignments, you can also tailor your homework to the mathematical level of your course. Problems that require mathematical manipulation are marked by an asterisk (*) in the end-of-chapter problem sets. The Mastering Astronomy Web site contains many additional resources for integrating mathematics into your course.

Sample Outline 1: One Semester, Solar System Emphasis

General Notes:

- This outline assumes that your course has no prerequisite. It is based on assuming approximately 3 contact hours per week over a 14-week semester.
- This outline maximizes coverage of the solar system chapters by minimizing coverage of the sky and sky phenomena. If you wish to spend more time on the sky, see Sample Outline 2.

Week	Reading	Suggested Coverage
1	Chapter 1	Begin your course with the overview of modern astronomy in Chapter 1.
2	Chapters 2 and 3	Cover Chapter 2 briefly, emphasizing "basics" such as seasons and phases of the Moon. In Chapter 3, focus on the Copernican revolution and the hallmarks of science.
3	Chapter 4	Coverage of Newton's laws, conservation laws, and gravity.
4	Chapters 5 and 6	Cover light and matter in some depth in Chapter 5; limited coverage of telescopes in Chapter 6.
5	Chapters 7 and 8	You should be able to cover Chapter 7 in one or two hours of class, then begin coverage of Chapter 8.
6	Chapter 9	Finish covering Chapter 8. Then begin full coverage of Chapter 9; you probably will get less than halfway through Chapter 9 in this week.
7	—	Finish covering Chapter 9.
8	Chapter 10	Begin full coverage of Chapter 10.
9	—	Finish covering Chapter 10.
10	Chapter 11	Begin full coverage of Chapter 11.
11	—	Finish covering Chapter 11.
12	Chapter 12	Full coverage of Chapter 12.
13	Chapter 13	Begin full coverage of Chapter 13.
14	Chapter 24	Finish covering Chapter 13. Optional: Cover life in the universe with Chapter 24.

Sample Outline 2: One Semester, Solar System/Sky Emphasis

General Notes:

- This outline is based on assuming approximately 3 contact hours per week over a 14-week semester.
- This outline includes more coverage of the sky and sky phenomena than Sample Outline 1. Inorder to incorporate this coverage, you will need to compress the coverage of other areas, as described in the "suggested coverage" column. This outline works best if you have access to a planetarium for teaching the sky phenomena.
- This outline assumes that your course has no prerequisite. If your course has a prerequisite that covered concepts in early chapters of the book, you can reduce the time spent on these chapters and increase the time spent on other chapters.
- This outline is also ideally suited to "honors" courses for more advanced students who can move more quickly through the course material.

Week	Reading	Suggested Coverage
1	Chapter 1	Begin your course with the overview of modern astronomy in Chapter 1. You may also wish to begin teaching students the major constellations in your local sky.
2	Chapter 2	Cover Chapter 2 in depth.
3	Chapter 3	Full coverage of this chapter.
4	Chapter S1	Full coverage of this chapter, ideally with the help of one or more planetarium visits.
5	Chapter 4	Coverage of Newton's laws, conservation laws, and gravity.
6	Chapters 5 and 6	Cover light and matter in some depth in Chapter 5; limited coverage of telescopes in Chapter 6.
7	Chapters 7 and 8	You should be able to cover Chapter 7 in one or two hours of class, then begin coverage of Chapter 8.
8	Chapter 9	Complete Chapter 8, and begin on Chapter 9, which will probably require a week and a half.
9	Chapter 10	Complete Chapter 9 and begin on Chapter 10.
10	—	Complete coverage of Chapter 10. Students will now have a solid understanding of planetary geology and atmospheres.
11	Chapter 11	You should be able to cover Chapter 11 in one week, perhaps with some selective emphasis.
12	Chapter 12	You should be able to cover Chapter 12 in one week, again perhaps with some selective emphasis.
13	Chapter 13	You should be able to cover Chapter 13 in one week.
14	Chapter 24	Wrap up with discussion of life in the universe in Chapter 24.

Sample Outline 3: One Semester, Stars/Galaxies/Cosmology Emphasis

General Notes:

- This outline is based on assuming approximately 3 contact hours per week over a 14-week semester.
- This outline assumes that your course has no prerequisite. We recommend Sample Outline 4 if your course has a prerequisite that covered concepts in early chapters of the book.
- This outline does not cover Part 4 on relativity and quantum ideas; see Sample Outline 4 if you wish to cover these chapters.

Week	Reading	Suggested Coverage
1	Chapter 1	Begin your course with the overview of modern astronomy in Chapter 1.
2	Chapters 3 and 4	In Chapter 3, emphasize ideas of modeling, the Copernican revolution, and the hallmarks of science. Begin Chapter 4 coverage of Newton's laws, conservation laws, and gravity.
3	Chapter 5	Complete coverage of Chapter 4, then continue on to light and matter in Chapter 5.
4	Chapter 6, 14	Spend perhaps one class period on telescopes in Chapter 6, then move to the Sun in Chapter 14.
5	Chapter 15	Full coverage of this chapter.
6	Chapter 16	Full coverage of this chapter.
7	Chapter 17	Full coverage of this chapter.
8	Chapter 18	Full coverage of this chapter.
9	Chapter 19	Full coverage of this chapter.
10	Chapter 20	Full coverage of this chapter.
11	Chapter 21	Full coverage of this chapter.
12	Chapter 22	Full coverage of this chapter.
13	Chapter 23	Full coverage of this chapter.
14	Chapter 24	Use this week as needed to complete coverage described above. If you have time, cover Chapter 24 briefly.

Sample Outline 4: One Semester, Stars/Galaxies/Cosmology Emphasis with Relativity

General Notes:

- This outline is based on the assumption that Parts 1 and 2 of the text are essentially review for your students, as would be the case if either:
  - Your course has a prerequisite in which Parts 1 and 2 were already covered.
  - You have "honors" students with strong backgrounds in high school science.
- This outline is based on assuming approximately 3 contact hours per week over a 14-week semester.

Week	Reading	Suggested Coverage
1	Review: Chapters 1–6	Use this first week for a review as needed of the overview of the universe, the nature of science, and of key physical concepts.
2	Chapter S2	Introduce special relativity with Chapter S2.
3	Chapter S3	Finish covering Chapter S2, and begin covering Chapter S3.
4	—	Finish covering Chapter S3.
5	Chapter S4	Full coverage of this chapter.
6	Chapters 14 and 15	Full coverage of these chapters. May require continuation in Week 7.
7	Chapter 16 and 17	Full coverage of this chapter. May require continuation in Week 8.
8	Chapter 18	Full coverage of this chapter. This chapter should take less than a full week, since much will be review of concepts introduced in Chapters S2–S4. Thus, you can also use this week as needed to complete coverage described above.
9	Chapter 19	Full coverage of this chapter.
10	Chapter 20	Full coverage of this chapter.
11	Chapter 21	Full coverage of this chapter.
12	Chapter 22	Full coverage of this chapter.
13	Chapter 23	Full coverage of this chapter.
14	Chapter 24	Wrap up with discussion of life in the universe in Chapter 24.

Sample Outline 5: One Semester, "Everything" Course

General Notes:

- This outline assumes your course covers all of astronomy (i.e., both solar system and stars/galaxies/cosmology) in a single, 14-week semester (approx. 3 contact hours per week).
- Trying to cover all of astronomy in a single semester necessarily means making some compromises; this sample outline is only one of many possible options for such a course.
- Note: If you are teaching this type of one-semester "everything" course, you may wish to consider using the condensed version of our book—called *The Essential Cosmic Perspective*—which is expressly designed for one-semester "everything" courses.

Week	Reading	Suggested Coverage
1	Chapter 1	Begin your course with the overview of modern astronomy in Chapter 1.
2	Chapters 2 and 3	Cover Chapter 2 briefly, emphasizing seasons and phases of the Moon. In Chapter 3, focus on the Copernican revolution and the hallmarks of science.
3	Chapters 4 and 5	Cover these chapters as fully as possible in one week.
4	Chapters 7 and 8	Cover Chapter 7 in one class period, skipping the material on spacecraft. Then cover the formation of the solar system in Chapter 8; de-emphasize the material on the heavy bombardment and radiometric dating,
5	Chapters 9 and 10	In Chapter 9, emphasize the four geological processes and the geological tours of the terrestrial worlds. In Chapter 10, emphasize the greenhouse effect, sources/losses of atmospheric gas, and the comparative histories of Mars, Venus, and Earth.
6	Chapters 11 and 12	In Chapter 11, emphasize the overview of jovian worlds and the major satellites and rings. In Chapter 12, emphasize Pluto and cosmic collisions.
7	Chapter 13	Cover extrasolar planets, and use any extra time to complete your coverage of the solar system.
8	Chapter 14	Start your classes on stars with this chapter on the Sun, emphasizing the concepts of gravitational contraction and gravitational equilibrium, nuclear fusion and study of the solar interior.
9	Chapters 15 and 16	In Chapter 15, emphasize the H–R diagram. Then move on to Chapter 16 on star birth.
10	Chapter 17	Finish coverage of star birth and cover stellar evolution in Chapter 17; de-emphasize close binaries.
11	Chapters 18 and 19	In Chapter 18, move quickly through white dwarfs and neutron stars to allow more time for black holes (a topic of high student interest). In Chapter 19, emphasize galactic structure and motion.
12	Chapters 20 and 21	In Chapter 20, move quickly through galaxy types to allow more time for the distance scale and the age of the universe. In Chapter 21, move quickly through general galaxy evolution; place emphasis on evidence for supermassive black holes.
13	Chapters 22 and 23	In Chapter 22, emphasize the role of dark matter in the fate of the universe and the possibility of an accelerating expansion. Cover Chapter 23 briefly, emphasizing the key evidence that supports the Big Bang theory and how recent discoveries tie back to ideas of the accelerating cosmos.
14	Chapter 24	Wrap up with discussion of life in the universe in Chapter 24.

Sample Outline 6: One Semester, "Everything" Course (Solar System Last)

General Notes:

- This outline covers essentially the same material as Sample Outline 5, but rearranged to cover stars/galaxies/cosmology before the solar system.
- NOTE: If you are teaching this type of one-semester "everything" course, you may wish to consider using the condensed version of our book—called *The Essential Cosmic Perspective*—which is expressly designed for one-semester "everything" courses.

Week	Reading	Suggested Coverage
1	Chapter 1	Begin your course with the overview of modern astronomy in Chapter 1.
2	Chapters 2 and 3	Cover Chapter 2 briefly, emphasizing seasons and phases of the Moon. In Chapter 3, focus on the Copernican revolution and the hallmarks of science.
3	Chapters 4 and 5	Cover these chapters as fully as possible in one week.
4	Chapter 14	Start your classes on stars with this chapter on the Sun, emphasizing the concepts of gravitational contraction and gravitational equilibrium, nuclear fusion and study of the solar interior.
5	Chapters 15 and 16	In Chapter 15, emphasize the H–R diagram. Then move on to Chapter 16 on star birth.
6	Chapter 17	Finish coverage of star birth and cover stellar evolution in Chapter 17; de-emphasize close binaries.
7	Chapters 18 and 19	In Chapter 18, move quickly through white dwarfs and neutron stars to allow more time for black holes (a topic of high student interest). In Chapter 19, emphasize galactic structure and motion.
8	Chapters 20 and 21	In Chapter 20, move quickly through galaxy types to allow more time for the distance scale and the age of the universe. In Chapter 21, move quickly through general galaxy evolution; place emphasis on evidence for supermassive black holes.
9	Chapters 22 and 23	In Chapter 22, emphasize the role of dark matter in the fate of the universe and the possibility of an accelerating expansion. Cover Chapter 23 briefly, emphasizing the key evidence that supports the Big Bang theory and how recent discoveries tie back to ideas of the accelerating cosmos.
10	Chapters 7 and 8	Cover Chapter 7 in one class period, skipping the material on spacecraft. Then cover the formation of the solar system in Chapter 8; de-emphasize the material on the heavy bombardment and radiometric dating,
11	Chapters 9 and 10	In Chapter 9, emphasize the four geological processes and the geological tours of the terrestrial worlds. In Chapter 10, emphasize the greenhouse effect, sources/losses of atmospheric gas, and the comparative histories of Mars, Venus, and Earth.
12	Chapters 11 and 12	In Chapter 11, emphasize the overview of jovian worlds and the major satellites and rings. In Chapter 12, emphasize Pluto and cosmic collisions.
13	Chapter 13	Cover extrasolar planets, and use any extra time to complete your coverage of the solar system.
14	Chapter 24	Wrap up with discussion of life in the universe in Chapter 24.

Sample Outline 7: One Quarter, Solar System Emphasis

General Notes:

- This outline assumes a 10-week quarter. It covers essentially the same material as Sample Outline 1, but in one 10-week quarter instead of one 14-week semester. As a result, it covers more material each week. If you have 4 contact hours per week (e.g., 3 hours of lecture and 1 hour of recitation), you should be able to cover everything in the same depth as in the one-semester course. If you have only 3 contact hours per week, you will need to choose areas of emphasis within each chapter.
- If you wish to add coverage of the sky and sky phenomena as in Sample Outline 2, you shouldadd 1 week on Chapter S1 by compressing the coverage of Chapters 8–14 into 5 weeks instead of 6.

Week	Reading	Suggested Coverage
1	Chapter 1	Begin your course with the overview of modern astronomy in Chapter 1.
2	Chapters 2 and 3	Cover Chapter 2 briefly, emphasizing basics such as seasons and phases of the Moon. In Chapter 3, focus on the Copernican revolution and the hallmarks of science.
3	Chapter 4	Coverage of Newton's laws, conservation laws, and gravity.
4	Chapters 5 and 6	Cover light and matter in some depth in Chapter 5; limited coverage of telescopes in Chapter 6.
5	Chapters 7 and 8	Full coverage of these chapters. If necessary, skip spacecraft in Chapter 7 and limit class time spent on radiometric dating in Chapter 8.
6	Chapter 9	Full coverage of this chapter.
7	Chapter 10	Full coverage of this chapter.
8	Chapter 11	Full coverage of this chapter.
9	Chapter 12	Full coverage of this chapter.
10	Chapters 13, 24	Full coverage of Chapter 13, and optional coverage of Chapter 24 if time allows.

Sample Outline 8: One Quarter, Stars/Galaxies/Cosmology Emphasis

General Notes:

- This outline assumes a 10-week quarter. It covers essentially the same material as Sample Outline 3, but in one 10-week quarter instead of one 14-week semester. As a result, it covers more material each week. If you have 4 contact hours per week (e.g., 3 hours of lecture and 1 hour of recitation), you should be able to cover everything in the same depth as in the one-semester course. If you have only 3 contact hours per week, you will need to choose areas of emphasis within each chapter.
- This outline assumes that your course has no prerequisite. We recommend Sample Outline 9 if your course has a prerequisite that covered concepts in early chapters of the book.
- This outline does not cover Part 4 on relativity and quantum ideas; see Sample Outline 9 if you wish to cover these chapters.

Week	Reading	Suggested Coverage
1	Chapter 1	Begin your course with the overview of modern astronomy in Chapter 1.
2	Chapters 3 and 4	In Chapter 3, emphasize ideas of modeling, the Copernican revolution, and the hallmarks of science. Begin Chapter 4 coverage of Newton's laws, conservation laws, and gravity.
3	Chapter 5 and 6	Cover chapter 5 as fully as possible, and spend limited time on telescopes as available.
4	Chapters 14 and 15	In Chapter 14, emphasize the first two sections. In Chapter 15, emphasize the H–R diagram.
5	Chapters 16 and 17	Full coverage of these chapters.
6	Chapter 18	Full coverage of this chapter.
7	Chapter 19	Full coverage of this chapter.
8	Chapters 20 and 21	Full coverage of these chapters. May require continuation in Week 9.
9	Chapter 22	Finish covering Chapter 21. Fully cover Chapter 22.
10	Chapters 23 and 24	Cover Chapter 23 as fully as possible in the final week. Wrap up with Chapter 24 if time allows.

Sample Outline 9: One Quarter, Stars/Galaxies/Cosmology Emphasis with Relativity

General Notes:

- This outline is based on the assumption that you do not need to spend class time on Parts I and II of the text, as would be the case if either:
  - Your course has a prerequisite in which Parts I and II were already covered.
  - You have "honors" students with strong backgrounds in high school science.
- This outline is also suitable for a course in relativity and cosmology if you de-emphasize the coverage of the H–R diagram and basic properties of stars and galaxies.

Week	Reading	Suggested Coverage
1	Chapter S2	Begin relativity, with full coverage of Chapters S2 and S3, to be spread out over weeks 1 through 3.
2	Chapter S3	Continue relativity.
3	—	Continue and conclude relativity.
4	Chapter S4	Quantum mechanics, with full coverage of this chapter.
5	Chapters 14 and 15	In Chapter 14, emphasize the first two sections. In Chapter 15, emphasize the H–R diagram.
6	Chapters 16 and 17	Full coverage of these chapters, to the extent possible.
7	Chapters 18 and 19	Full coverage of these chapters, to the extent possible.
8	Chapters 20 and 21	Full coverage of these chapters, to the extent possible.
9	Chapters 22 and 23	Full coverage of these chapters, to the extent possible.
10	Chapter 24	Finish discussion of Chapter 23. Wrap up with Chapter 24.

The Pedagogical Approach of *The Cosmic Perspective*

Our modern understanding of the cosmos is both broad and deep, making it impossible to cover all of what we know today in a one- or two-term introductory course. After all, professional astronomers typically take years to master just one subdiscipline of astronomical research, and even then the star specialists may be unfamiliar with much of planetary astronomy, and vice versa. Thus, success in an introductory astronomy course requires a carefully chosen set of key concepts presented with a coherent and integrated pedagogy. This is precisely what we have tried to offer in *The Cosmic Perspective*. We believe that you will be able to make more effective use of our text if you understand the how and why of its pedagogical approach. We therefore discuss this approach in some detail in this section.

Themes of *The Cosmic Perspective*

The Cosmic Perspective offers a broad survey of modern understanding of the cosmos and of how we have gained that understanding. Such a survey can be presented in a number of different ways. We have chosen to present it by interweaving a few key themes throughout the narrative, all chosen to help make the subject more appealing to students who may never have taken any formal science courses and who may enter the course with little understanding of how science works. The following are five key themes around which we built our book:

> *Theme 1: We are a part of the universe, and thus can learn about our origins by studying the universe.* This is the overarching theme of *The Cosmic Perspective*, as we continually emphasize that learning about the universe helps us understand what has made our existence possible. Studying the intimate connections between human life and the cosmos gives students a reason to care about astronomy and also deepens their appreciation of the unique and fragile nature of our planet and its life.
>
> *Theme 2: The universe is comprehensible through science that can be understood by anyone.* The universe is comprehensible because the same physical laws appear to be at work in every aspect, on every scale, and in every age of the universe. Moreover, while the laws have generally been discovered by professional scientists, they can be understood by anyone. Students can learn enough in one term of astronomy to comprehend many phenomena they see around them—from seasonal changes and phases of the Moon to the most esoteric astronomical images that appear in the news.
>
> *Theme 3: Science is not a body of facts but rather a process through which we seek to understand the world around us.* Many students assume that science is just a laundry list of facts. The long history of astronomy can show them that science is a process through which we learn about our universe—a process that is not always a straight line to the "truth." That is why our ideas about the cosmos sometimes change as we learn more, as they did dramatically when we first recognized that Earth is a planet going around the Sun rather than the center of the universe. In this book, we continually emphasize the nature of science so that students can

understand how and why modern theories have gained acceptance and why these theories may still change in the future.

Theme 4: A course in astronomy is the beginning of a lifelong learning experience. Building upon the prior themes, we emphasize that what students learn in their astronomy course is not an end but a beginning. By remembering a few key physical principles and understanding the nature of science, students can follow astronomical developments for the rest of their lives. We therefore seek to motivate students enough that they will have the desire to continue to participate in the ongoing human adventure of astronomical discovery.

Theme 5: Astronomy affects each of us personally with the new perspectives it offers. We all conduct the daily business of our lives with reference to some "world-view"—a set of personal beliefs about our place and purpose in the universe—that we have developed through a combination of schooling, religious training, and personal thought. This world-view shapes both our beliefs and many of our actions. Although astronomy should not be construed to support any particular world-view, it certainly provides the foundations from which world-views are built. For example, a person's world-view would likely be quite different if they believed the Earth to be the center of the universe, instead of a tiny and fragile world in a vast cosmos. In many respects, the role of astronomy in shaping world-views represents the deepest connection between the universe and the everyday lives of humans.

Pedagogical Principles of *The Cosmic Perspective*

No matter how you choose to teach your course, it is very important to present material according to a clear set of pedagogical principles. The following list summarizes the major pedagogical principles that we focused on in the writing of every chapter, section, paragraph, and sentence in the book:

- *Stay focused on the "big picture."* No matter how well we teach, within a relatively short time (a year, say) our students will inevitably forget most the specific facts they learn in any course. Therefore it is critical that even as we delve into details, we always stay focused on the "big picture" ideas that we hope our students will retain long after the course is over. For us, these key big picture ideas are the themes we have outlined above. Throughout the text, you will find that we have made decisions on what details to include—and what to leave out—by asking whether they help support the big picture themes that we want students to take away from their course.
- *Always provide context first.* We all learn new material more easily when we understand why we are learning it. In essence, this is simply the idea that it is easier to get somewhere when you know where you are going. (As Yogi Berra said, "If you don't know where you're going, you'll probably end up someplace else.") Because most students enter introductory courses with little prior knowledge of astronomy, it is critical that we give them a mental framework for the subject before they begin. We therefore begin the book (Chapter 1) with a broad overview of modern understanding of the cosmos—our cosmic address and origins, the scale of the universe, and motion of the universe. This allows us to give students at least a general understanding of such things as the levels of structure in the cosmos, how

vastly the scale of the solar system and galaxy differ (and remember that students often confuse the two terms), what we mean by a beginning in a Big Bang, and what we mean by the idea that the universe is expanding. Once students have this basic overview in their heads, it is much easier for them to make sense of all the details that follow. We maintain this "context first" approach in the rest of the book as well. For example, we always begin chapters by telling students what we hope to learn and why, and we always review the context at the end of each chapter (in the end-of-chapter sections called "The Big Picture").

- *Make the material relevant.* It's human nature to be more interested in subjects that seem relevant to our lives.[1] Thus, while there's nothing wrong with emphasizing that science is often driven by inherent curiosity, we are more likely to reach our students if we emphasize the many connections between astronomy and their personal concerns. Indeed, this idea of relevance lies behind the choice of our themes listed above. It also has practical implications to the way we structure the text. For example, when studying the solar system we emphasize how learning about other planets helps us understand our own (since Earth is clearly relevant to everyone); when studying stars we emphasize that we are "star stuff" (to quote Carl Sagan) in the sense that the atoms of our body were forged in stars; and when studying galaxies we emphasize that we are also "galaxy stuff" in the sense that galaxies are necessary for the cosmic recycling that makes possible multiple generations of stars.
- *Emphasize conceptual understanding over "stamp collecting" of facts.* All too often, astronomy is presented as a large set of facts to remember. As discussed earlier, students inevitably forget these facts after the exams are over. If we want students to retain something of value from their astronomy course, we must emphasize just a few key conceptual ideas that we use over and over again. For example, students encounter the laws of conservation of energy and conservation of angular momentum over and over throughout the book, and students learn about how terrestrial planets are shaped by just a few basic geological processes. Students retain such learning far better than when they are asked to memorize facts.
- *Proceed from the more familiar and concrete to the less familiar and abstract.* It's well known that children learn best by starting with concrete ideas and then generalizing to abstractions later. In fact, the same is true for all of us at any age. To the extent possible, when introducing new ideas we should always seek "bridges to the familiar" (a term coined by our colleague Jeff Goldstein of the Challenger Center for Space Science Education). For example, in our chapter on motion and gravity, we begin with "everyday" experiences that can be expanded upon and generalized as we discuss the much more abstract physical laws of the cosmos. Similarly, in our chapter on terrestrial planets, we begin with ideas that are familiar from everyday observations of our own world. We can carry this idea forward even with topics like galaxies; for example, we start our chapter on galaxies beyond the Milky Way (Chapter 20) by showing the Hubble Deep Field, because it allows students to use their own senses to see that galaxies have different shapes, sizes, and colors. Once they experience these concrete ideas for themselves, we can begin to ask more abstract questions about why galaxies differ.

---

[1] As an example that may carry weight for many instructors, consider investments in the stock market. Twenty-five years ago, few faculty knew anything about the market, but today nearly all of us follow the markets and have a fairly deep understanding of how they work. What changed? The development of retirement plans for which we must make choices ourselves made the markets relevant and gave us a reason to learn about them.

- *Use plain language.* Surveys have found that the number of new terms in typical introductory astronomy books is larger than the number of words taught in many first courses in foreign languages. In essence, this means the books are teaching astronomy in what looks to students like a foreign language. Clearly, it will be much easier for students to understand key astronomical concepts if these concepts can be explained in plain English, without resort to unnecessary jargon. We have therefore gone to great lengths to eliminate jargon as much as possible, or at minimum to replace standard jargon with terms that will be easier to remember in the context of the subject matter. For example, instead of "lunar regolith" (a term you'll find in nearly all texts but which even most astronomers don't know), we refer to "powdery lunar soil"—which is in essence what the regolith is. As an example of a case where standard jargon can be replaced with something more meaningful, consider the two basic mechanisms for supernovae. One mechanism involves an explosion at the end of the life of a massive star, which is generally called a Type II supernova (but also includes Type Ib and Ic supernovae). The other mechanism involves white dwarfs in binary systems, and is known to astronomers as a Type Ia supernova. Most books use these Type I/II terms, but we find students understand and remember them much better when we refer to the two basic types simply as massive star supernovae and white dwarf supernovae.[2]
- *Recognize and address student misconceptions.* Students do not arrive as blank slates. Most students enter our courses not only lacking the knowledge we hope to teach, but often holding misconceptions about astronomical ideas. Therefore, in order to teach correct ideas, we must also help students recognize the paradoxes in their prior misconceptions. This recognition creates the conditions under which they can undergo "conceptual change" from a misconception to a correct conceptual understanding. We address this issue in a number of ways in our text, the most obvious being the presence of our many "common misconceptions" boxes. These summarize commonly held misconceptions and explain why they cannot be correct.

Note: The above pedagogical ideas are discussed in more detail (and with a slightly different presentation) in the article "Strategies for Teaching Astronomy," by Jeffrey Bennett, in Mercury, Nov/Dec 1999.[3] The article is also posted in the instructor's area of the MasteringAstronomy website (www.masteringastronomy.com) and on Jeff's personal web page (www.jeffreybennett.com).

[2] The Type I/II designations are based on spectra (Type II have hydrogen lines and Type I do not), which is why there is not a one-to-one correspondence with the presumed mechanisms (white dwarf or massive star undergoing supernova). In the past, before we had a good idea of the mechanisms, it made sense to teach from the directly observable phenomena of spectra. Today, however, we think it makes far more sense to teach based on the mechanisms.

[3] Jeff also has a talk for science faculty based on these ideas that he has given at numerous colleges and universities. If you are interested in having him present the talk at your school, contact him directly: jeff@bigkidscience.com or jeffrey.bennett@comcast.net.

The Topical Structure of *The Cosmic Perspective*

In order to implement the themes and pedagogical principles described above, we had to make decisions about how we would treat the major topical areas of astronomy. We decided to organize the book broadly into seven major topical areas, referred to as Parts I through VII in the table of contents. In this section, we describe each of the seven major parts, with a short description of its pedagogical approach and how it may differ from the approach you may have used if you taught in the past from different texts.

Part I—Developing Perspective (Chapters 1–3, S1)

Basic Content. As the Part title suggests, the three chapters here are designed to give students a general perspective that we will build upon throughout the rest of the book. Chapter 1 offers an overview of modern understanding of the cosmos, thereby giving students perspective on the entire universe. Chapter 2 provides an introduction to basic sky phenomena, including seasons and phases of the Moon—perspective on how phenomena we experience every day are tied to the broader cosmos. Chapter 3 discusses the history and nature of science, offering a historical perspective on the development of science and giving students perspective on how science works and how it differs from nonscience. Note that this chapter includes discussion of the Copernican revolution and Kepler's laws.

Pedagogical Approach—The Big Picture, The Process of Science, and Integrated History. The basic pedagogical approach in these chapters is designed to make sure students get their "big picture" overview and context for the rest of the book, and to make sure they develop an appreciation for the process of science and how science has developed through history. In addition, please note that:

- Chapter 1 gives a fairly comprehensive overview of modern astronomy, including an introduction to ideas of the Big Bang, the expanding universe, and the mystery of dark matter. In a sense, this chapter covers nearly all of the important ideas that we want students to take away from their course, so that the rest of the book is simply reinforcement and filling in the details.
- We believe that historical ideas of astronomy are much more meaningful when taught in context rather than as a separate "track" in an astronomy course. We therefore introduce the accomplishments of many ancient cultures in the context of how they relate to what we consider science today, and introduce Greek ideas of astronomy through the context of the modeling (e.g., the transition from the geocentric model to Copernican model) that has become the foundation of modern science. Note also that we have taken extraordinary care to make sure that our historical discussions are based on the most current understanding of history; to help ensure accuracy, noted historian of science Owen Gingerich reviewed our historical material.

How and Why Is Our Approach Different? Traditionally, most introductory astronomy textbooks have begun with the sky as viewed by the ancients, and then gradually worked through astronomical topics in a fairly historical progression, so that ideas of the expanding universe and modern cosmology come only at the very end of the book. If these books offer any "big picture" overview at all in the beginning, it has typically been limited to structural hierarchy and perhaps a bit of scale. Our approach of introducing all the major cosmological ideas in Chapter 1 is clearly different from this traditional approach. To understand why we believe our approach is more successful with students, it's useful to consider both the potential advantages and drawbacks of the traditional approach. The potential advantage of the traditional approach is this: if students fully absorb it, they recreate for themselves the thought patterns and paradigm shifts that led from ancient superstitions about the sky to modern understanding of the universe. Clearly, that would lead to a fairly deep understanding of the scientific process. Unfortunately, we believe that this potential advantage is almost impossible to realize in practice, for two major reasons.

- First, the traditional approach in essence asks students to recreate the thinking produced over thousands of years by some of the greatest minds in history—all with just a few hours of study per week for a few months. We do not think this is realistic, and we believe that students who seem to succeed at it are almost always those students who enter the course with prior knowledge of the types of big picture ideas that we cover in our Chapter 1. Thus, we believe the potential advantage instead becomes a practical unfairness, because it puts students with prior knowledge at a competitive advantage over those for whom a course with no prerequisites is truly designed.
- Second, because the traditional approach leaves all mention of cosmology to the very end of the course, it leaves this material at risk of being lopped off due to time constraints. That is, in the very common situation that you must slow your pace of coverage, the traditional approach could mean that you never get to talk at all about such things as the expanding universe, dark matter, and the Big Bang. Given that these topics permeate virtually all of modern astronomy, we think it almost criminal to risk the possibility that students will leave an introductory astronomy class without at least having been introduced to them.

We authors began our teaching careers following the traditional approach, and it was its failure for so many students that convinced us to change. Our experience, and that of others who have made the same change, tells us that our "big picture" approach will allow a far greater fraction of your students to learn and retain key astronomical ideas than will the traditional approach.

Part II—Key Concepts for Astronomy (Chapters 4–6)

Basic Content. These chapters lay the groundwork for understanding astronomy through what is sometimes called the "universality of physics"—the idea that a few key principles governing matter, energy, light, and motion explain both the phenomena of our daily lives and the mysteries of the cosmos. Chapter 4 covers the laws of motion, the crucial conservation laws of angular momentum and energy, and the universal law of gravitation. Chapter 5 covers the nature of light and matter, including the formation of spectra, the laws of thermal radiation, and the Doppler effect. Chapter 6 covers telescopes and astronomical observing techniques.

Pedagogical Approach—Bridges to the Familiar. We approach this material by following the principle of building "bridges to the familiar." Each chapter makes connections between science and the phenomena of everyday life. We then build on this "everyday knowledge" to help students learn the formal principles of physics needed for the rest of their study of astronomy.

How and Why Is Our Approach Different? All introductory astronomy textbooks include discussion of key physical ideas, but our approach differs from what you may have found in other books in at least two key ways:

- Most introductory texts tend to start with the abstract laws (such as Newton's laws) and then give examples of them in action. Following our principle of building from the concrete to the abstract, we always begin with everyday experiences first. We believe this helps students learn the physical laws in a way they are more likely to understand and retain.
- We place much more emphasis on conservation laws and their implications than most other texts. In particular, our section 4.3 on conservation laws in astronomy shows students why energy and angular momentum conservation are so useful to understanding physical processes throughout the universe, and we refer back to these laws frequently in the rest of the book.

Part III—Learning From Other Worlds (Chapters 7–13)

Basic Content. This is our section about our planetary system. This set of chapters begins with a broad overview of the solar system in Chapter 7, including a 10-page tour that highlights some of the most important and interesting features of the Sun and each of the nine planets in turn. Chapter 8 uses these concrete features to build student understanding of the current theory of solar system formation. Chapters 9 and 10 focus on the terrestrial planets, covering key ideas of geology and atmospheres, respectively, and showing how these ideas apply on each of the terrestrial worlds. Chapter 11 covers the jovian planets and their moons and rings. Chapter 12 covers small bodies in the solar system—asteroids, comets, and Pluto and other large objects of the Kuiper belt. It also covers cosmic collisions, including the impact linked to the extinction of the dinosaurs and a discussion of how seriously we should take the ongoing impact threat. Finally, Chapter 13, which is new to this edition, explores recent discoveries about other planetary systems. Note that Part III is essentially independent of Parts IV through VI, and thus can be covered either before or after them.

Pedagogical Approach—True Comparative Planetology. We offer a comparative planetology approach, in which the discussion emphasizes the processes that shape the planets, rather than the "stamp collecting" of facts about the individual planets. As always, we begin with the concrete and familiar. For example, Chapter 7 presents the very concrete ideas of the solar system's basic layout, while Chapter 8 then explains how that layout came to be. Similarly, in the other chapters we always begin with things that may be familiar from Earth or that are at least easy to see on other planets, and then move on to explain how these things came to be.

How and Why Is Our Approach Different? Most other introductory textbooks cover the planets in a "march of the planets," covering them individually rather than through our process-oriented comparative planetology approach. Why did we choose to do something so different? There are four major reasons:

- First and foremost, we are convinced that students learn and retain much more from the comparative planetology approach. The traditional approach teaches lots of facts about the planets—facts that tend to be quickly forgotten after exams (e.g., where is Caloris Basin located again?). In contrast, our emphasis on processes means students only need to learn a few key ideas that they can then apply over and over to phenomena on Earth or other worlds.

- Of almost equal importance, our approach helps students see the relevance of planetary science to their own lives by enabling them to gain a far deeper appreciation of our own unique world. With the traditional approach, students can legitimately ask, "why should I care about the names of volcanoes on Mars?" With our approach, students see how the study of volcanoes on Mars and elsewhere allows us to learn more about Earth and the conditions under which life is possible.
- We sincerely believe that the traditional approach is now outdated. Covering planets individually made perfect sense when we did not know a lot about them. But as more spacecraft show us the planets in more and more detail, and as we discover more and more planets in other solar systems, the idea of simply adding more detailed coverage of the planets quickly becomes untenable. The comparative planetology approach allows students to see how rapidly accumulating new discoveries fit into our overall understanding of how planets work—or, in some cases, how they force us to modify our theories of planetary processes.[4]
- Finally, with our emphasis on how and why planets work, students' minds are better prepared for the future of planetary science. What do we still need to learn about planets in our own solar system? What should we expect from discoveries of planets around other stars? Are the conditions for life in the universe rare and random, or do they follow the same principles operating in our solar system? Students using this textbook will be ready to understand and appreciate planetary science for the rest of their lives.

Part IV—A Deeper Look at Nature (Chapters S2–S4)

Basic Content. *These chapters are labeled "supplementary" because coverage of them is optional.* The three chapters of Part IV cover special relativity (Chapter S2), general relativity (Chapter S3), and key astronomical ideas of quantum mechanics (Chapter S4). Covering them will give your students a deeper understanding of the topics that follow on stars, galaxies, and cosmology, but the later chapters are self-contained so that they may be covered without having covered Part IV at all.

Pedagogical Approach—Ideas of Relativity and Quantum Mechanics are Accessible to Anyone. These chapters all begin with sections summarizing the basic ideas that we will cover, then proceed to explain the ideas, then discuss the evidence and implications of the ideas. The main thrust throughout is to demystify relativity and quantum mechanics by convincing students that they are capable of understanding the key ideas despite the reputation of these subjects for being hard or counterintuitive.

How and Why Is Our Approach Different? Well, the big difference is that no other introductory text devotes full chapters to relativity and quantum ideas as we do. So why do we do it? Here are three major reasons:

- Many of the most important and most interesting ideas of modern astronomy really cannot be understood unless students have a solid foundation in relativity and quantum mechanics. For example, while nearly every book offers a rubber sheet analogy for the curvature of spacetime near a black hole, without much

[4] Not to toot our own horn too loudly, but: Properly incorporating the recent discovery of "Planet X" — an object in the Kuiper belt larger than Pluto — required very little rewriting in our book, because it had already been anticipated by our comparative planetology organization.

more discussion students will completely miss the point of the analogy and may even take it literally—picturing a black hole as a funnel-shaped structure! Similarly, when introducing the important meaning of the observable universe, we expect students to take it for granted that faster-than-light travel is not possible—but of course students want to know why. (Worse yet, many students assume that saying you can't go faster than light is like past engineers saying you can't break the sound barrier—which leads them to conclude that scientists are closed-minded and cynical!) Newsworthy ideas of the overall geometry of the universe, such as what we mean when we say that new WMAP results support a flat universe, also require relativity. Quantum ideas are no less important, and not only because of the role they play in understanding light and spectra. For example, quantum ideas are needed to understand the degeneracy pressure that supports brown dwarfs, white dwarfs, and neutron stars, as well as to understand cosmological ideas such as inflation in the very early universe.

- We want to demystify these topics and show students that they can indeed understand these important astronomical ideas. There's a common myth that relativity and quantum ideas are too hard for the average person. (Consider the still-persistent urban legend that only 12 people in the world understand relativity.) As in our theme 2 above, we want students to feel that all of science is accessible to them, if it is presented simply and clearly. Thus, taking the time to do relativity and quantum ideas "right" seems well worth the effort.
- Students love it. Nearly all students have at least heard of things like the prohibition on faster-than-light travel, curvature of spacetime, and the uncertainty principle. But few (if any) enter an introductory astronomy course with any idea of what these things mean. They are naturally curious, and get great satisfaction from discovering that they can actually understand these topics. Indeed, in our courses in which we've covered these topics, end-of-course surveys have consistently found that a plurality of students chooses relativity as their favorite part of their astronomy course.

Part V—Stellar Alchemy (Chapters 14–18)

Basic Content. These are our chapters on stars and stellar lifecycles. Chapter 14 covers the Sun in depth, so that it can serve as our concrete model for building an understanding of other stars. Chapter 15 describes the general properties of other stars, how we measure these properties, and how we classify stars with the HR diagram. Chapters 16 and 17 cover stellar evolution, with the first chapter focusing on star birth and the second on the birth-to-death lives of both low- and high-mass stars. Chapter 18 covers the end points of stellar evolution: white dwarfs, neutron stars, and black holes.

Pedagogical Approach—We are Intimately Connected to the Stars. Perhaps we can best describe our approach here as "less is more." Today, we know so much about stars that the primary challenge in teaching them is deciding what NOT to cover. We therefore have chosen to focus on those aspects of stars that support our themes—especially those aspects that reveal our intimate connections to the stars (e.g., nucleosynthesis). In order to make sure these themes stand out clearly, we deliberately leave out many other "details" that students would surely forget soon anyway.

How and Why Is Our Approach Different? The basic structure of our chapters on stars is the same as that used in nearly all textbooks. We believe this similarity simply reflects the fact that the basics of stellar evolution have been well-understood for decades, so that by now everyone has converged on an optimal order of presentation. Nevertheless, you'll still find at least two important differences between our coverage of stars and that in most other textbooks:

- Our "less is more" approach, as discussed above. If you want students to memorize facts about Herbig-Haro objects or the many different types of binary star systems—or if you want to quiz them on standard jargon like Type I and II supernovae, asymptotic-giant branch stars, or hydrostatic equilibrium—then we may not be the book for you. But if you want students to truly understand the connections between stars and human existence, and to be amazed by the bizarre objects left behind when stars die, then we think you'll like our approach.
- Most books include substantial discussion of the interstellar medium along with their discussion of stars and star birth. We choose instead to leave most of our discussion of the interstellar medium to a later chapter—Chapter 19 on the Milky Way. Why? Because aside from the crucial ideas of how and why interstellar clouds sometimes contract under gravity, none of the other details of the interstellar medium are needed to understand star formation. In contrast, the workings of the interstellar medium are crucial to understanding galactic "ecology," and thus belong with the discussion of the Milky Way.

Part VI—Galaxies and Beyond (Chapters 19–23)

Basic Content. These chapters cover galaxies and cosmology. Chapter 19 presents the Milky Way, emphasizing that it is a gravitationally confined system that gradually turns gas into stars while continually enriching it with heavier elements. Chapter 20 presents the variety of galaxies, and how we determine key parameters such as galactic distances and age. Chapter 21 presents current understanding of galaxy evolution. Chapter 22 focuses on dark matter and dark energy and their role in the fate of the universe. Chapter 23 covers the theory of the Big Bang.

Pedagogical Approach—Present Galaxy Evolution in a Way That Parallels the Teaching of Stellar Evolution, and Integrate Cosmological Ideas In the Places Where They Most Naturally Arise. Given the success and general agreement on the pattern of organization for teaching stars, it seems wise to use a parallel structure for galaxies. For example:

- Chapter 19 presents the Milky Way as a paradigm for galaxies in much the same way that Chapter 14 uses the Sun as a paradigm for stars.
- Chapter 20 presents the classification of galaxies and discusses how we determine key parameters such as galactic distances and ages, making it somewhat analogous to the Chapter 15 presentation of stellar parameters.
- Chapter 21 discusses the current state of knowledge regarding galaxy evolution, much as Chapters16 and 17 cover stellar evolution.
- The relationship between Chapters 22–23 (dark matter/energy and the Big Bang) and Chapter 18 (stellar corpses) is somewhat looser, but similarities include showing where astronomy makes contact with the frontiers of physics and discussing some of the most exciting ideas in modern astronomy—the ideas that draw many students into astronomy courses in the first place.

How and Why Is Our Approach Different? Once again (as in Part III), we have an approach that is fundamentally different from that used in nearly all other introductory astronomy textbooks. The traditional approach found in other books goes something like this: morphology and structure of the Milky Way; "normal galaxies" discussed via the traditional Hubble tuning fork classification; "active galaxies" discussed as distinct set of entities from "normal" galaxies; and, finally, cosmology, including Hubble's Law and the Big Bang. Quite frankly, we believe this traditional approach is outdated in light of discoveries made over the past three decades or so. For example:

- The traditional approach gives students very little reason to care about the Milky Way, since it is basically full of facts about its shape and structure. In contrast, our emphasis is on the Milky Way as a cosmic recycling plant, making it possible for generations of stars to live, die, and chemically enrich the interstellar medium—a clear prerequisite to our own existence.
- The traditional emphasis on Hubble's galaxy classification scheme made sense decades ago when astronomers hoped that the tuning fork diagram might turn out to be as important to understanding galaxy evolution as the H–R diagram is to understanding stellar evolution. But it has not turned out that way. Instead, the most fundamental distinction between galaxy types is between those with gas-rich disks (spirals) and those without them (ellipticals), and we are rapidly building an understanding of what factors play a role in a galaxy's type. These are the types of issues that we emphasize with our evolutionary approach to galaxies.
- The traditional distinction between "normal" and "active" galaxies just does not hold water any more. Many galaxies that would have formerly been called "active" are actually quite ordinary except for their unusually bright galactic nuclei. Indeed, the correlation between bulge mass and AGN mass suggests that all galaxies may have a black hole of some kind in their core—it is possible that at some point in time, all galaxies were "active." Other so-called "active galaxies" are now clearly seen as galaxies that may previously have been quite normal but are now in disarray for some reason or other. We therefore discuss galaxy evolution as a single, coherent subject in which various forms of "activity" are seen in context as transient events that occur in the lives of some galaxies.
- The traditional approach keeps cosmology fairly distinct from the study of galaxies. But the formation and evolution of galaxies is so inextricably linked to initial conditions (e.g., seeds of structure present in the cosmic microwave background) and the overall evolution of the universe (e.g., ongoing expansion even as large-scale structures may still be forming) that we do not believe they should be separated. Indeed, to our minds, trying to teach galaxies without cosmology is like trying to teach biology without the theory of evolution—it leaves out the key element of the modern paradigm. We have therefore woven galaxy evolution and cosmic evolution together, and we believe this approach gives students a deeper understanding of both.
- The traditional approach leaves little room for a coherent discussion of the two things now thought to dominate the mass-energy of the universe: dark matter and dark energy. As a result, these topics usually are presented in a way that leads to an incomplete and fragmented understanding and a lot of unnecessary mystification. By devoting an entire chapter (Chapter22) to dark matter and dark energy, we are able to give students a solid understanding of the greatest mysteries in modern astronomy. Indeed, this pedagogical structure allows the integration of ideas of universal acceleration and dark energy so smoothly that we did not even have to make any major changes to our text to accommodate these recent discoveries.

Part VII— Life on Earth and Beyond (Chapter 24)

Basic Content. This Part consists of just a single chapter. It may be considered optional, to be used as time allows. Those who wish to teach a more detailed course on astrobiology may wish to consider the text *Life in the Universe*, by Bennett, Shostak, and Jakosky.

Pedagogical Approach—Bridges to the Familiar. We begin with a discussion of life on Earth, then discuss the search for life elsewhere.

How and Why Is Our Approach Different? Many books end with a chapter on life in the universe, so our approach is not fundamentally different. However, we believe you'll find greater emphasis on a few key ideas in our book:

- We place greater emphasis on understanding the key biological arguments that seem to favor life being easy to get, while at the same time presenting the alternative arguments behind the rare Earth hypothesis.
- We include some discussion of the challenge of interstellar travel, because we believe this is necessary in order for students to understand why scientists give so little credence to claims of alien visitations on Earth.

Chapter Structure in *The Cosmic Perspective*

So far, we have given you a fairly extensive description of the pedagogical rationale for The Cosmic Perspective. But on a day-to-day basis, what really matters is the structure of the presentation in the text, since that is how your students will encounter all of our pedagogy. If you thumb through the book, you'll quickly notice that each chapter follows a similar structure, designed to make it easy both for students to study and for you to choose what topics you wish to emphasize in your course. Here is a brief overview of the chapter structure and how it should help you and your students.

Basic Structural Elements

All chapters in the book follow the same basic structural flow, which consists of the following elements:

- Learning Goals: These are presented as key questions at the start of each chapter. They are designed to give students context for the material to come and to help students focus their attention on the most important concepts ahead.
- Chapter Opening Quotation: Every chapter begins with a short quotation, usually from a well-known historical or cultural personality, designed to draw connections between the chapter topic and other human endeavors.[5]
- Chapter Introduction: Prior to the first numbered section (e.g. Section 3.1), each chapter has two to three paragraphs that describe why the chapter material is important and interesting and offer a brief road map of how we will cover it.
- Numbered Sections (Chapter Narrative): The bulk of each chapter comes with the numbered sections that present the narrative and all of the inter-narrative features (such

[5] It's perhaps worth noting that the quotations tend to be much appreciated by students whose primary interest is in social sciences, humanities, or the arts, because it shows them that science is not isolated from these other pursuits.

as Common Misconceptions and Special Topics boxes, and icons to indicate when and where students can find relevant tutorials on the Mastering Astronomy website).

- Learning Goal Headings: The Learning Goal questions from the beginning of the chapter reappear as subheadings throughout the narrative, thereby making it easier for students to find material that they need to study in the text.
- The Big Picture: This end-of-chapter feature helps students put what they've learned in the chapter into the context of the overall goal of gaining a new perspective on ourselves and our planet.
- Chapter Summary: Coming at the end of the chapter narrative (just before the problem set), these return to the questions in the learning goals with answers that help reinforce student understanding of key concepts presented in the chapter.
- End-of-Chapter Questions: Each chapter ends with several sets of questions that can be used for study, discussion, or assignment (answers and solutions for most of them are included in the chapter-by-chapter sections of this Instructor's Guide). They are organized into subcategories as follows:
  - Review Questions: These questions are designed primarily for self-study, since they can generally be answered simply by looking back at the relevant portions of the text.
  - "Does it Make Sense?" Questions: These questions ask the students to evaluate a simple statement that is designed to help them focus on important concepts from the chapter. They should be quite easy for students who understand the concept, but nearly impossible for students who don't understand it—thus pointing those students to their need for further study. Note that the "Does it Make Sense?" heading is sometimes replaced with a slight variation, such as "True or False?" or "Surprising Discoveries?," but in all cases the basic idea is for students to evaluate a statement and explain their reasoning clearly. Note also that, in at least some cases, it is possible to make a case for either answer (e.g., sensible or nonsense), so that the most important part of a student's answer is how well they defend their choice.
  - Quick Quiz: A short multiple choice quiz for students to check their progress.
  - Short Answer/Essay Questions: These go beyond the review questions in asking for conceptual interpretation.
  - Quantitative Problems: These problems require some mathematics, usually based on topics covered in the Mathematical Insight boxes.
  - Discussion Questions: These questions are meant to be particularly thought-provoking and generally do not have objective answers. As such, they are ideal for in-class discussion or extended essays.
- Media Explorations: Each chapter ends with a section or page of "Media Exploration," which is designed to highlight some of the many media resources available to aid students in studying the chapter material.
  - For those chapters in which MasteringAstronomy™ tutorials are relevant, we have included short sets of sample questions that can be assigned along with the tutorial to focus on the issues relevant to that chapter.
  - Every Media Explorations section concludes with one or more "Web Projects" designed for independent research. The projects typically ask students to learn more about a topic of relevance to the chapter, such as a current or planned mission. Students can find useful links for all the Web projects on the MasteringAstronomy™ website.
- Online Quizzes: Finally, although they do not appear in the printed book, the online quizzes for each chapter should be considered an integral part of the chapter.

They are the best way for students to check that they have really grasped the key chapter concepts.

Additional "Feature" Elements

Within the main narrative of each chapter, you'll find a number of pedagogical elements that appear whenever they can help illuminate important topics. These features include all of the following:

- Think About It: This feature, which appears as short questions integrated into the narrative, gives students the opportunity to reflect on important new concepts. It also serves as an excellent starting point for classroom discussions. Answers or discussion points for all theThink About It questions can be found in the chapter-by-chapter sections of this Instructor'sGuide.
- Common Misconceptions: These boxes address and correct popularly held but incorrect ideas related to the chapter material. In a few cases in which the boxes focus on misconceptions perpetrated by television or movies, the boxes carry the title "Movie Madness."
- Mathematical Insights: These boxes contain most of the mathematics used in the book and can be covered or skipped, depending on the level of mathematics that you wish to include in your course.
- Special Topic Boxes: These boxes contain supplementary discussion of topics related to the chapter material but not prerequisite to the continuing discussion.
- Wavelength/Observatory Icons: For astronomical photographs (or art that might be confused with photographs), we include simple icons that identify the wavelength band of the photo or identify the figure as an art piece or computer simulation. Alongside the wavelength icon for photos, we also include an icon to indicate whether the image came from ground-based or space-based observations.
- Cross-References: When we discuss a concept that is covered in greater detail elsewhere in the book, we include a cross-reference in brackets. For example, "[Section 5.2]" means that the concept being discussed is covered in greater detail in Section 5.2.
- Glossary: Although not part of the chapter flow, the glossary should be considered another pedagogical element tied to the chapters, since it makes it easy for students to look up important terms.

Getting the Most Out of Each Chapter

The chapter structure is designed so that it is easy for students to study effectively and efficiently. Nevertheless, it may be useful to suggest the following study plan to your students:

- Begin by reading the Learning Goals to make sure you know what you will be learning about in this chapter.
- Before reading in-depth, start by skimming the chapter and focusing only on the headings, callouts, and illustrations. Study each illustration and read the captions so that you will get an overview of the key chapter concepts.
- Next, read the chapter narrative. Try to answer the Think About It questions as you go along, but you may save the other boxed features (Common Misconceptions, Special Topics) to read later.

- After reading the chapter once, go back through and read the boxed material. Also look for the tutorial icons that tell you when there is a relevant tutorial on the Mastering Astronomy website. If you are having difficulty with a concept, be sure you try the tutorial.
- Study the Chapter Summary by first trying to answer the Learning Goals questions for yourself, then checking your understanding against the answers given in the summary.
- Check your understanding by trying some of the end-of-chapter problems and the online quizzes.

Preparing Your Course: Suggestions for First-Time Astronomy Teachers

If you've taught astronomy before, then you know how much work you must put into course preparation in order for the course to run smoothly and seamlessly for your students. This section is intended to help reduce that workload by giving you some suggestions that we've found useful in preparing our own courses. It is primarily intended for those of you who are new to teaching astronomy; if you've been teaching for many years, they you probably already have your own good systems for course preparation.

What is Teaching All About?

Forgive us for pontificating, but before we get into details of course preparation we'd like to make a general point about teaching:

- This may sound strange, but if you'll think about it you'll realize that you can't "teach" anyone anything. The only way that any of us ever learn is by learning for ourselves. Your job as a teacher is to make learning possible. If you simply try to "pour" facts into student heads, it won't work. But if you motivate them, encourage them to work at learning, and answer their questions, you will be a great teacher.

General Notes on Course Preparation

The broad range of material covered in introductory astronomy courses can be intimidating to teach, especially if your primary scientific training was not in astronomy. (FYI: A large fraction of the people teaching introductory astronomy courses have degrees in physics rather than astronomy.) Here are a few general suggestions that will help you stay ahead of your students and keep your course running smoothly. We apologize if they seem obvious, but for some new instructors they will be useful.

- Be Clear About Your Expectations of Students. Perhaps the single most important thing you can do to help your students achieve success in science is to lay out clearly what it takes for them to succeed in your class. Thus, before your course begins, you should decide what you intend to ask of your students—for example, how much homework you will assign (see p. 37), when exams are scheduled, whether work with Mastering Astronomy tutorials is optional or required, and how you will determine final grades. Once you have made your decisions, you should communicate them clearly to students. This communication can be done in at least three ways:
  - Prepare a syllabus that lays out all your expectations and grading policies clearly. This syllabus can then serve as the single reference for students whenever they have questions about class logistics or if they begin to struggle or fall behind. Appendix 4 offers a sample syllabus from one of our own courses, which you can use or modify as you wish.
  - Spend some time on the first day of class going over your syllabus and making sure that students understand that they should take your expectations seriously.

- Throughout the course, continue to emphasize your expectations, especially to students who are not meeting them. We strongly believe that the vast majority of students will rise to meet your expectations.
  - One caveat: don't be inflexible. You will inevitably encounter a wide variety of issues with different students, and sometimes the expectations you set before the course begins may prove to be too high or too low. Be willing to make adjustments in your expectations if necessary, and remember that your job is not to be perfect but rather to do the best that is possible with the teaching situation into which you are thrown.
- Read the Chapters and Try the Assignments You Give to Students. Although it may seem obvious that you should do what you ask of your students, we've found a depressing fraction of college professors (across all subjects) who, for example, assign reading that they've never read themselves or assign problems that they haven't tried to solve themselves. Clearly, you cannot anticipate student questions or understand the problems they run into if you don't know what they've been reading and working. Ideally, you would read all the assigned chapters before your course begins. Realistically, as some of you may have very little prep time available, you can succeed by skimming the chapters before the course begins and reading them in depth a day or two ahead of when you expect students to have read them.
- Try to Be Consistent With the Text (Especially with Jargon). Students learn best when the reading, assignments, class lectures, and discussions all reinforce one another. This is not to say that you cannot deviate at all from the textbook presentation; it only means that you should strive for as much consistency as possible and that you should acknowledge it when you choose to do things differently. For example:
  - In many cases, it may work quite well if you present a topic in class in a different way than we present it in the book or include a topic that we don't cover at all in the book. However, since students are (hopefully) reading the book, it's important that you let them know when and why you are doing something differently, so that they don't become confused (e.g., asking themselves, "who's right, the book or the prof?"). In other words, it's fine to say, "the authors do it this way, but I prefer this other way," but not fine to simply do the other way without acknowledging that it's different from what they are reading.
  - Consistency of jargon is particularly important. Remember that we have worked hard to reduce jargon in The Cosmic Perspective. If you then use the standard jargon in class, it will only confuse students. For example, if you talk about Type I and II supernovae when we talk about white dwarf and massive star supernovae, you'll lose the intended benefit of the jargon reduction.
- Become Familiar With the Study Resources Available To Your Students. No matter how well you teach or how good our book, many of your students will still have difficulty in your course. Thus, one of your most important jobs as a teacher is to help steer these students to resources that will help them overcome their difficulties. Obviously, you cannot steer them well unless you know what resources are available. In particular, you should familiarize your self with all of the following:
  - The study resources on the MasteringAstronomy™ website, including the tutorials and the on-line quizzes. For many students, making good use of these resources will be all they need to overcome their difficulties with course material.
  - The Addison-Wesley Astronomy Tutor Center. It would be nice in principle if you could hold enough office hours to accommodate all student needs, but in reality it's probably impossible. The Tutor Center, when used effectively, can

be a big help by giving students a way to get help when you are not available. (Remember that your students can have free access to the Tutor Center, but only if you order your books bundled with it. See www.aw.com/tutorcenter or talk to your local Addison Wesley sales representative for more information about the Tutor Center.)
    - Campus resources for students, such as tutoring centers, study skills workshops, and counseling services. Most college campuses offer at least some resources to help students who are having difficulty. Some campuses have tutor centers staffed by undergraduate or graduate students. Many campuses offer study skills workshops for students who don't know how to study efficiently. Most campuses offer counseling services, which may help students whose difficulties stem from issues that may not be directly related to your course. Sadly, the students who most need these services are usually the ones who are least likely to know that they exist. Therefore, you can do your students a great service if you are sufficiently familiar with the available campus resources to point your students in the right direction.
- Stay Current With Astronomical News. Astronomy generates frequent headlines in the news media, and we believe that a major goal of an astronomy course should be to get students sufficiently interested in the subject that they become lifelong followers of the astronomical news. One of the best ways to help generate this enthusiasm is for you to bring current astronomical news into your class. You should follow the news closely for reports of discoveries or missions that may be relevant to your class. It's well worth taking a few minutes out of your planned lecture to report on an exciting new discovery, even if it's in a different area of astronomy than the area you are covering at that moment. Note: The lead author (Bennett) sends out an occasional e-mail with updates on sky events and recent astronomy news. You and your students may subscribe to it by sending an e-mail to: subscribe-spacenews_bennett@lists.awl.com. Put the word "subscribe" (without the quotes) in the body of the message.
- Share Resources and Ideas With Other Instructors. Just as your students should never feel helpless when they run into difficulties, you should never feel like you are alone as a teacher. There are lots of other astronomy teachers out there with lots of great ideas, and there's always some way to share the load. For example:
    - If you teach at a school where multiple instructors are teaching the same course from the same textbook, try to share resources, teaching ideas, and plans for homework assignments or exams. If possible, hold a weekly meeting of all the instructors where you can simply discuss your experiences so that you all get the benefit of each others' insights.
    - If you are the only person teaching the course at your school, seek out others who may have taught the course in the past or teaching colleagues at other schools. Again, it is always valuable to share ideas. Meetings of the American Astronomical Society (AAS) and American Association of Physics Teachers (AAPT) often have sessions on the teaching of astronomy, and you may also find these well worth attending.
    - If you need supplements or have questions or difficulties with any materials for The Cosmic Perspective, you can always contact your local Addison Wesley sales representative. And, again, you can also feel free to e-mail the authors with your questions or comments (see the first page of the IG for e-mail address).

Setting Your Grading Policy

A key part of setting student expectations is making sure that you have a clearly stated grading policy. Setting such a policy can be more difficult than it sounds, and depends in part on the time you have available for grading work. Nevertheless, as a general guideline, we recommend basing grades on some combination of the following:

1. Homework sets consisting of selections from the end-of-chapter problems, assignable work from the tutorials on the MasteringAstronomy™ website, or activities based on the SkyGazer software. We generally recommend homework scheduled on a regular basis, such as weekly or every other week, because it makes it possible for students to budget their time in this and other courses. Of course, homework requires grading, which may limit how much you can assign; please see "Selective Homework Grading" (p. 37) for suggestions on making the grading work manageable.
2. Some set of quizzes and/or exams; e.g., you might choose short weekly quizzes (or completion at home of some of the on-line quizzes), and one or two longer midterms and a final exam.
3. One or more projects, selected from the Web projects on the Media Explorations page of each chapter.

   Note: Be very clear about your policy regarding late assignments and make-ups. For a large class, you may wish to say "no make-ups" of quizzes or exams and simply drop one or two lowest scores—otherwise you might find yourself spending a lot of time giving make-up tests. For homework, you might penalize late assignments with a one grade penalty for each day late. That way students have some leeway for occasional catastrophes, but the penalty is stiff enough to discourage procrastination.

Weighting the Components of the Final Grade

Once you decide what to assign, you must assign a weight to each component in the final grade. Again, the weight will depend on your personal style and resources for grading. However, we have some personal opinions that we'll now share with you:

- In science, we have a tendency to base grades strictly on content knowledge. For example, in a physics course for majors, those who can solve the problems get high grades and those who can't get low grades. However, we believe that in a course for nonscience majors, grades should reflect effort at least as much as content knowledge. The reason is simple: because an introductory astronomy course has no prerequisites, students enter with a wide range of prior content knowledge. Thus, if you base your grades strictly on content knowledge, you may effectively be grading students as much on what they knew before your course as you are on what they know after it. To take an extreme example, we know of many instances in which science or engineering majors have enrolled in introductory astronomy courses, even though these courses are not really intended for them. Not surprisingly, these students can often "ace" the exams, even without studying very much, because they already are adept at science and may have learned much of this material in high school. Clearly, letting these students "skew the curve" would be unfair to those students for whom the course is really intended. By making effort a part of the grade, we not only level the playing field, but we also make sure that even these advanced students must do enough work so that they actually learn something new.

- The best way to reward effort is to make homework and projects count as a substantial fraction of the final grade. This is particularly true of essay-type questions, in which students must write clearly and demonstrate an ability to defend their answers, as opposed to simply getting an answer right or wrong. We like to further reward effort by making the exams closely tied to the homework, so that those who really understood what they were doing on the homework are likely to get good grades on the exams because the questions cover similar concepts.
- Regarding exams: again, in science we have a tendency to ask exam questions that check whether students can "go beyond" the concepts learned in class. This is a great way to pick out those most likely to succeed among our science majors, but we urge you to resist the temptation to ask such questions in nonscience classes. The basic work of learning a science is challenging enough for nonscience students, and we should be quite satisfied if they learn the key concepts presented in the book or in class. This is not say that you should avoid challenging questions—only that the questions should be challenging in the sense of making students think hard about something that they studied, rather than in the sense of asking them to think of something entirely new that was not covered in class.

You can choose for yourself how to balance these ideas in weighting your grades. A policy that we have used is shown on the sample syllabus in Appendix 4.

The Grading Scale

The primary question in setting a grading scale is whether to use a curve or a "straight" scale. There's really no right answer to this question, but our own preferences lean toward a straight scale for one key reason: We feel an obligation to combat the rampant grade inflation on college campuses. With a curve, it's too easy to succumb to student pressure (and these days parent pressure, too) and decide that "average" should be a B or an A– rather than a C. In contrast, if you set a policy that a weighted average of 90% on all assignments constitutes an A–, it is very difficult for anyone to argue with you.

An added advantage of the straight grading scale is that it allows you tell the students straight-faced (no pun intended) that you are willing to give each and every one of them an A—all they have to do is work hard and learn the material. Moreover, if you follow our guidelines above about rewarding effort, it really does become possible for even the most science-phobic students to rise up and get an A (we've seen it happen many times!). Don't worry about this causing grade inflation, though—sadly, not all of your students will put in the necessary effort. You'll end up with an average course grade that makes you look like one of the toughest instructors on campus, while your students will still come away feeling that you gave them a fair chance to earn whatever grade they were aiming for.

Setting Your Homework Policy

As discussed above, we suggest assigning at least some homework unless a lack of grading resources makes it completely impossible. The key issue is deciding precisely what and how much to assign. You have a lot of choices. The book alone has far more end-of-chapter problems than you could reasonably assign to your students, and on top of that you have the option of assignments based on the web tutorials, SkyGazer, and other projects. A good way to decide what to assign is read through the various problems and projects and then decide which ones you'd like to emphasize. When selecting particular problems for assignment, you may wish to consider the following:

- The "How to Succeed" section in the Preface recommends that students in a 3-credit class expect to spend about 2–3 hours per week on homework (i.e., on working problems; not counting time for reading the text or preparing for exams). Thus, we suggest you seek to assign homework that you believe the average student can complete within this 2–3-hour-per-week guideline.
- As a guideline to help you estimate the appropriate length for a homework assignment, we've found that homework will generally take an average student roughly 4 times as long as it takes you. For example, if you can do a set of problems in 1 hour, the same set will take an average student about 4 hours. Of course, this guideline may vary depending on the level of your students and your own speed of work, so adjust accordingly from experience.
- We recommend a mix of easier and harder problems. For example, from a particular chapter you might assign three or four of the "Does it Make Sense?" questions, two or three more extended problems, and one longer essay or Web project.
- Although weekly homework is a good idea in principle, you may choose to give homework somewhat less often so that, for example, homework is not due in the same week as a major midterm.

Finally, to ensure that students turn in their homework in an appropriate fashion (e.g., not done in crayon!), we suggest you make use of our guidelines on "Presenting Homework and Writing Assignments," which appear in Appendix 5.

Selective Homework Grading

How are you going to grade all that homework? This is one of the most challenging questions in teaching, since we rarely have enough resources to grade all the work that we'd like to assign. We therefore recommend that rather than shortening the assignments, you instead use a selective grading strategy. That way students still get the benefit of the learning that homework entails (and homework really is the best way to learn), while you are able to assign grades in a practical and fair way. Our recommended strategy goes like this:

- Tell students in advance that you will grade only a few of the problems that you assign. Although some students will complain about turning in "extra" work that you do not grade, most will recognize that you are assigning this work for their benefit.
- In addition to grading a few problems in detail on each assignment, also assign part of their grade based on a skim of its completeness and presentation. A quick scan is usually all you need to figure out who really put in effort and did a good

job and who didn't. For example, if you say that a homework assignment is worth 10 points, reserve 2 points for this overall "presentation" score while grading 4 other problems for 2 points each. (On a 10-point homework scale, we usually award half-points as needed; thus, the 2 points can essentially become 2, 1.5, 1, 0.5, or 0, which corresponds to A, B, C, D, F.) While the 2 points for "presentation" may seem rather small given that it actually represents most of the work the students turned in (since you grade in detail only a small fraction of the problems), it is a strong motivator for success. For example, even if students guessed correctly which problems you were going to grade in detail, getting a presentation score of 0 for not doing the rest would mean a maximum possible score of 8 on the homework assignment (which is a B– on our straight scale). Thus, students who want a good grade will be forced to work hard on the entire assignment. Moreover, those who put in the effort necessary to get the full 2 points for presentation usually have done all their work carefully and well, so these 2 points help to reward effort (which, as noted earlier, we think is very important).

- Hand out or post detailed solutions to all the problems you assign so that students can check for themselves whether they did ungraded exercises correctly.
- Try to make at least some exam questions directly test concepts from the homework. In that way, students will feel that working the exercises was beneficial because it helped them do better on the exam. This, along with the laws of statistics, will ensure that your homework grading policy is fair even though you do not grade everything.

Homework Help

We encourage you to find ways to provide your students with plenty of homework help on request. Again, because of the varying science backgrounds of students in introductory courses, this is the only way to ensure that those with weak backgrounds still have a chance at success. We have found that one good approach is to offer students unlimited help—with the caveat that they'll get this help only if they have already tried the problems themselves. We go so far as to tell students that, if necessary, we'll lead them step-by-step through everything they must do to write down a correct solution (in reality, it very rarely comes to this point). This policy not only makes students feel confident that they can succeed in your class if they are willing to put in the effort, but also has the effect of helping prevent procrastination. After all, students who wait to start on their homework until the night before it is due will have no time left to take advantage of your offer of help. (Don't forget the availability of the Addison Wesley Astronomy Tutor Center as a source of homework help.)

Setting Your Testing Policy

Studying for tests actually helps students consolidate their learning. Thus, it is a good idea to give quizzes and exams even if you assign extensive homework. Here, we discuss a few considerations that concern testing policy.

Multiple-Choice or Essay?

It's widely acknowledged that short-answer or essay-type questions are a better diagnostic of student understanding of particular concepts than are multiple choice questions. However, this advantage must be weighed against two drawbacks to essay-type exams: (1) They take much longer to grade than machine-readable multiple-choice exams; and (2) because the questions take longer to answer, an essay-type exam cannot be as comprehensive as a multiple-choice exam with the same time limit. Different instructors will of course weigh the pros and cons of essay-type exams differently, and much will depend on the number of students in your class and the availability of grading resources. Again, there is no right answer, and you will come to your own conclusions about what works best for you.

However, we wish to make one general recommendation: If you have only enough grading resources (e.g., just yourself!) for either homework or essay-type exams but not both, we believe that the grading resources are better spent on the homework, which means going with machine gradable multiple-choice exams. Our reason is simple: While essay-type questions provide a better diagnostic of what students have already learned, homework is a way of getting them to learn in the first place. Thus, if you must choose, the option that forces students to do homework is far more likely to result in greater student learning.

Should You Give Short Quizzes?

We have found that, because of the rapid pace at which students encounter new concepts in astronomy courses, it is useful to give frequent quizzes. There are two basic ways in which you can do this:

- Using MasteringAstronomy, you can receive scores when your students complete the on-line quizzes. Because there are quizzes for each chapter in the book, this will ensure that your students are continually studying enough to consolidate their learning. A good strategy is to require that students complete the basic and/or visual quiz for a chapter before you cover that chapter in class, then complete the conceptual quiz for the chapter a day or two after you complete your class coverage of the chapter.
- Alternatively, you might give a weekly multiple-choice quiz designed to take, say, 15–20 minutes. The drawback to this idea compared to use of the on-line quizzes is that it uses up some of your class time. The advantages are that you can cover more than one chapter at once, you can choose your own questions, and you can monitor against potential cheating. Note: if you go this route, be sure to make answers available immediately after the quiz so that students can think about their answers while the quiz is fresh in their minds. (With the on-line quizzes, the feedback is immediate after each question.)

How About Exams?

Quizzing will help your students a great deal, but it still tests only small chunks of material at a time. To really ensure that your students put everything they are learning together, we suggest that you also give at least one longer midterm and a final exam. Personally, we tend to favor two midterms and the final exam, mainly to give students more "practice" at studying. Ideally, the exams should count enough toward the final grade so that students will be forced to study and learn, but no so much as to put overwhelming pressure on those students who may have fears of in-class testing. A few other notes that may be of interest:

- Remember that a midterm is a learning experience while a final is a testing experience. That is, students can learn from their mistakes on a midterm, but when the final is past they are unlikely to ever look at the material again.
- Given the above statement, we have sometimes implemented strategies designed to maximize the learning that students get from their midterm mistakes. For example, we may promise to repeat difficult questions on the final (thereby giving students the message that they really must go back and see what they did wrong).
- A more labor-intensive strategy for midterm learning is to offer what we call "exam rebates." After the graded exam is returned, we offer the students the opportunity to earn back up to half of the points they lost on the exam. To get these "rebate" points, they must write a short essay for each question that they missed in which they not only explain the correct answer but also explain what they now realize was wrong with their original answer. For students, going through this process is a great way to figure out what they have been misunderstanding and to get themselves onto a better learning track. We have had many cases in which students fail the first midterm, but after going through this rebate process they have improved their studying so much that they get A's on subsequent exams. Note that the "up to half back" usually also gives such students an opportunity to overcome their poor scores. For example, a student who fails with a score of 50 (out of 100) can use the rebates to raise their score to 75—making it much easier for them to get a B or A average overall when they do well for the rest of the term. (The question sometimes arises as to whether this offer should be open even to the A students; e.g., do we allow a student who gets a score of 96 to use the rebates to raise their score to 98? Our answer is yes, though we don't go out of our way to encourage it since such a student is clearly doing well anyway.)
- If you are feeling especially generous, you might even offer to replace midterm scores with final scores for students who do better on the final. We don't recommend doing this every term, since otherwise word will get out and students won't study hard for your midterm, but it can be a useful motivator if you happen to have a class that does much worse than you hoped on a midterm. (Note: Some instructors have offered a gambling variation on this theme, in which students tell you in advance (either before the final or just as they turn it in but before it is graded) whether they want to have their midterm score replaced. Those who choose the replacement option get the replacement score whether it is better or worse than the original. Personally, we aren't big fans of gambling in general, but some teachers think this is a good way to go.)

The Test Bank

It's not easy to write good test questions, so we encourage you to take advantage of the Test Bank available with this book (in both printed and computerized form). Why reinvent the wheel when others have already gone to great length to write exam questions? You may use questions from theTest Bank directly, modify them as you wish, or simply use it to get ideas for questions you write yourself.

- Note that all the questions that appear in the on-line quizzes also appear in the Test Bank. You may wish to focus on these if you want to "reward" those students have spent time with the on-line quizzes as a study aid.

The First Day of Class

OK, so you've set all your policies, made your syllabus, and now you are ready to face students on the first day of class. What should you do? We believe that your primary goals on the first day should be: (1) make sure students understand your expectations; and (2) motivate them by starting in on the big picture ideas of Chapter 1. A few notes:

- Be sure not only to go over your syllabus, but also to point students to the section of the book's Preface titled "How to Succeed in Your Astronomy Class." This section emphasizes that students should expect to study 2–3 hours outside class for each hour in class. We have found that most students are not accustomed to this much studying in their classes, but it is crucial to success in science. Thus, your students will be unaware that your class requires this much work unless you tell them so explicitly.
- Let's say the above another way: Tell you students that you can provide them with the key to their success in your course: They must put in real effort. If you go with our guidelines above on grading, students will understand that effort = success, while lack of effort = failure.
- If you have a small class, spent a little time learning student names, and perhaps going around the room so that students can introduce themselves. After all, one of the advantages of a small class is that it can lead to collaborative learning, but collaboration will only occur if the students get to know one another.
- Finally, we have found it very useful to assign something like the following "Assignment 1" on the first day of class. It will not only help students get "up and running" with the text website, but will also provide you with direct student contact and some understanding of what has motivated these students to be in your class.

Suggested First Day Assignment

Feel free to hand this out verbatim or modify as you wish:

Assignment 1 (due by second day of class)

Go to the text website, www.masteringastronomy.com. Spend at least 1 hour exploring the site. After you have completed this exploration, send a single e-mail message to your instructor (or TA). In your e-mail message, you should:

1. Make a brief, clearly organized list that includes your name, student ID number, a telephone number where you can be reached during the semester, your status in school (e.g., freshman, sophomore, etc.), and a list of any prior astronomy classes you have taken (including names of instructors, if you remember them).
2. Write one or two paragraphs telling us what you hope to get out of this class.
3. Write two or three paragraphs briefly describing what you learned from your "Web surfing." What was your favorite feature of the MasteringAstronomy Web site, and why? If you explored links to other sites, which site was your favorite, and why?

Structuring Class Sessions

After the first day, you can and should expect students to come to class prepared. The structure of each class period will depend on how frequently you meet and how long the class lasts. Nevertheless, we recommend a few general strategies:

- Remember that students may be coming to your class with other things on their mind. Help them get on track by spending a few minutes at the beginning of each class reviewing where you left off the last time. Then save the last couple of minutes of class to review the topics covered that day (which helps them consolidate what they've just learned) and to tell students what will be covered next time (which helps them be prepared for the next class).
- Expect your students to do the assigned reading, and show that you expect it by avoiding direct repetition of what is in the book during your lectures; focus your class time on the more difficult concepts. Note: You can reinforce the need for your students to come prepared by requiring them to complete the on-line basic quiz for a chapter before you begin covering the chapter in class. If you really want to encourage early studying, you might allow multiple attempts to get the best possible score—but only up to the time that class begins. This gives even weak students a chance to get a high score, as long as they start early enough.
- Consider trying new techniques to make your lectures more interactive. A few such techniques are discussed in the next section of this Instructor's Guide.
- Show your students that you care. Expect them to attend class and participate in any discussions or activities. Constantly remind them that help is available if they need it. If you have a student who misses class or an assignment, call the student to ask if everything is OK and, at least the first time, offer a way for them to make up the missed work. You'll be amazed at how quickly you can turn around some students with the simple act of a phone call that makes them feel that you care.

Evaluating Your Teaching

We recommend that teachers always ask their students for feedback on what is working and not working in their teaching, and this is especially important in science classes where material builds over the course of the term. There are many ways to get this feedback. Here are two evaluation strategies that have worked well for us.

Post-Lecture Evaluations

This evaluation exercise asks for anonymous feedback at the end of a class period. It is best tried after you've "hit your stride" in the class, but early enough in the term so that you can still incorporate what you learn from the feedback—e.g., you might try it after a class about a third of the way into your course, then repeat it around the two-thirds point. Here's how it works:

Stop your class a couple of minutes before the end of period and ask students to briefly answer the following three questions before they leave. You can either ask them to write on a piece of scratch paper or hand out a prepared form. Either way, be sure that the feedback is anonymous:

1. What constructive criticism do you have on how this class is taught?
2. What's good about the way this class is taught?
3. What's the main thing you learned in today's class?

Note: You might think that only the first question is important, but the second gives a more balanced view of your teaching, and the third offers a valuable insight into whether students got what you had intended out of that day's class.

Feedback on Homework

We have also found it very useful to ask students for feedback that they can turn in with their homework assignments. We do this by including the following optional question at the end of each homework assignment:

Question X. Comments (please answer, but this will not be graded): How long did this homework assignment take you? Please comment on the assignment and the class in general. For example: Do you feel you are understanding the material? Do you feel that what you are learning will be beneficial to you? Other comments or suggestions?

If you feel that anonymity is the only way to get honest responses, you can ask students to turn in their answers to this question separately from their graded papers. However, we've found that in many cases we can get more useful information by having the answers with the graded papers. For example, it's useful to know whether those who said the homework was easy were really the ones who did well. In addition, we've often found that we can write personal notes back to the students on their papers, thereby helping them understand specific difficulties that they called out in their comments. Moreover, we've found that that lack of anonymity is usually not a deterrent to honest comments when the comments are associated with homework—perhaps because they are already "exposing" themselves with their written assignment, students often are quite willing to tell you exactly how they feel about all the work they just completed.

Notes on Interpreting Evaluations

We all put a great deal of effort into teaching and would like our evaluations to show an appreciation for all we do for the students, so it can be very disappointing when student evaluations tell you that you're not the greatest teacher in the history of the planet—or something even worse! This can be especially true if you are new to teaching, and thus have not yet developed the thick skin of old-timers. We therefore offer a few suggestions on how to get the most out of your teaching evaluations:

- First and perhaps most important, remember these are comments on a task to which you have been assigned and NOT comments on you as a person. Especially if you are new to teaching, the personality that comes across in class may be very different from the personality that your friends see. If the students hate you, it just means you may want to do some things differently next time. It does not make you a bad person or even a bad teacher—just a teacher with room for improvement.
- Students will write all kinds of strange things on course evaluations, so use a filter to decide which comments truly represent constructive criticism and which comments can be safely ignored. For example, students may complain about class being too early in the morning, which is clearly not your fault. Or if ten students say you gave just the right amount of homework and one says you gave too much, the one may just have had too big a course load. (We've even seen complaints like "the professor made me think too much," which you might take as a compliment!) A good guideline can be the amount that students write—those who make the effort to give you extensive comments usually have something worthwhile to say.

- Develop your thick skin: No matter how well you teach, some students will write horrible things about you that you don't deserve. This may be because they just didn't "click" with you, or because they happened to be having a bad day when you handed out the course evaluation, or because they are just mean people. Regardless, remember that the point of the evaluations is to help you improve your teaching. If you can extract some constructive criticism from the negative comments, then great—but if they only make you feel bad, then you should feel free to ignore them.
- As a corollary to the above, remember that it's not easy to get positive comments, especially in large classes that students take only to fulfill a requirement. Students have a lot going on in their lives. If someone says that they loved your class, it means you have succeeded in breaking through the "noise" of everyday life to touch that person in a very deep way. If even a modest fraction of the class rates you highly, you've done a great job. (If you manage to get half your students saying they loved your class, you are the teacher of the year.) Allow yourself to feel good about the positive comments, even while you try to learn from the negative ones so that you can be even better the next time.

Suggestions on Making Your Lectures Interactive

We're now ready to discuss how you can attempt to break out of "lecture mode," changing learning from a passive to an active experience. This is particularly important in large classes, but the methods described below are also effective in smaller classes. Some of these ideas are formally developed pedagogical tools with published references, while others are just simple ideas that may keep your students engaged during class. It's not possible (or recommended) to implement many of these at once, nor will all suggestions match an individual instructor's teaching style. We encourage you to read through the ideas and pick out the ones that will help you most. We hope that they will make class time more enjoyable for you and your students.
Note: For additional ideas on the user of clickers, see *Clickers in the Astronomy Classroom*, by Douglas Duncan ISBN 0-8053-9616-0 (available from your Addison Wesley sales representative).

Basic Interactions

You may have noticed the following curious effect: The more students you have in your class, the fewer actively participate in basic question-and-answer interactions. Although you may ask "Are there any questions?" you've probably noticed that few take advantage of the opening. And when you ask a question, only a few hands come up, and they're always the same ones. Statistically speaking, they're not likely to be female or minority students. Here we offer a few suggestions on how to improve simple interactions without changing your class structure very much.

- First, note that success in all interactive teaching depends on keeping control over your classroom—it is important that students do not become rowdy or discourteous. To help you keep control over your classroom, we suggest that you emphasize the importance of the guidelines on common courtesy discussed in Appendix 4 with our sample syllabus.
- One simple way to involve the whole class is to convert each question into a yes/no or multiple-choice question and ask students to vote for their answer. For example, ask "Does the full moon rise at sunrise, noon, sunset, or midnight?" A vote (by show of hands, color-coded cards, or electronic transmitter) is a good way to tell whether students have understood a recent topic well enough for you to move on. It also forces the students to think and participate to some degree.
- Regarding the tally of votes as in the above bullet: There are also new technologies to allow students to interact in class with electronic transmitters. Student participation can thereby be recorded and graded, with very positive effects on participation, attendance and understanding. An example of this approach is described in a short article and set of view graphs compiled by TCP author Nick Schneider. These are posted and available for download at http://ganesh.colorado.edu/nick/TeachTech1.pdf and http://ganesh.colorado.edu/nick/TeachTech2.pdf.
- You can expand the discussion by calling on students who have indicated one opinion or another to explain their viewpoints. Another way of involving a larger

fraction of the class is calling on students instead of waiting for volunteers. If you use this method, be sure to read "Avoiding Intimidation" below.

- Even the physical action of raising hands is better than just listening. Try other techniques to involve the students physically. Ask them to stand for yes votes instead of raising hands. Have them do simple demonstrations whenever possible (e.g., demonstrate parallax by asking them to hold their hands out with an extended index finger at arm's length). If you are conducting a demonstration at the front of a large lecture hall, give them the opportunity to leave their seats and file by so they can see the demonstration up close.
- Nonscience majors are keenly aware that instructors are often looking for the one "right answer," and doubt that they actually know it. You can break the ice in classroom discussions by including topics that have more than one answer. Examples range from opinion questions (e.g., "Do you think there's other intelligent life in the universe?" "How much is it worth spending on space exploration?") to lists (e.g., "What can you learn about a star with telescopes?" "What kinds of geological features are present on Venus?"). With practice, general questions like these can extend into more challenging and interesting questions, such as the Discussion Questions provided at the end of each chapter.
- Perhaps the most important point to keep in mind is that good discussion requires good questions. Try to spend a few minutes before each class period coming up with good questions. The Think About It features in our textbook are often excellent starting points and have the advantage that students who have done the reading will already have some answers in mind.

Collaborative Learning

The unifying theme behind the suggestions that follow is challenging students to work through thought-provoking questions, to discuss them with their neighbors, and to gain confidence and practice in working with new ideas.

- You can make classroom questioning even more interactive by encouraging students to work together to determine the right answer. Allow students to break into discussion groups of two or three for a couple of minutes. This gets everyone talking, often arguing for their answer in an animated way. Students learn to discuss scientific issues and end up learning from their peers. Reconvene the class with some prearranged signal like dimming the lights. You can either call for another "vote" for the right answer or poll individual groups to get an explanation for the answer they chose. One version of this approach has been carefully designed and tested under the names ConcepTests and Peer Instruction; the book and website are listed below under "Resources on Interactive Teaching."
- A variation on this method uses index cards for student answers. Students can exchange the cards to learn from their neighbors. The instructor can then ask students if the card they are holding has an interesting or useful answer. The exchange of cards ensures anonymity (see "Avoiding Intimidation" below) and usually brings out more thoughtful (or amusing) responses than students will volunteer on their own. Alternatively, students can collaboratively answer questions on worksheets distributed in class.
- Another variant uses demonstrations as the focal point of discussion. Specifically, a well-designed, thought-provoking demonstration is set up and described—but not performed. (Galileo's classic test of dropping balls of different masses is a

good example.) Students then discuss in small groups what they think will happen and report their predictions. The demonstration is then performed, and students reconcile the results with their predictions (if necessary).

Avoiding Intimidation

Perhaps the biggest obstacle to classroom participation—particularly in large lecture classes—is intimidation. Even students who know the right answer or who have a genuinely intelligent question may not feel comfortable speaking in front of a few hundred fellow students. Many of the approaches described above are designed to eliminate this problem. The small group breakouts are particularly effective. First, almost all students will speak in small groups and have their point of view validated or corrected before being called on. Second, asking "What did your group conclude?" takes the burden off the individual—the credit or blame is shared with those around you. Third, frequent use of any of the techniques described here will show that participation and active learning are more important than looking smart. Here are a few other notes:

- Be aware that body language can also set the tone for discussions. Avoid looking impatient or disapproving. When calling on students to speak, use an open-handed inviting gesture instead of pointing with the index finger. Also be aware that student responses may not be audible throughout the classroom; repeat the main points of student responses if necessary.
- Make it clear that there is no such thing as a dumb question. Make some attempt to answer all questions; if a student asks a question that you don't understand, ask him/her to clarify the question in a polite way. If you still don't understand it, help students avoid embarrassment by suggesting that they come talk to you about their questions after class; say something like "I see what you are getting at, but I don't want to take class time right now. Can you talk to me about it after class?" Note: If you are short on time, it is better to say that you cannot take any more questions than to give unsatisfactory answers.
- On occasion you may ask students a question that they get utterly wrong. While they may be in good company, the embarrassment they feel in class may have a long-lasting effect. Some students may even stop attending class if they fear further embarrassment. It's therefore important to help students to a better ending, even if it takes a while. One of the best ways to take the embarrassment away from the student is to take the blame on yourself, the instructor. Statements like "Perhaps I worded my question in a roundabout way" or "That's the right approach if you just consider ..." will take the sting off the student. Then back up to a related question that you are certain the student will get right. Then, you can move forward with that student (if you think he/she can get the concept) or move on to another student. Emphasizing the importance of working out the right answer (as opposed to already knowing it) will help in the long run. The suggestions of the following section will also make you, the instructor, less threatening and more human.

Personalizing the Impersonal Classroom

Lecture mode tends to distance the instructor from the students, but there are a number of ways to bridge that gap.

- Do what you can to bring your audience close to you:

    - In large lecture halls, ask students to move forward so you don't have students concentrated in the back with lots of empty seats near the front.
    - In smaller classrooms, ask students to rearrange furniture to form a semicircle so they can see each other as well as you.
- In any lecture period there is probably at least one common experience that is relevant to the lecture topic. You can draw students into discussion by asking "Have you been watching the Moon recently? Did anybody notice what phase it's in?" Or "Have any of you ever traveled to the Southern Hemisphere?" Or "How many of you caught the latest picture from Hubble on the front page?" After you see a few nodding heads, ask someone to expand on what they have seen, read, or experienced. Even small interactions like these familiarize students with talking in class and demonstrate the relevance of their experience and interests to the class material.
- Most of us are so busy that we rush into class just before the period starts and rush out afterward. But the few minutes before and after class are often the times when students are most in need of your time—and not just for homework answers. If at all possible, arrive a few minutes before class and stay long enough to answer questions afterward. Striking up brief conversations not related to the course offers another way to set students at ease talking to you. (If you have a decent sound system in your lecture hall, you might try playing some music in the few minutes before class; it will help students relax, and the end of the music will signal the beginning of class.)
- Many instructors are surprised at how much help students need in their course, but how little help they seek. If you wish more people took advantage of your office hours, try renaming them "review sessions" on your syllabus. Consider holding one of your office hour periods on "neutral ground" in a cafeteria or common area.
- Students often feel isolated in large classes, which makes them less likely to participate in discussions. Encourage them to learn their neighbors' names early in the semester. (The collaborative learning method described above offers a natural means.) Help students organize study groups, for example, by encouraging interested students to meet in front after the lecture. If feasible, have your students write or e-mail you a few sentences about themselves, their interests, and what they hope to get out of the class. (See "A Suggested First-Day Assignment," page 41.)
- Our final suggestion involves the daunting task of learning student names. While there are legendary instructors who have memorized classes of hundreds, most of us cannot. But chances are that you can learn a lot more names than you think if the information is presented to you in a usable form: names and faces together. One convenient method requires a digital camera and just a few minutes of class time: Have your students write their names on pieces of paper with a magic marker. Parade your students past the camera and take their pictures while they hold their papers in front of them. With more than one camera, you can enlist student help to take pictures even faster. Most cameras come with software that allows you to print many small pictures per page. You can also combine the pictures into a movie file that you can step through frame by frame. If you don't have a digital camera, a camcorder can be used in a similar way. After only a few times looking over the pictures, you'll be surprised at how many students you know by name. It's possible to learn classes of up to 100 students with this method with less than an hour's total effort. For smaller classes, you can take

group photos with an instant camera and have the students sign their names. Some instructors prefer to use a seating chart. Try whatever method works for you. Calling on students by name has a remarkably positive effect on the tone of the class.

Resources on Interactive Teaching

Clickers in the Astronomy Classroom, by Douglas Duncan, published by Addison Wesley, ISBN 0-8053-9616-0.

Peer Instruction, by Eric Mazur, published by Prentice Hall, ISBN 0-13-565441-6. http://galileo.harvard.edu/.

Mysteries of the Sky: Activities for Collaborative Groups, by Adams and Slater, Kendall Hunt, ISBN 0-7872-5126-7. http:/www/montana.edu/~wwwph/research/phys_ed.html.

Cooperative Learning Activities in Introductory Astronomy for Non-science Majors, by Deming, Miller, and Trasco, available from the author. Contact Grace Deming, Dept. of Astronomy, U. Maryland, College Park, MD 20742.

Science Teaching Reconsidered, National Research Council, National Academy Press, ISBN 0-309-05498-2.

Chapter-by-Chapter Guide

Part I: Developing Perspective

The remainder of this Instructor's Guide goes through the book chapter by chapter. Within each chapter, it is organized as follows:

- A brief introduction with general comments about the chapter.
- "What's New in the Fourth Edition That Will Affect My Lecture Notes?" This short section is aimed at those who may have notes from teaching with past editions of our book.
- Teaching Notes. Organized section by section for the chapter, these are essentially miscellaneous notes that may be of use to you when teaching your course.
- Answers/Discussion Points for Think About It Questions.
- Solutions to End-of-Chapter Problems.

Chapter 1. Our Place in the Universe

The purpose of this first chapter is to provide students with the contextual framework they need to learn the rest of the course material effectively: a general overview of our cosmic address and origins (Section 1.1), an overview of the scale of space and time (Section 1.2), and an overview of motion in the universe (Section 1.3). We often tell students that, after completing this first chapter, they have essentially learned all the major ideas of astronomy, and the rest of their course will be building the detailed scientific case for these general ideas.

As always, when you prepare to teach this chapter, be sure you are familiar with the relevant media resources (see the complete, section-by-section resource grid in Appendix 3 of this Instructor's Guide) and the online quizzes and other study resources available on the Mastering Astronomy Web site.

What's New in the Fourth Edition That Will Affect My Lecture Notes?

As everywhere in the book, we have added learning goals that we use as subheadings, rewritten to improve the text flow, improved art pieces, and added new illustrations. The art changes, in particular, will affect what you wish to show in lecture. We have not made any substantial content or organizational changes to this chapter.

- The Mathematical Insights for Chapter 1 are significantly different from those of the prior edition. This chapter now contains four Mathematical Insights boxes that, together, provide students with an overview of basic mathematical techniques that they can use throughout the book: an overview of problem solving; an introduction to unit analysis, scientific notation, and order of magnitude estimation; and the use of simple formulas.

Teaching Notes (By Section)

Section 1.1 Our Modern View of the Universe

This section provides a brief overview of our modern view of the universe, including the hierarchical structure of the universe (our cosmic address) and the history of the universe (our cosmic origins).

- We urge you to pay special attention to the two full-page paintings (Figures 1.1 and 1.2). These pieces should help your students keep our cosmic address and origins in context throughout the course, and you may wish to refer back to them often.
- Note the box on "Basic Astronomical Objects, Units, and Motion": Although some of the terms in this box are not discussed immediately, having them here in the beginning of the book should be helpful to students. All these terms also appear in the Glossary, but they are so basic and important that we want to emphasize them here in Chapter 1.
- Note that we've chosen to use *light-years* rather than *parsecs* as our primary unit for astronomical distances for the following three reasons:
  1. We have found that light-years are more intuitive than parsecs to most students because light-years require only an understanding of light travel times, and not of the more complex trigonometry of parallax.
  2. Lookback time is one of the most important concepts in astronomy, and use of light-years makes it far easier to think about lookback times (e.g., when a student hears that a star is 100 light-years away, he/she can immediately recognize that we're seeing light that left the star 100 years ago).
  3. Fortuitously, 1 light-year happens to be very close to 10^{13} kilometers (9.46×10^{12} km), making unit conversions very easy—this helps students remember that light-years are a unit of distance, not of time.
- FYI: The 2.5-million-light-year distance to the Andromeda Galaxy is based on results reported by K. Stanek and P. Garnavich in *Astrophysical Journal Letters*, 20 August 1998 (503, L131). They give the distance to Andromeda as 784 kpc, with a statistical error of ±13 and a systematic error of ±17. This distance is based on Hipparcos distances of red clump (helium core–burning) stars in the Milky Way and Hubble observations of red-clump stars in Andromeda.
- We give the age of the universe as "about 14 billion years" based on the recent WMAP results (http://map.gsfc.nasa.gov/). The WMAP results are consistent with an age of 13.7 billion years with a 1 sigma error bar of 0.2 billion years.

Section 1.2 The Scale of the Universe

We devote this section to the scale of space and time because our teaching experience tells us that this important topic generally is underappreciated by students. Most students enter our course without any realistic view of the true scale of the universe. We believe that it is a disservice to students to teach them all about the content and physics of the universe without first giving them the large-scale context.

- The "walking tour of the solar system" uses the 1-to-10-billion scale of the Voyage scale model solar system in Washington, D.C., a project that was proposed by *The Cosmic Perspective* author Bennett. Voyage replicas are being developed for other science centers; if you are interested in learning more about how to get a Voyage replica in your town, please contact the author. (The same scale is also used in the Colorado Scale Model Solar System in Boulder.)
- With regard to the count to 100 billion, it can be fun in lecture to describe what happens when you ask children how long it would take. Young children inevitably say they can count much faster than one per second. But what happens when they get to, say, "twenty-four billion, six hundred ninety-seven million, five hundred sixty-two thousand, nine hundred seventy-seven . . ."? How fast can they count now? And can they remember what comes next?
- Regarding our claim that the number of stars in the observable universe is roughly the same as the number of grains of sand on all the beaches on Earth, here are the assumptions we'vemade:
  - We are using 10^{22} as the number of stars in the universe. Assuming that grains of sand typically have a volume of 1 mm^3 (correct within a factor of 2 or 3), 10^{22} grains of sand would occupy a volume of 10^{22} mm^3, or 10^{13} m^3.
  - We estimate the depth of sand on a typical beach to be about 2–5 meters (based on beaches we've seen eroded by storms) and estimate the width of a typical beach at 20–50 meters; thus, the cross-sectional area of a typical beach is roughly 100 m^2.
  - With this 100 m^2 cross-sectional area, it would take a length of 10^{11} meters, or 10^8 kilometer, to make a volume of 10^{13} m^3. This is almost certainly greater than the linear extent of sandy beaches on Earth.
- The idea of a "cosmic calendar" was popularized by Carl Sagan. Now that we've calibrated the cosmic calendar to a cosmic age of 14 billion years, note that 1 average month = 1.17 billion years.
- You may want to make note of the new Special Topic box that discusses the "Planet X" discovery. It provides a good example of how rapidly things can change in astronomy, and also gets students thinking early about the definition of *planet.*

Section 1.3 Spaceship Earth

This section completes our overview of the "big picture" of the universe by focusing on motion in the context of the motions of the Earth in space, using R. Buckminster Fuller's idea of *spaceshipEarth* .

- There are several different ways to define an average distance between the Earth and the Sun (e.g., averaged over phase, over time, etc.). In defining an AU, we use the term *average* to mean (perihelion + aphelion)/2, which is equivalent to the semimajor axis. This has advantages when it comes to discussing Kepler's third law, as it is much easier for students to think of a in the equation $p^2 = a^3$ as *average* than as *semimajor axis.*
- We use the term *tilt* rather than *obliquity* as part of our continuing effort to limit the use ofjargon.

- We note that universal expansion generally is not discussed until very late in other books. However, it's not difficult to understand through the raisin cake analogy; most students have heard about it before (though few know what it means); and it's one of the most important aspects of the universe as we know it today. Given all that, why wait to introduce it?

Section 1.4 The Human Adventure of Astronomy

Although the philosophical implications of astronomical discoveries generally fall outside the realm of science, most students enjoy talking about them. This final section of Chapter 1 is intended to appeal to that student interest, letting them know that philosophical considerations are important to scientists as well.

- FYI: Regarding the Pope's apology to Galileo, the following is a quotation from *Time* magazine, December 28, 1992:

  > Popes rarely apologize. So it was big news in October when John Paul II made a speech vindicating Galileo Galilei. In 1633 the Vatican put the astronomer under house arrest for writing, against church orders, that the earth revolves around the sun. The point of the papal statement was not to concede the obvious fact that Galileo was right about the solar system. Rather, the Pope wanted to restore and honor Galileo's standing as a good Christian. In the 17th century, said the Pope, theologians failed to distinguish between belief in the Bible and interpretation of it. Galileo contended that the Scriptures cannot err but are often misunderstood. This insight, said John Paul, made the scientist a wiser theologian than his Vatican accusers. More than a millennium before Galileo, St. Augustine had taught that if the Bible seems to conflict with "clear and certain reasoning," the Scriptures obviously need reinterpretation.

Answers/Discussion Points for Think About It Questions

The Think About It questions are not numbered in the book, so we list them in the order in which they appear, keyed by section number.

Section 1.1

- (p. 2) This question is, of course, very subjective, but can make for a lively in-class debate.
- (p. 8) If people are looking from the Andromeda Galaxy at the Milky Way, they would see a spiral galaxy looking much like their galaxy looks to us. They would see our galaxy as it was about 2.5 million years ago (due to light travel time) and thus could not know that our civilization exists here today.

Section 1.2

- (p. 9) This is another very subjective question, but it should get students thinking about the size of Earth in the cosmos. At the least, most students are very surprised at how small our planet seems in relation to the solar system. For most students, it makes Earth seem a little more fragile, and often makes them think more about how we can best take care of our planet.
- (p. 14) This question also can be a great topic of debate. We've found that most students tend to think it is inconceivable that we could be the only intelligent beings. However, some religious students will assume we are alone on grounds of

their faith. In both cases, it can generate discussion about how science goes only on evidence. For example, we don't assume there are others because we have no evidence that there are, and we don't assume we are alone for the same reason.

Section 1.3

- (p. 16) As we authors understand it, the only real reason that globes are oriented with north on top is because most of the early globe makers lived in the Northern Hemisphere. In any case, it is certainly equally correct to have the globe oriented in any other way.
- (p. 18) This question is easy to discuss if you refer to the 1-to-10-billion scale model developed earlier. On this scale, entire star systems are typically only a few hundred meters in diameter (including all their planets), while they are separated from other systems by thousands of kilometers (at least in our vicinity of the galaxy).

Solutions to End-of-Chapter Problems (Chapter 1)

1. A geocentric universe is one in which the Earth is assumed to be at the center of everything. In contrast, our current view of the universe suggests that Earth is a rather ordinary planet orbiting a rather ordinary star in an ordinary galaxy, and there is nothing "central" about Earth at all.
2. The largest scale is the universe itself, which is the sum total of all matter and energy. The largest-known organized structures are superclusters of galaxies, then clusters and groups of galaxies, and then the roughly 100 billion individual galaxies, most of which are many thousands of light-years across. Each galaxy contains billions of stars, and many or most stars may be orbited by planets.
3. When we say that the universe is expanding, we mean that the average distance between galaxies is increasing with time. If the universe is expanding, then if we imagine playing time backward, we'd see the universe shrinking. Eventually, if we went far enough back in time, the universe would be compressed until everything were on top of everything else. This suggests that the universe may have been very tiny and dense at some point in the distant past and has been expanding ever since. This beginning is what we call the Big Bang.
4. Most of the atoms in our bodies (all the elements except for hydrogen, since our bodies generally do not contain helium) were made by stars well after the Big Bang. So most of the stuff in our bodies was once part of stars.
5. Light travels at 300,000 kilometers per second. A light-year is the distance that light travels in 1 year, which is about 9.46 trillion kilometers.
6. Because light travels at a fixed speed, it takes time for it to go between two points in space. Although light travels very quickly, the distances in the universe are so large that the time for light to travel between stars is years or longer. The farther away we look, the longer it takes light to have traveled to us from the objects. So the light we see from more distant objects started its journey longer ago. This means that what we see when we look at more distant objects is how they looked longer ago in time. So looking farther away means looking further back in time.

7. The observable universe is the portion of the entire universe that we can, in principle, see; it is presumably about 14 billion light-years in radius, because light from more than 14 billion light-years away could not yet have reached us during the 14 billion years since the Big Bang. Scientists currently think that the entire universe is larger than the observable universe.
8. On the 1-to-10-billion scale, the Sun is about the size of a grapefruit and the planets are the sizes of marbles or smaller. The distances between the planets are a few meters for the inner solar system to many tens of meters in the outer solar system. On the same scale, the nearest stars are thousands of kilometers away.
9. One way to understand the size of our galaxy is to note that if the Milky Way were the size of a football field, then the distance to the nearest star would be about 4 millimeters. One way to get a sense of the size of the observable universe is to note that the number of stars in it is comparable to the number of grains of sand on all of the beaches on the entire planet Earth.
10. One thing that you could tell your friend to give him or her a sense of the age of the universe compared to the time that humans have been around is that if the entire history of the universe were compressed into a single year, modern humans would have evolved only 2 minutes ago and that the pyramids would have been built only 11 seconds ago.
11. Astronomical Unit: The average distance between the Earth and Sun, which is about 1.496×10^8 kms.

    Ecliptic Plane: The two-dimensional plane in which Earth orbits around the Sun. Most of the other planets orbit nearly in this same plane.

    Axis Tilt: The amount that a planet's rotation axis is tipped relative to a line *perpendicular* to the ecliptic plane.
12. The Milky Way Galaxy is a spiral galaxy, which means that it is disk-shaped with a large bulge in the center. The galactic disk includes a few large spiral arms. Our solar system is located about 28,000 light-years from the center of the galaxy, or about halfway out to the edge of the galaxy. Our solar system orbits about the galactic center in a nearly circular orbit, making one trip around every 230 million years.
13. The disk of the galaxy is the flattened area where most of the stars, dust, and gas reside. The halo is the large, spherical region that surrounds the entire disk and contains relatively few stars and virtually no gas or dust. Dark matter resides primarily in the halo.
14. Edwin Hubble discovered that most galaxies are moving away from our galaxy, and the farther away they are located, the faster they are moving away. While at first this might seem to suggest that we are at the center of the universe, a little more reflection indicates that this is not the case. If we imagine a raisin cake rising, we can see that every raisin will move away from every other raisin. So each raisin will see all of the others moving away from it, with more distant ones moving faster—just as Hubble observed galaxies to be moving. Thus, just as the raisin observations can be explained by the fact that the raisin cake is expanding, Hubble's galaxy observations tell us that our universe is expanding.
15. *Our solar system is bigger than some galaxies.* This statement does not make sense, because all galaxies are defined as collections of many (a billion or more) star systems, so a single star system cannot be larger than a galaxy.

16. *The universe is billions of light-years in age.* This statement does not make sense because it uses the term "light-years" as a time, rather than as a distance.
17. *It will take me light-years to complete this homework assignment.* This statement does not make sense, because it uses the term "light-years" as a time, rather than as a distance.
18. *Someday, we may build spaceships capable of traveling a light-year in only a decade.* This statement is fine. A light-year is the distance that light can travel in 1 year, so traveling this distance in a decade would require a speed of 10% of the speed of light.
19. *Astronomers recently discovered a moon that does not orbit a planet.* This statement does not make sense, because a moon is defined to be an object that orbits a planet.
20. *NASA soon plans to launch a spaceship that will photograph our Milky Way Galaxy from beyond its halo.* This statement does not make sense, because of the size scales involved: Even if we could build a spaceship that traveled close to the speed of light, it would take tens of thousands of years to get high enough into the halo to photograph the disk, and then tens of thousands of years more for the picture to be transmitted back to Earth.
21. *The observable universe is the same size today as it was a few billion years ago.* This statement does not make sense, because the universe is growing larger as it expands.
22. *Photographs of distant galaxies show them as they were when they were much younger than they are today.* This statement makes sense, because when we look far into space we also see far back in time. Thus, we see distant galaxies as they were in the distant past, when they were younger than they are today.
23. *At a nearby park, I built a scale model of our solar system in which I used a basketball to represent the Earth.* This statement does not make sense. On a scale where Earth is the size of a basketball, we could not fit the rest of the solar system in a local park. (A basketball is roughly 200 times the diameter of Earth in the Voyage model described in the book. Since the Earth-Sun distance is 15 meters in the Voyage model, a basketball-size Earth would require an Earth-Sun distance of about 3 kilometers, and a Sun-Pluto distance of about 120 kilometers.)
24. *Because nearly all galaxies are moving away from us, we must be located at the center of the universe.* This statement does not make sense, as we can tell when we think about the raisin cake model. Every raisin sees every other raisin moving away from it, so in this sense no raisin is any more central than any other. (Equivalently, we could say that every raisin—or galaxy—is the center of its own observable universe, which is true but very different from the idea of an absolute center to the universe.)
25. a; 26. b; 27. c; 28. b; 29. c;
30. b; 31. a; 32. a; 33. b; 34. a.
35. This is a short essay question. Key points should include the fact that we are made of elements forged in past generations of stars and that those elements were able to be brought together to make our solar system because of the recycling that occurs within the Milky Way Galaxy.

36. This is a short essay question. Key points should include discussion of the difference in scale between interstellar travel and travel about our own world, so that students recognize that alien technology would have to be far more advanced than our own to allow them to visit us with ease.

37. There is no danger of a collision between our star system and another in the near future. Such collisions are highly improbable in any event—remember that our Sun is separated from the nearest stars like grapefruits spaced thousands of miles apart. Moreover, we can observe the motions of nearby stars, and none of them are headed directly our way.

38. a. The diagrams should be much like Figure 1.16, except that the distances between raisins in the expanded figure will be 4 centimeters instead of 3 centimeters.

b.

Distances and Speeds of Other Raisins as Seen from the Local Raisin			
Raisin Number	**Distance Before Baking**	**Distance After Baking (1 hour later)**	**Speed**
1	1 cm	4 cm	3 cm/hr
2	2 cm	8 cm	6 cm/hr
3	3 cm	12 cm	9 cm/hr
4	4 cm	16 cm	12 cm/hr
M	M	M	M
10	10 cm	40 cm	30 cm/hr
M	M	M	M

c. As viewed from any location inside the cake, more distant raisins appear to move away at faster speeds. This is much like what we see in our universe, where more distant galaxies appear to be moving away from us at higher speeds. Thus, we conclude that our universe, like the raisin cake, is expanding.

39. a. This problem asks students to draw a sketch. Using the scale of 1 cm = 100,000 light-years, the sketches should show that each of the two galaxies is about 1 centimeter in diameter and that the Milky Way and M 31 are separated by about 25 centimeters.

b. The separation between the Milky Way and M 31 is only about 25 times their respective diameters—and other galaxies in the Local Group lie in between. In contrast, the model solar system shows that, on a scale where stars are roughly the size of grapefruits, a typical separation is thousands of kilometers (at least in our region of the galaxy). Thus, while galaxies can collide relatively easily, it is highly unlikely that two individual stars will collide. *Note:* Stellar collisions are more likely in places where stars are much closer together, such as in the galactic center or in the centers of globular clusters.

40. This is a subjective essay question. Grade should be based on clarity of the essay.

41. a. A light-second is the distance that light travels in 1 second. We know that light travels at a speed of 300,000 km/s, so a light-second is a distance of 300,000 kilometers.

b. A light-minute is the speed of light multiplied by 1 minute:

$$\begin{aligned}1\text{ light-minute} &= (\text{speed of light}) \times (1\text{ min})\\ &= 300{,}000\frac{\text{km}}{\cancel{\text{s}}} \times 1\ \cancel{\text{min}} \times \frac{60\ \cancel{\text{s}}}{1\ \cancel{\text{min}}}\\ &= 18{,}000{,}000\text{ km}\end{aligned}$$

That is, "1 light-minute" is just another way of saying "18 million kilometers."

c. Following a similar procedure, we find that 1 light-hour is 1.08 billion kilometers; and

d. 1 light-day is 2.59×10^{10} kms, or about 26 billion kilometers.

42. Recall that

$$\text{speed} = \frac{\text{distance traveled}}{\text{time of travel}}$$

We can rearrange this with only a little algebra to solve for time:

$$\text{time of travel} = \frac{\text{distance traveled}}{\text{speed}}$$

The speed of light is 3×10^5 km/s according to Appendix A. (We choose the value in km/s rather than m/sec because looking ahead we see that the distances in Appendix E are in kilometers.)

a. According to Appendix E, the Earth-Moon distance is 3.844×10^5 km. Using this distance and the equation above for travel time, we get

$$\text{time of travel} = \frac{3.844 \times 10^5\ \cancel{\text{km}}}{3.00 \times 10^5\ \cancel{\text{km}}/\text{s}} = 1.28\text{ s}$$

Light takes 1.28 seconds to travel from the Moon to Earth.

b. Appendix E also tells us that the distance between the Earth and the Sun is 1.496×10^8 km. So we calculate:

$$\text{time of travel} = \frac{1.496 \times 10^8\ \cancel{\text{km}}}{3.00 \times 10^5\ \cancel{\text{km}}/\text{s}} = 499\text{ s}$$

But most people don't really know how long 499 seconds is. It would be more useful if this number were in a more appropriate time unit. So we start by trying to convert this to minutes:

$$499\ \cancel{\text{sec}} \times \frac{1\text{ min}}{60\ \cancel{\text{sec}}} = 8.32\text{ min}$$

Since 8 minutes is 480 seconds $\left(8\text{ min} \times \frac{60\text{ s}}{1\text{ min}} = 480\text{ s}\right)$, 499 seconds is also equivalent to 8 minutes and 19 seconds. Thus, light takes 8 minutes and 19 seconds to travel from the Sun to Earth.

43. We will use the speed-time-distance relationship from the prior problem:

$$\text{time of travel} = \frac{\text{distance traveled}}{\text{speed}}$$

We also need the speed of light, which is 3×10^5 km/s; the problem tells us that the distance from Earth to Mars varies from 56 million kilometers to 400 million kilometers. (Or 5.6×10^7 km and 4×10^8 km in scientific notation.)

a. Mars at its nearest is 56 million kilometers away, so the light travel time is:

$$\text{time of travel} = \frac{5.6 \times 10^7 \cancel{\text{km}}}{3.00 \times 10^5 \cancel{\text{km}}/\text{s}} = 187 \text{ s}$$

We would prefer this answer in minutes, so converting:

$$187 \cancel{\text{s}} \times \frac{1 \text{ min}}{60 \cancel{\text{s}}} = 3.11 \text{ min}$$

It takes a little over 3 minutes each way to communicate with a spacecraft on Mars at closest approach.

b. At the most distant, Mars is 400 million kilometers from Earth, so we can compute the travel time for light:

$$\text{time of travel} = \frac{4 \times 10^8 \cancel{\text{km}}}{3.00 \times 10^5 \cancel{\text{km}}/\text{s}} = 1{,}330 \text{ s}$$

We would again prefer this in minutes, so we convert:

$$1{,}330 \cancel{\text{s}} \times \frac{1 \text{ min}}{60 \cancel{\text{s}}} = 22.2 \text{ min}$$

It takes a little over 22 minutes each way to communicate with a spacecraft on Mars when Mars is at its farthest from Earth.

44. We will again use the speed-time-distance relationship:

$$\text{time of travel} = \frac{\text{distance traveled}}{\text{speed}}$$

In this case, we seek the time it takes light to travel from Earth to Pluto. According to Appendix E, Earth's average distance from the Sun is 1.496×10^8 km and Pluto's average distance is 5.916×10^9 km. If we assume the two problems are lined up on the same side of the Sun at their average distances, the distance between them is:

$$\begin{aligned}\text{Earth to Pluto distance} &= 5.916 \times 10^9 \text{ km} - 1.496 \times 10^8 \text{ km} \\ &= 5.766 \times 10^9 \text{ km}\end{aligned}$$

Using this value, the light travel time is:

$$\text{time of travel} = \frac{5.766 \times 10^9 \cancel{\text{km}}}{3.00 \times 10^5 \cancel{\text{km}}/\text{s}} = 1.922 \times 10^4 \text{ s}$$

This is nearly 20,000 seconds, so let's try changing to more comprehensible units. We'll start by converting to minutes:

$$1.922 \times 10^4 \cancel{\text{s}} \times \frac{1 \text{ min}}{60 \cancel{\text{s}}} = 320 \text{ min}$$

That's a lot better, but it's still a lot more than 1 hour, so let's convert to hours:

$$320 \text{ min} \times \frac{1 \text{ hr}}{60 \text{ min}} = 5.33 \text{ hr}$$

It would take light 5.33 hours, or 5 hours and 20 minutes, to travel from Earth to Pluto under the alignment conditions and average distances we've assumed.

45. Since this question asks "how many times greater," the easiest way to solve it is with a ratio. We are asked how many times bigger the Earth to Alpha Centauri distance is than the Earth-Moon distance, so we'll set this up with Earth-Alpha Centauri distance on the top of the ratio:

$$\text{ratio} = \frac{\text{Earth to Alpha Centauri distance}}{\text{Earth to Moon distance}}$$

Now all we need are the two distances. From Appendix E, the Earth-Moon distance is 3.844×10^5 kms. Both the problem and Appendix E tell us that the Earth-Alpha Centauri distance is 4.4 light-years, but we need that in kilometers in order to set up the ratio properly since the units have to be the same on the top and on the bottom. To find out how to make this conversion, we will look in Appendix A where we find that 1 light-year = 9.46×10^{12} kms. So converting:

$$4.4 \text{ light-years} \times \frac{9.46 \times 10^{12} \text{ km}}{1 \text{ light-year}} = 4.2 \times 10^{13} \text{ km}$$

Our ratio becomes

$$\text{ratio} = \frac{4.2 \times 10^{13} \text{ km}}{3.844 \times 10^{5} \text{ km}} = 1.1 \times 10^{8}$$

We have found that Alpha Centauri is 110 million times farther away than the Moon.

46. We are asked to find how many times larger the Milky Way Galaxy is than the planet Saturn's rings. We are told that Saturn's rings are about 270,000 kilometers across and that the Milky Way is 100,000 light-years. Clearly, we'll have to convert one set of units or the other. Let's change light-years for kilometers. In Appendix A we find that 1 light-year = 9.46×10^{12} km, so we can convert:

$$100{,}000 \text{ light-years} \times \frac{9.46 \times 10^{12} \text{ km}}{1 \text{ light-year}} = 9.46 \times 10^{17} \text{ Km}$$

We can now find the ratio of the two diameters:

$$\text{ratio} = \frac{\text{diameter of Milky Way}}{\text{diameter of Saturn's rings}}$$

$$= \frac{9.46 \times 10^{17} \text{ km}}{2.7 \times 10^{5} \text{ km}} = 3.5 \times 10^{12}$$

The diameter of the Milky Way Galaxy is about 3.5 trillion times as large as the diameter of Saturn's rings!

47. We are given the distance to Alpha Centauri as 4.4 light-years, but those units aren't very useful to us. So let's convert to kilometers. From Appendix A we see that 1 light-year = 9.46×10^{12} km. So the distance to Alpha Centauri is

$$4.4 \text{ light-years} \times \frac{9.26 \times 10^{12} \text{ km}}{1 \text{ light-year}} = 4.2 \times 10^{13} \text{ km}$$

At a scale of 1-to-10^{19}, the distance to Alpha Centauri on this scale is

$$\frac{4.2 \times 10^{13} \text{ km}}{10^{19}} = 4.2 \times 10^{-6} \text{ km}$$

As numbers go, that one's not very easy to picture. So let's convert it to some smaller units. Since there are a thousand (10^3) meters to a kilometer and a thousand (10^3) millimeters to a meter, there are (10^6) millimeters to a kilometer. Those units seem about right for this distance, so let's convert:

$$4.2 \times 10^{-6} \text{ km} \times \frac{10^6 \text{ mm}}{1 \text{ km}} = 4.2 \text{ mm}$$

From Appendix E, the Sun is 7.0×10^5 km in radius, so the diameter is twice this, or 1.4×10^6 km. On the 1-to-10^{19} scale, this is

$$\frac{1.4 \times 10^6 \text{ km}}{10^{19}} = 1.4 \times 10^{-13} \text{ km}$$

We're asked to compare this to the size of an atom, 10^{-10} meter, so we had better convert one of the numbers. We'll convert from kilometers to meters:

$$1.4 \times 10^{-13} \text{ km} \times \frac{1000 \text{ m}}{1 \text{ km}} = 1.4 \times 10^{-10} \text{ m}$$

Note that this is about the same size as a typical atom. In summary, we've found that on a scale on which the Milky Way Galaxy would fit on a football field, the distance from the Sun to Alpha Centauri is only about 4.2 millimeters—smaller than the width of a finger—and the Sun itself becomes as small as an atom.

48. a. First, we need to work out the conversion factor to use to go between the real universe and our model. We are told that the Milky Way Galaxy is to be about 1 centimeter in the new model. We know from this chapter that the Milky Way Galaxy is about 10^5 light-years across. We could convert this to kilometers, but looking at what we're asked to convert, it appears that all the numbers we'll be using will be given in light-years anyway. So our conversion factor is:

$$1 \text{ cm} = 10^5 \text{ light-years}$$

So for the first part, we need to figure out how far the Andromeda Galaxy (also known as M 31) is from the Milky Way. The chapter tells us that their actual separation is 2.5 million light-years. So the distance on our scale is:

$$2.5 \times 10^6 \text{ light-years} \times \frac{1 \text{ cm}}{10^5 \text{ light-years}} = 25 \text{ cm}$$

The Andromeda Galaxy would be 25 centimeters away from our galaxy on this scale.

b. From Appendix E, we learn that Alpha Centauri is 4.4 light-years from the Sun. So using the conversion factor we developed in part (a), we convert this to the scale model distance:

$$4.4 \text{ light-years} \times \frac{1 \text{ cm}}{10^5 \text{ light-years}} = 4.4 \times 10^{-5} \text{ cm}$$

The separation between Alpha Centauri and our Sun is about 4.4×10^{-5} cm on this scale. As we will see in Chapter 6, this is comparable to the wavelength of blue light.

c. The observable universe is about 14 billion light-years in radius, so let's use that distance as an approximation to the distance of the most distant observable galaxies. Converting this distance to our model's scale:

$$14 \times 10^9 \text{ light-years} \times \frac{1 \text{ cm}}{10^5 \text{ light-years}} = 1.4 \times 10^6 \text{ cm}$$

That's more than a million centimeters, but it's difficult to visualize how far that is offhand. Let's convert that number to kilometers. We'll convert to meters first since we know how to go from centimeters to meters and meters to kilometers, but it's difficult to remember how to go straight from centimeters to kilometers:

$$1.4 \times 10^6 \text{ cm} \times \frac{1 \text{ m}}{100 \text{ cm}} \times \frac{1 \text{ km}}{1000 \text{ m}} = 14 \text{ km}$$

On a scale where our entire Milky Way Galaxy is the size of a marble, the most distant galaxies in our observable universe would be located some 14 kilometers away.

49. a. The circumference of the Earth is $2\pi \times 6{,}380$ km = 40,087 km. At a speed of 100 km/hr, it would take:

$$40{,}087 \text{ km} \div 100 \text{ km/hr} = 40{,}087 \text{ km} \times \frac{1 \text{ hr}}{100 \text{ km}} \times \frac{1 \text{ day}}{24 \text{ hr}} = 16.7 \text{ days}$$

to drive around the Earth. That is, a trip around the equator at 100 km/hr would take a little under 17 days.

b. We find the time by dividing the distance to the planet from the Sun by the speed of 100 km/hr. It would take about 170 years to reach the Earth and about 6,700 years to reach Pluto (at their mean distances).

c. Similarly, it would take 6,700 years to drive to Pluto at 100 km/hr. FYI: The following table shows the driving times from the Sun to each of the planets at a speed of 100 km/hr.

Planet	Driving Time
Mercury	66 years
Venus	123 years
Earth	170 years
Mars	259 years
Jupiter	888 years
Saturn	1,630 years
Uranus	3,300 years
Neptune	5,100 years
Pluto	6,700 years

d. We are given the distance to Alpha Centauri in light-years; converting to kilometers, we get:

$$4.4 \text{ light-years} \times \frac{9.46 \times 10^{12} \text{ km}}{1 \text{ light-year}} = 41.6 \times 10^{12} \text{ km}$$

At a speed of 100 km/hr, the travel time to Proxima Centauri would be about:

$$4.16 \times 10^{13} \text{ km} \div 100 \frac{\text{km}}{\text{hr}} = 4.16 \times 10^{13} \text{ km} \times \frac{1 \text{ hr}}{100 \text{ km}} \times \frac{1 \text{ day}}{24 \text{ hr}} \times \frac{1 \text{ yr}}{365 \text{ day}} = 4.7 \times 10^{7} \text{ yr}$$

It would take some 47 million years to reach Proxima Centauri at a speed of 100 km/hr.

50. a. To reach Alpha Centauri in 100 years, you would have to travel at 4.4/100 = 0.044 of the speed of light, which is about 13,200 km/s or nearly 50 million km/hr.

b. This is about 1,000 times the speed of our fastest current spacecraft.

51. The average speed of our solar system in its orbit of the Milky Way is the circumference of its orbit divided by the time it takes for one orbit:

$$v = \frac{2\pi(28{,}000 \text{ ly})}{2.3 \times 10^{8} \text{ yr}} = \frac{1.76 \times 10^{5} \text{ ly} \times 9.46 \times 10^{12} \frac{\text{km}}{\text{ly}}}{2.3 \times 10^{8} \text{ yr} \times 365 \frac{\text{day}}{\text{yr}} \times 24 \frac{\text{hr}}{\text{day}}} \approx 8.3 \times 10^{5} \frac{\text{km}}{\text{hr}}$$

We are racing around the Milky Way Galaxy at about 830,000 km/hr, or about 510,000 mi/hr.

52. We'll use the relationship that says that

$$\text{speed} = \frac{\text{distance traveled}}{\text{time}}$$

To compute each speed, we'll find the distance a person travels around the Earth's axis in 1 day (24 hours). Using the hint from the problem, we can find the radius of the circle that a person at that latitude travels is $R_{Earth} \times \cos(\text{latitude})$, where R_{Earth} = 6,378 km from Appendix E.

a. The radius of the circle that a person at latitude 30° travels is:

$$(6{,}378 \text{ km}) \times \cos(30°) = 5{,}534 \text{ km}$$

To get the distance traveled, we use the fact that a circle's circumference is given by $c = 2\pi r$, so in this case:

$$\text{distance} = 2\pi\ 5{,}524 \times \text{km} = 34{,}700 \text{ km}$$

Using the relationship between speed, time, and distance given above, the speed is

$$\text{speed} = \frac{34{,}700 \text{ km}}{24 \text{ hr}} = 1{,}446 \text{ km/hr}$$

A person at 30° latitude would be traveling at 1,446 km/hr around Earth's axis because of Earth's rotation.

b. The radius of the circle that a person at 60° travels is

$$(6{,}378 \text{ kms}) \times \cos(60°) = 3{,}189 \text{ km}$$

The distance traveled is

$$\text{distance} = 2\pi(3{,}189 \text{ km}) = 20{,}040 \text{ km}$$

The speed is

$$\text{speed} = \frac{20{,}040 \text{ km}}{24 \text{ hr}} = 834 \text{ km/hr}$$

A person at 60° latitude travels around the axis at 834 km/hr due to Earth's rotation.

c. Answers will vary depending on location.

Chapter 2. Discovering the Universe for Yourself

This chapter introduces major phenomena of the sky, with emphasis on:

- The concept of the celestial sphere.
- The basic daily motion of the sky, and how it varies with latitude.
- The cause of seasons.
- Phases of the Moon and eclipses.
- The apparent retrograde motion of the planets, and how it posed a problem for ancient observers.

As always, when you prepare to teach this chapter, be sure you are familiar with the relevant media resources (see the complete, section-by-section resource grid in Appendix 3 of this Instructor's Guide) and the online quizzes and other study resources available on the Mastering Astronomy Web site.

What's New in the Fourth Edition That Will Affect My Lecture Notes?

As everywhere in the book, we have added learning goals that we use as subheadings, rewritten to improve the text flow, improved art pieces, and added new illustrations. The art changes, in particular, will affect what you wish to show in lecture. We have not made any substantial content or organizational changes to this chapter.

Teaching Notes (By Section)

Section 2.1 Patterns in the Night Sky

This section introduccs thc concepts of constellations and of the celestial sphere, and introduces horizon-based coordinates and daily and annual sky motions.

- Stars in the daytime: You may be surprised at how many of your students actually believe that stars disappear in the daytime. If you have a campus observatory or can set up a small telescope, it's well worth offering a daytime opportunity to point the telescope at some bright stars, showing the students that they are still there.
- In class, you may wish to go further in explaining the correspondence between the Milky Way Galaxy and the Milky Way in our night sky. Tell your students to imagine being a tiny grain of flour inside a very thin pancake (or crepe!) that bulges in the middle and a little more than halfway toward the outer edge. Ask, "What will you see if you look toward the middle?" The answer should be "dough." Then ask what they will see if they look toward the far edge, and they'll give the same answer. Proceeding similarly, they should soon realize that they'll see a band of dough encircling their location, but that if they look away from the plane, the pancake is thin enough that they can see to the distant universe.
- Sky variation with latitude: Here, the intention is only to give students an overview of the idea and the most basic rules (such as latitude = altitude of NCP). Those instructors who want their students to be able to describe the sky in detail should cover Chapter S1, which covers this same material, but in much more depth.
- Note that in our jargon-reduction efforts, we do not introduce the term *asterism*, instead speaking of patterns of stars in the constellations. We also avoid the term *azimuth* when discussing horizon-based coordinates. Instead, we simply refer to *direction* along the horizon (e.g., south, northwest). The distinction of "along the horizon" should remove potential ambiguity with direction on the celestial sphere (where "north" would mean toward the north celestial pole rather than toward the horizon).

Section 2.2 The Reason for Seasons

This section focuses on seasons and why they occur.

- In combating misconceptions about the cause of the seasons, we recommend that you follow the logic in the Common Misconceptions box. That is, begin by asking your students what they think causes the seasons. When many of them suggest it is linked to distance from the Sun, ask how seasons differ between the two hemispheres. They should then see for themselves that it can't be distance from the Sun, or seasons would be the same globally rather than opposite in the two hemispheres.

- As a follow-up on the above note: Some students get confused by the fact that season diagrams (such as our Figure 2.15) cannot show the Sun-Earth distance and size of the Earth to scale. Thus, unless you emphasize this point (as we do in the figure and caption), it might actually look like the two hemispheres are at significantly different distances from the Sun. This is another reason why we believe it is critical to emphasize ideas of scale throughout your course. In this case, use the scale model solar system as introduced in Section 1.2, and students will quickly see that the two hemispheres are effectively at the same distance from the Sun at alltimes.
- Note that we do not go deeply into the physics that causes precession, as even a basic treatment of this topic requires discussing the vector nature of angular momentum. Instead, we include a brief motivation for the cause of precession by analogy to a spinning top.
- FYI regarding Sun signs: Most astrologers have "delinked" the constellations from the Sun signs. Thus, most astrologers would say that the vernal equinox still is in Aries—it's just that Aries is no longer associated with the same pattern of stars as it was in A.D. 150. For a fuller treatment of issues associated with the scientific validity (or, rather, the lack thereof) of astrology, see Section 3.5.

Section 2.3 The Moon, Our Constant Companion

This section discusses the Moon's motion and its observational consequences, including the lunar phases and eclipses.

- For what appears to be an easy concept, many students find it remarkably difficult to understand the phases of the Moon. You may want to do an in-class demonstration of phases by darkening the room, using a lamp to represent the Sun, and giving each student a Styrofoam ball to represent the Moon. If your lamp is bright enough, the students can remain in their seats and watch the phases as they move the ball around their heads.
- Going along with the above note, it is virtually impossible for students to understand phases from a flat figure on a flat page in a book. Thus, we have opted to eliminate the "standard" Moon phases figure that you'll find in almost every other text, which shows the Moon in eight different positions around the Earth—students just don't get it, and the multiple moons confuse them. Instead, our Figure 2.22 shows how students can conduct a demonstration for themselves that will help them understand the phases. The Phases of the Moon tutorial on the Mastering Astronomy Web site has also proved very successful at helping students understand phases.
- When covering the causes of eclipses, it helps to demonstrate the Moon's orbit. Keep a model "Sun" on a table in the center of the lecture area; have your left fist represent the Earth, and hold a ball in the other hand to represent the Moon. Then you can show how the Moon orbits your "fist" at an inclination to the ecliptic plane, explaining the meaning of the nodes. You can also show eclipse seasons by demonstrating the Moon's orbit (with fixed nodes) as you walk around your model Sun: The students will see that eclipses are possible only during two periods each year. If you then add in precession of the nodes, students can see why eclipse seasons occur slightly more often than every 6 months.

- The *Moon Pond* painting in Figure 2.24 should also be an effective way to explain what we mean by *nodes* of the Moon's orbit.
- FYI: We've found that even many astronomers are unfamiliar with the saros cycle of eclipses. Hopefully our discussion is clear, but some additional information may help you as an instructor: The nodes of the Moon's orbit precess with an 18.6-year period; note that the close correspondence of this number to the 18-year 11-day saros has no special meaning (it essentially is a mathematical coincidence). The reason that the same type of eclipse (e.g., partial versus total) does not recur in each cycle is because the Moon's line of apsides (i.e., a line connecting perigee and apogee) also precesses—but with a different period (8.85 years).
- FYI: The actual saros period is 6,585.32 days, which usually means 18 years 11.32 days, but instead is 18 years 10.32 days if 5 leap years occur during this period.

Section 2.4 The Ancient Mystery of the Planets

This section covers the ancient mystery of planetary motion, explaining the motion, how we now understand it, and how the mystery helped lead to the development of modern science.

- We have chosen to refer to the westward movement of planets in our sky as *apparent* retrograde motion, in order to emphasize that planets only appear to go backward but never really reverse their direction of travel in their orbits. This makes it easy to use analogies—e.g., when students try the demonstration in Figure 2.28, they never say that their friend really moves backward as they pass by, only that the friend appears to move backward against thebackground.
- You should emphasize that apparent retrograde motion of planets is noticeable only by comparing planetary positions over many nights. In the past, we've found a tendency for students to misinterpret diagrams of retrograde motion and thereby expect to see planets moving about during the course of a single night.
- It is somewhat rare among astronomy texts to introduce stellar parallax so early. However, it played such an important role in the historical debate over a geocentric universe that we feel it must be included at this point. Note that we do *not* give the formula for finding stellar distances at this point; that comes in Chapter 15.

Answers/Discussion Points for Think About It Questions

The Think About It questions are not numbered in the book, so we list them in the order in which they appear, keyed by section number.

Section 2.1

- (p. 29) The simple answer is no, because a galaxy located in the direction of the galactic center will be obscured from view by the dust and gas of the Milky Way. Note, however, that this question can help you root out some student misconceptions. For example, some students might wonder if you could see the galaxy "sticking up" above our own galaxy's disk—illustrating a misconception about how angular size declines with distance. They might also wonder if a telescope would make a difference, illustrating a misconception about telescopes' being able to "see through" things that our eyes cannot see through. Building on this idea, you can also foreshadow later discussions of nonvisible light by pointing out that while no telescope can help the problem in visible light, we CAN penetrate the interstellar gas and dust in some other wavelengths.

- (p. 30) No. We can only describe angular sizes and distances in the sky, so physical measurements do not make sense. This is a difficult idea for many children to understand, but hopefully comes easily for college students!
- (p. 31) Yes, because it is Earth's rotation that causes the rising and setting of all the objects in the sky. *Note:* Many instructors are surprised that this question often gives students trouble, but the trouble arises from at least a couple misconceptions harbored by many students. First, even though students can recite the fact that the motion of the stars is really caused by the rotation of Earth, they haven't always absorbed the idea and therefore don't automatically apply it to less familiar objects like galaxies. Second, many students have trouble visualizing galaxies as fixed objects on the celestial sphere like stars, perhaps because they try to see them as being "big" and therefore have trouble fitting them onto the sphere in their minds. Thus, this simple question can help you address these misconceptions and thereby make it easier for students to continue their progress in the course.
- (p. 35, left) This question is designed to make sure students understand basic ideas of the sky. Answers are latitude-dependent. Sample answer for latitude 40°N: The north celestial pole is located 40° above the horizon, due north. You can see circumpolar stars by looking toward the north, anywhere between the north horizon and altitude 80°. The lower 40° of the celestial sphere is always below your horizon.
- (p. 35, right) Depends on the time of year; this question really just checks that students can properly interpret Figure 2.14. Sample answer for September 21: The Sun appears to be in Virgo, which means you'll see the opposite zodiac constellation—Pisces—on your horizon at midnight. After sunset, you'll see Libra setting in the western sky, since it is east of Virgo and therefore follows it around the sky.

Section 2.2

- (p. 37) Jupiter does not have seasons because of its lack of appreciable axis tilt. Saturn, with an axis tilt similar to Earth, does have seasons.
- (p. 40) In 2,000 years, the summer solstice will have moved about the length of one constellation along the ecliptic. Since the summer solstice was in Cancer a couple thousand years ago (as you can remember from the Tropic of Cancer) and is in Gemini now, it will be in Taurus in about 2,000 years.

Section 2.3

- (p. 43, left) A quarter moon visible in the morning must be third-quarter, since third-quarter moon rises around midnight and sets around noon.
- (p. 43, right) About 2 weeks each. Because the Moon takes about a month to rotate, your "day" would last about a month. Thus, you'd have about 2 weeks of daylight followed by about 2 weeks of darkness as you watched Earth hanging in your sky and going through its cycle of phases.
- (p. 48) Remember that each eclipse season lasts a few weeks. Thus, if the timing of the eclipse season is just right, it is possible for two full moons to occur during the same eclipse season, giving us two lunar eclipses just a month apart. In such cases the eclipses will almost always be penumbral, because the penumbral

shadow is much larger than the umbral shadow; thus, it's far more likely that the Moon will pass twice in the same eclipse season through the large penumbral shadow than through the much smaller umbral shadow.

Section 2.4

- (p. 52) Opposite ends of the Earth's orbit are about 300 million kilometers apart, or about 30 meters on the 1-to-10-billion scale used in Chapter 1. The nearest stars are tens of trillions of kilometers away, or thousands of kilometers on the 1-to-10-billion scale, and are typically the size of grapefruits or smaller. The challenge of detecting stellar parallax should now be clear.

Solutions to End-of-Chapter Problems (Chapter 2)

1. A constellation is a section of the sky, like a state within the United States. They are based on groups of stars that form patterns that suggested shapes to the cultures of the people who named them. The official names of most of the constellations in the Northern Hemisphere came from ancient cultures of the Middle East and the Mediterranean, while the constellations of the Southern Hemisphere got their official names from seventeenth-century Europeans.
2. If we were making a model of the celestial sphere on a ball, we would definitely need to mark the north and south celestial poles, which are the points directly above the Earth's poles. Halfway between the two poles we would mark the great circle of the celestial equator, which is the projection of Earth's equator out into space. And we definitely would need to mark the circle of the ecliptic, which is the path that the Sun appears to make across the sky. Then we could add stars and borders of constellations.
3. No, space is not really full of stars. Because the distance to the stars is very large and because stars lie at different distances from Earth, stars are not really crowded together.
4. The local sky looks like a dome because we see half of the full celestial sphere at any one time.

   Horizon—The boundary line dividing the ground and the sky.

   Zenith—The highest point in the sky, directly overhead.

   Meridian—The semicircle extending from the horizon due north to the zenith to the horizon due south.

   We can locate an object in the sky by specifying its altitude and its direction along the horizon.
5. We can measure only angular size or angular distance on the sky because we lack a simple way to measure distance to objects just by looking at them. It is therefore usually impossible to tell if we are looking at a smaller object that's near us or a more distant object that's much larger.

   Arcminutes and arcseconds are subdivisions of degrees. There are 60 arcminutes in 1 degree, and there are 60 arcseconds in 1 arcminute.
6. Circumpolar stars are stars that never appear to rise or set from a given location, but are always visible on any clear night. From the North Pole, every visible star is circumpolar, as all circle the horizon at constant altitudes. In contrast, a much

smaller portion of the sky is circumpolar from the United States, as most stars follow paths that make them rise and set.

7. Latitude measures angular distance north or south of Earth's equator. Longitude measures angular distance east or west of the Prime Meridian. The night sky changes with latitude, because it changes the portion of the celestial sphere that can be above your horizon at any time. The sky does not change with changing longitude, however, because as Earth rotates, all points on the same latitude line will come under the same set of stars, regardless of their longitude.
8. The zodiac is the set of constellations in which the Sun can be found at some point during the year. We see different parts of the zodiac at different times of the year because the Sun is always somewhere in the zodiac and so we cannot see that constellation at night at that time of the year.
9. If Earth's axis had no tilt, Earth would not have significant seasons because the intensity of sunlight at any particular latitude would not vary with the time of year.
10. The summer solstice is the day when the Northern Hemisphere gets the most direct sunlight and the southern hemisphere the least direct. Also, on the summer solstice the Sun is as far north as it ever appears on the celestial sphere. On the winter solstice, the situation is exactly reversed: The Sun appears as far south as it will get in the year, and the Northern Hemisphere gets its least direct sunlight while the Southern Hemisphere gets its most direct sunlight.

    On the equinoxes, the two hemispheres get the same amount of sunlight, and the day and night are the same length (12 hours) in both hemispheres. The Sun is found directly overhead at the equator on these days, and it rises due east and sets due west.
11. The direction in which the Earth's rotation axis points in space changes slowly over the centuries and we call this change "precession." Because of this movement, the celestial poles and therefore the pole star changes slowly in time. So while Polaris is the pole star now, in 13,000 years the star Vega will be the pole star instead.
12. The Moon's phases start with the new phase when the Moon is nearest the Sun in our sky and we see only the unlit side. From this dark phase, one side of the Moon's visible face slowly becomes lit, moving to the first-quarter phase, when we see a half-lit moon. During the time when the Moon's illuminated fraction is increasing, we say that the Moon is waxing. When the entire visible face of the Moon is lit up and the Moon is visible all night long, we say that the Moon is in its full phase. The process then occurs in reverse over the second half of the month as the Moon's lit fraction decreases, through third-quarter when it is half-lit, back to new again. During the second half of the month when the Moon's illuminated fraction is decreasing, we say that the Moon is waning.

    We can never see a full moon at noon because for the Moon to be full, it and the Sun must be on opposite sides of the Earth. So as the full moon rises, the Sun must be setting and when the Moon is setting, the Sun is rising. (*Exception*: At very high latitudes, there may be times when the full moon is circumpolar, in which case it could be seen at noon—but would still be 180° away from the Sun's position.)

13. When we say that the Moon displays synchronous rotation, we mean that the Moon's spin period and its orbital period around the Earth are the same. So from the Earth, we always see the same side of the Moon and someone on the Moon always sees the Earth in the same place in her local sky.
14. While the Moon must be in its new phase for a solar eclipse or in its full phase for a lunar eclipse, we do not see eclipses every month. This is because the Moon usually passes to the north or south of the Sun during these times, because its orbit is tilted relative to the ecliptic plane.
15. The apparent retrograde motion of the planets refers to the planets' behaviors when they sometimes stop moving eastward relative to the stars and move westward for a while. While the ancients had to resort to complex systems to explain this behavior, our Sun-centered model makes this motion a natural consequence of the fact that the different planets move at different speeds as they go around the Sun. We see the planets appear to move backward because we are actually overtaking them in our orbit (if they orbit farther from the Sun than Earth) or they are overtaking us (if they orbit closer to the Sun than Earth).
16. Stellar parallax is the apparent movement of some of the nearest stars relative to the more distant ones as Earth goes around the Sun. This is caused by our slightly changing perspective on these stars through the year. However, the effect is very small because Earth's orbit is much smaller than the distances to even the closest stars. Because the effect is so small, the ancients were unable to observe it. However, they correctly realized that if the Earth is going around the Sun, they should see stellar parallax. Since they could not see the stars shift, they concluded that the Earth does not move.
17. *The constellation of Orion didn't exist when my grandfather was a child.* This statement does not make sense, because the constellations don't appear to change on the time scales of human lifetimes.
18. *When I looked into the dark lanes of the Milky Way with my binoculars, I saw what must have been a cluster of distant galaxies.* This statement does not make sense, because we cannot see through the band of light we call the Milky Way to external galaxies; the dark fissure is gas and dust blocking our view.
19. *Last night the Moon was so big that it stretched for a mile across the sky.* This statement does not make sense, because a mile is a physical distance, and we can measure only angular sizes or distances when we observe objects in the sky.
20. *I live in the United States, and during my first trip to Argentina I saw many constellations that I'd never seen before.* This statement makes sense, because the constellations visible in the sky depend on latitude. Since Argentina is in the Southern Hemisphere, the constellations visible there include many that are not visible from the United States.
21. *Last night I saw Jupiter right in the middle of the Big Dipper. (Hint: Is the Big Dipper part of the zodiac?)* This statement does not make sense, because Jupiter, like all the planets, is always found very close to the ecliptic in the sky. The ecliptic passes through the constellations of the zodiac, so Jupiter can appear to be only in one of the 12 zodiac constellations—and the Big Dipper (part of the constellation Ursa Major) is not among these constellations.

22. *Last night I saw Mars move westward through the sky in its apparent retrograde motion.* This statement does not make sense, because the apparent retrograde motion is noticeable only over many nights, not during a single night. (Of course, like all celestial objects, Mars moves from east to west over the course of EVERY night.)
23. *Although all the known stars appear to rise in the east and set in the west, we might someday discover a star that will appear to rise in the west and set in the east.* This statement does not make sense. The stars aren't really rising and setting; they only appear to rise in the east and set in the west because the EARTH rotates.
24. *If Earth's orbit were a perfect circle, we would not have seasons.* This statement does not make sense. As long as Earth still has its axis tilt, we'll still have seasons.
25. *Because of precession, someday it will be summer everywhere on Earth at the same time.* Thisstatement does not make sense. Precession does not change the tilt of the axis, only its orientation in space. As long as the tilt remains, we will continue to have opposite seasons in the two hemispheres.
26. *This morning I saw the full moon setting at about the same time the Sun was rising.* This statement makes sense, because a full moon is opposite the Sun in the sky.
27. c; 28. a; 29. a; 30. a; 31. a;
32. b; 33. b; 34. b; 35. a; 36. b.
37. The planet will have seasons because of its axis tilt, even though its orbit is circular. Because its 35° axis tilt is greater than Earth's 23.5° axis tilt, we'd expect this planet to have more extreme seasonal variations than Earth.
38. Answers will vary with location; the following is a sample answer for Boulder, Colorado.
    a. The latitude in Boulder is 40°N and the longitude is about 105°E.
    b. The north celestial pole appears in Boulder's sky at an altitude of 40°, in the direction duenorth.
    c. Polaris is circumpolar because it never rises or sets in Boulder's sky. It makes a daily circle, less than 1° in radius, around the north celestial pole.
39. a. When you see a full earth, people on Earth must have a new moon.
    b. At full moon, you would see new earth from your home on the Moon. It would be daylight at your home, with the Sun on your meridian and about a week until sunset.
    c. When people on Earth see a waxing gibbous moon, you would see a waning crescent earth.
    d. If you were on the Moon during a total lunar eclipse (as seen from Earth), you would see a total eclipse of the Sun.
40. You would not see the Moon go through phases if you were viewing it from the Sun. You would always see the sunlit side of the Moon, so it would always be "full." In fact, the same would be true of Earth and all the other planets as well.
41. If the Moon were twice as far from the Earth, its angular size would be too small to create a total solar eclipse. It would still be possible to have annular eclipses, though the Moon would cover only a small portion of the solar disk.
42. If the Earth were smaller in size, solar eclipses would still occur in about the same way, since they are determined by the Moon's shadow on the Earth.
43. This is an observing project that will stretch over several weeks.

44. This is a literary essay that requires reading the Mark Twain novel.

45. a. There are $360 \times 60 = 21{,}600$ arcminutes in a full circle.

b. There are $360 \times 60 \times 60 = 1{,}296{,}000$ arcseconds in a full circle.

c. The Moon's angular size of 0.5° is equivalent to 30 arcminutes or $30 \times 60 =$ 1,800 arcseconds.

46. a. We know that circumference = $2 \times \pi \times$ radius, so we can compute the circumference of the Earth:

$$\begin{aligned}\text{circumference} &= 2 \times \pi \times (6370 \text{ km}) \\ &= 40{,}000 \text{ km}\end{aligned}$$

b. There are 90° of latitude between the North Pole and the equator. This distance is also one-quarter of Earth's circumference. Using the circumference from part (a), this distance is

$$\begin{aligned}\text{equator to pole distance} &= \frac{\text{circumference}}{4} \\ &= \frac{40{,}000 \text{ km}}{4} \\ &= 10{,}000 \text{ km}\end{aligned}$$

So if 10,000 kilometers is the same as 90° of latitude, then we can convert 1° into kilometers:

$$1^\circ \times \frac{10{,}000 \text{ km}}{90^\circ} = 111 \text{ km}$$

So 1° of latitude is the same as 111 kilometers on the Earth.

c. There are 60 arcminutes in a degree. So we can find how many arcminutes are in a quarter-circle:

$$90^\circ \times \frac{60 \text{ arcminutes}}{1^\circ} = 5{,}400 \text{ arcminutes}$$

Doing the same thing as in part (b):

$$1 \text{ arcminute} \times \frac{10{,}000 \text{ km}}{5400 \text{ arcminutes}} = 1.85 \text{ km}$$

Each arcminute of latitude represents 1.85 kilometers.

d. We cannot provide similar answers for longitude, because lines of longitude get closer together as we near the poles, eventually meeting at the poles themselves. So there is no single distance that can represent 1° of longitude everywhere on Earth.

47. a. We start by recognizing that there are 24 whole degrees in this number. So we just need to convert the 0.3° into arcminutes and arcseconds. So first converting to arcminutes:

$$0.3^\circ \times \frac{60 \text{ arcminutes}}{1^\circ} = 18 \text{ arcminutes}$$

Since there is no fractional part left to convert into arcseconds, we are done. So 24.3° is the same as 24° 18′ 0″.

b. Leaving off the whole degree, we convert the 0.59° to arcminutes:

$$0.59^\circ \times \frac{60 \text{ arcminutes}}{1^\circ} = 35.4 \text{ arcminutes}$$

So we have 35 whole arcminutes and a fractional part of 0.4 arcminute that we need to convert into arcseconds:

$$0.4 \text{ arcminute} \times \frac{60 \text{ arcseconds}}{1 \text{ arcminute}} = 24 \text{ arcseconds}$$

So 1.59° is the same as 1° 35′ 24″.

c. We have 0 whole degrees, so we convert the fractional degree into arcminutes:

$$0.1^\circ \times \frac{60 \text{ arcminutes}}{1^\circ} = 6 \text{ arcminutes}$$

Since there is no fractional part to this, we do not need any arcseconds to represent this number. So 0.1° is the same as 0° 6′ 0″.

d. We again have no whole degrees, so we start by converting 0.01° to arcminutes:

$$0.01^\circ \times \frac{60 \text{ arcminutes}}{1^\circ} = 0.6 \text{ arcminute}$$

There are no whole arcminutes here, either, so we have to convert 0.6 arcminute into arcseconds:

$$0.6 \text{ arcminute} \times \frac{60 \text{ arcseconds}}{1 \text{ arcminute}} = 36 \text{ arcseconds}$$

So 0.01° is the same as 0° 0′ 36″.

e. We again have no whole degrees, so we start by converting 0.001° to arcminutes:

$$0.001^\circ \times \frac{60 \text{ arcminutes}}{1^\circ} = 0.06 \text{ arcminute}$$

There are no whole arcminutes here, either, so we have to convert 0.06 arcminute into arcseconds:

$$0.06 \text{ arcminute} \times \frac{60 \text{ arcseconds}}{1 \text{ arcminute}} = 3.6 \text{ arcseconds}$$

So 0.01° is the same as 0° 0′ 3.6″.

48. a. We will start by converting the 42 arcseconds into arcminutes:

$$42 \text{ arcseconds} \times \frac{1 \text{ arcminute}}{60 \text{ arcseconds}} = 0.7 \text{ arcsecond}$$

So now we have 7° 38.7′. Converting the 38.7 arcminutes to degrees:

$$38.7\ \cancel{\text{arcminutes}} \times \frac{1^\circ}{60\ \cancel{\text{arcminutes}}} = 0.645^\circ$$

So 7° 38′ 42″ is the same as 7.645°.

b. We will start by converting the 54 arcseconds into arcminutes:

$$54\ \cancel{\text{arcseconds}} \times \frac{1\ \text{arcminute}}{60\ \cancel{\text{arcseconds}}} = 0.9\ \text{arcminute}$$

So now we have 12.9 arcminutes. Converting this to degrees:

$$12.9\ \cancel{\text{arcminutes}} \times \frac{1^\circ}{60\ \cancel{\text{arcminutes}}} = 0.215^\circ$$

So 12′ 54″ is the same as 0.215°.

c. We will start by converting the 59 arcseconds into arcminutes:

$$59\ \cancel{\text{arcseconds}} \times \frac{1\ \text{arcminute}}{60\ \cancel{\text{arcseconds}}} = 0.9833\ \text{arcminute}$$

So now we have 1° 59.9833′ arcminutes. Converting this to degrees:

$$59.9833\ \cancel{\text{arcminutes}} \times \frac{1^\circ}{60\ \cancel{\text{arcminutes}}} = 0.9997^\circ$$

So 1° 59′ 59″ is the same as 1.9997°, very close to 2°.

d. In this case, we need only convert 1 arcminute to degrees:

$$1\ \cancel{\text{arcminute}} \times \frac{1^\circ}{60\ \cancel{\text{arcminutes}}} = 0.017^\circ$$

So 1′ is the same as 0.017°.

e. We can convert this from arcseconds to degrees in one step since there are no arcminutes to add in:

$$1\ \cancel{\text{arcsecond}} \times \frac{1\ \cancel{\text{arcminute}}}{60\ \cancel{\text{arcseconds}}} \times \frac{1^\circ}{60\ \cancel{\text{arcminutes}}} = 2.78 \times 10^{-4\circ}$$

So 1″ is the same as $2.78 \times 10^{-4\circ}$

49. From Appendix E, the Moon's orbit has a radius of 384,400 kilometers. The distance that the Moon travels in one orbit is the circumference of the orbit:

$$\begin{aligned} \text{distance traveled} &= 2 \times \pi \times \text{radius} \\ &= 2 \times \pi \times (384{,}400\ \text{km}) \\ &= 2{,}415{,}000\ \text{km} \end{aligned}$$

To find the Moon's speed in kilometers per hour, we also need to find how many hours are in the Moon's $27\frac{1}{3}$-day orbit:

$$27.3 \cancel{\text{days}} \times \frac{24 \text{ hr}}{1 \cancel{\text{day}}} \approx 656 \text{ hr}$$

The speed is the distance over the time,

$$\begin{aligned} \text{speed} &= \frac{\text{distance traveled}}{\text{time}} \\ &= \frac{2{,}415{,}000 \text{ km}}{656 \text{ hr}} \\ &\approx 3{,}680 \text{ km/hr} \end{aligned}$$

The Moon orbits Earth at a speed of 3,680 km/hr.

50. Starting with the size of the Moon, we convert to the scale model distance by dividing by 10 billion:

$$\frac{3500 \text{ km}}{10^{10}} = 3.5 \times 10^{-7} \text{ km}$$

This number is pretty hard to understand, so we should convert it to something more useful. Judging by the sizes of other objects in the model, let's convert from kilometers to millimeters:

$$3.5 \times 10^{-7} \cancel{\text{km}} \times \frac{1000 \cancel{\text{m}}}{1 \cancel{\text{km}}} \times \frac{1000 \text{ mm}}{1 \cancel{\text{m}}} = 0.35 \text{ mm}$$

The Moon's size on this scale is 0.35 millimeter.

We perform the same conversion to get to the Moon's scale distance:

$$\frac{380{,}000 \text{ km}}{10^{10}} = 3.8 \times 10^{-5} \text{ km}$$

Just as above, this number is hard to understand. We'll also convert it to millimeters:

$$3.8 \times 10^{-5} \cancel{\text{km}} \times \frac{1000 \cancel{\text{m}}}{1 \cancel{\text{km}}} \times \frac{1000 \text{ mm}}{1 \cancel{\text{m}}} = 38 \text{ mm}$$

The distance to the Moon on this scale is 38 millimeters. Since there are 10 millimeters to 1 centimeter, we can convert this to centimeters:

$$38 \cancel{\text{mm}} \times \frac{1 \text{ cm}}{10 \cancel{\text{mm}}} = 3.8 \text{ cm}$$

The Moon's scaled distance is 3.8 centimeters, which is less than 2 inches. It also means that the Moon's orbit is about half the size of the ball of the Sun. The ball of the Sun was the size of a grapefruit in this scale model, so sticking with fruit, we could say that the Moon's orbit has the diameter of a medium-size orange or an apple.

51. Following the method of Example 1 in Mathematical Insight 2.1, we can use the angular separation formula to find the Sun's actual diameter from its angular diameter of 0.5° and its distance of 1.5×10^8 km:

$$\text{physical diameter} = \text{angular diameter} \times \frac{2\pi \times \text{distance}}{360°}$$
$$= 0.5° \times \frac{2\pi(1.5 \times 10^8 \text{ km})}{360°}$$
$$\approx 1.3 \times 10^6 \text{ km}$$

Using this very approximate value of 0.5° for the Sun's angular size, we find that the Sun's diameter is about 1.3 million kilometers—fairly close to the actual value of 1.39 million kilometers.

52. To solve this problem, we turn to Mathematical Insight 2.1, where we learn that the physical size of an object, its distance, and its angular size are related by the equation:

$$\text{physical size} = \frac{2\pi \times (\text{distance}) \times (\text{angular size})}{360°}$$

We are told that the Sun is 0.5° in angular diameter and is about 150,000,000 kilometers away. So we put those values in:

$$\text{physical size} = \frac{2\pi \times (150{,}000{,}000 \text{ km}) \times (0.5°)}{360°}$$
$$= 1{,}310{,}000 \text{ km}$$

For the values given, we estimate the size to be about 1,310,000 kilometers. We are told that the actual value is about 1,390,000 kilometers. The two values are pretty close and the difference can probably be explained by the Sun's actual diameter not being exactly 0.5° and the distance to the Sun not being exactly 150,000,000 kilometers.

53. To solve this problem, we use the equation relating distance, physical size, and angular size given in Mathematical Insight 2.1:

$$\text{physical size} = \frac{2\pi \times (\text{distance}) \times (\text{angular size})}{360°}$$

In this case, we are given the distance to Betelgeuse as 427 light-years and the angular size as 0.044 arcsecond. We have to convert this number to degrees (so that the units in the numerator and denominator cancel), so:

$$0.044 \text{ arcsecond} \times \frac{1 \text{ arcminute}}{60 \text{ arcseconds}} \times \frac{1°}{60 \text{ arcminutes}} = 1.22 \times 10^{-5}°$$

We can leave the distance in light-years for now. So we can calculate the size of Betelgeuse:

$$\text{physical size} = \frac{2\pi \times (427 \text{ light-years}) \times (1.22 \times 10^{-5}°)}{360°}$$
$$= 9.1 \times 10^{-5} \text{ light-years}$$

Clearly, we've chosen to express this in the wrong units: lights-years are too large to be convenient for expressing the size of stars. So we convert to kilometers using the conversion factor found in Appendix A:

$$9.1 \times 10^{-5} \text{ light-years} \times \frac{9.46 \times 10^{12} \text{ km}}{1 \text{ light-year}} = 8.6 \times 10^{8} \text{ km}$$

(Note that we could have converted the distance to Betelgeuse to kilometers before we calculated Betelgeuse's size and gotten the diameter in kilometers out of our formula for physical size.)

The diameter of Betelgeuse is about 860 million kilometers, which is more than 600 times the Sun's diameter of 1.39×10^6 km. It is also almost six times the distance between the Earth and Sun (1.5×10^8 km, from Appendix E).

54. a. Using the small-angle formula given in Mathematical Insight 2.1, we know that:

$$\text{angular size} = \text{physical size} \times \frac{360°}{2\pi \times \text{distance}}$$

We are given the physical size of the Moon (3,476 kilometers) and the minimum orbital distance (356,400 kilometers), so we can compute the angular size:

$$\text{angular size} = (3{,}476 \text{ km}) \times \frac{360°}{2\pi \times (356{,}400 \text{ km})} = 0.559°$$

When the Moon is at its most distant, it is 406,700 kilometers, so we can repeat the calculation for this distance:

$$\text{angular size} = (3{,}476 \text{ km}) \times \frac{360°}{2\pi \times (406{,}700 \text{ km})} = 0.426°$$

The Moon's angular diameter varies from 0.426° to 0.559° (at its farthest from Earth and at its closest, respectively).

b. We can do the same thing as in part (a), except we use the Sun's diameter (1,390,000 kilometers) and minimum and maximum distances (147,500,000 kilometers and 152,600,000 kilometers) from Earth. At its closest, the Sun's angular diameter is:

$$\text{angular size} = (1{,}390{,}000 \text{ km}) \times \frac{360°}{2\pi \times (147{,}500{,}000 \text{ km})} = 0.540°$$

At its farthest from Earth, the Sun's angular diameter is:

$$\text{angular size} = (1{,}390{,}000 \text{ km}) \times \frac{360°}{2\pi \times (152{,}600{,}000 \text{ km})} = 0.522°$$

The Sun's angular diameter varies from 0.522° to 0.540°.

c. When both objects are at their maximum distances from Earth, both objects appear with their smallest angular diameters. At this time, the Sun's angular diameter is 0.522° and the Moon's angular diameter is 0.426°. The Moon's angular diameter under these conditions is significantly smaller than the Sun's, so it could *not* fully cover the Sun's disk. Since it cannot completely cover the Sun, there can be no total eclipse under these conditions. There can be only an annular or partial eclipse under these conditions.

Chapter 3. The Science of Astronomy

Most students do not really understand how science works, and our aim in this chapter is to edify them in an interesting and multicultural way. If you are used to teaching from other textbooks, you may be surprised that we have chosen to wait until Chapter 3 to introduce this material. However, we have found that students are better able to appreciate the development of science and how science works after they first have some idea of what science has accomplished. Thus, we find that covering the development of science at this point is more effective than introducing it earlier.

- If your course focuses on the solar system, you may wish to emphasize this material heavily in class, and perhaps supplement it with your favorite examples of ancient astronomy.
- If your course focuses on stars, galaxies, or cosmology, you may decide not to devote class time to this chapter at all, or to concentrate only on Section 3.4 on the nature of science. However, you may still wish to have your students read this chapter, as it may prove useful in later discussions about the nature of science.

As always, when you prepare to teach this chapter, be sure you are familiar with the relevant media resources (see the complete, section-by-section resource grid in Appendix 3 of this Instructor's Guide) and the online quizzes and other study resources available on the Mastering Astronomy Web site.

What's New in the Fourth Edition That Will Affect My Lecture Notes?

Those who have taught from previous editions of *The Cosmic Perspective* should be aware of the following organizational or pedagogical changes to this chapter (i.e., changes that will influence the way you teach) from the third edition:

- We have substantially revised our discussion of archaeoastronomy in Section 3.1, and updated our examples with new research results in this field. Also note that we have learned that some of our examples from prior editions have either been discredited or are subject to more controversy than we had been aware of; for example, we now show the Wyoming Medicine Wheel as an example of apparent misinterpretation of data, although some archaeoastronomers still think the Medicine Wheel is a valid example.
- We have further revised our discussion in Section 3.4 of the nature of science to make the distinction between science and nonscience even clearer to students. In particular, note the new subsection on "Verifiable Observations" on page 80.

Teaching Notes (By Section)

Section 3.1 The Ancient Roots of Science

This section introduces students to the development of astronomy by discussing how ancient observations were made and used by different cultures. We stress that these ancient observations helped lay the groundwork for modern science. The particular examples cited were chosen to give a multicultural perspective on ancient astronomy; instructors may wish to add their own favorite examples of ancient observations. In teaching from this section, you can take one of two basic approaches, depending on how much time you have available: (1) If you have little time to discuss the section in class, you can focus on the examples generally without delving into the observational details; or (2) if you have more time available, you can emphasize the details of how observations allowed determination of the time and date, and of how lunar cycles are used to make lunar calendars.

Section 3.2 Ancient Greek Science

This section focuses on the crucial role of the ancient Greeks in the development of science. We focus on the idea of creating scientific models through the example of the gradual development of the Ptolemaic model of the universe. The section concludes with discussion of the Islamic role in preserving and expanding upon Greek knowledge, setting the stage for discussion of the Copernican revolution in the next section.

- The flat earth: There's a good article about the common misconception holding that medieval Europeans thought the Earth to be flat in *Mercury*, Sept/Oct 2002, page 34.

Section 3.3 The Copernican Revolution

With the background from the previous two sections, students now are capable of understanding how and why the geocentric model of the universe was abandoned. We therefore use this section to discuss the unfolding of the Copernican revolution by emphasizing the roles of each of the key personalities.

- Note that Kepler's laws are introduced in this section, in their historical context.
- Note that we present Galileo's role by focusing on how he overcame remaining objections to the Copernican model. This is a particularly good example of the working of science, since it shows both that old ideas were NOT ridiculous while also showing how new ideas gained acceptance.

Section 3.4 The Nature of Science

The historical background of the previous sections has students ready to discuss just what science really is. A few notes:

- We emphasize that the traditional idea of a "scientific method" is a useful idealization, but that science rarely proceeds so linearly.
- The most important part of this section is the "hallmarks of science." We have developed these three hallmarks through extensive discussions with both scientists and philosophers of science, and we believe they represent a concise summary of the distinguishing features of science.
- One of the key reasons that the hallmarks are useful is that they make it relatively easy for students to distinguish between science and nonscience.
- We include only a brief discussion of the idea of scientific paradigms; you may wish to supplement this discussion with your favorite examples of paradigm shifts.

Section 3.5 Astrology

Public confusion between astronomy and astrology is well known. To address this confusion, we end this chapter with a discussion designed to help students distinguish between the two. We have tried to avoid direct criticism of astrology and astrologers even while pointing out that it is clearly not a science. Nevertheless, we suggest that you treat this topic carefully. A fair number of students are hard-core believers in astrology, and an attempt to dissuade them may backfire by making them dislike you and/or your course. If you can at least get such students to ask a few questions of themselves and their beliefs, you will have achieved a great deal.

Answers/Discussion Points for Think About It Questions

The Think About It questions are not numbered in the book, so we list them in the order in which they appear, keyed by section number.

Section 3.1

- (p. 58) This question simply asks students to think about the process of learning by trial and error. If you use this question for in-class discussion, you should encourage students to think about how this process is similar to the process of thinking used in science.
- (p. 63) This question often generates interesting discussion, particularly if some of your students have read the claims that the Nazca lines have alien origins. We hope students will recognize that such claims shortchange the people who lived there by essentially claiming that they weren't smart enough to have created the lines and patterns themselves. In that way, students usually conclude that the arguments favoring alien origins do not make much sense.

Section 3.2

- (p. 69) The intent of this question is to help students gain appreciation for the accomplishments of ancient Greece. In class, this question can lead to further discussion of how much was lost when the Library of Alexandria was lost and also to discussion of whether the knowledge of our own civilization might suffer a similar fate.

Section 3.3

- (p. 74) Kepler's third law tells us that orbital period depends only on average distance, so the comet with an average distance of 1 AU would orbit the Sun in the same time that Earth orbits the Sun: 1 year. Kepler's second law tells us that the comet would move fast near its perihelion and slow near its aphelion, so it would spend most of its time far from the Sun, out near the orbit of Mars.

Section 3.4

- (p. 82) When someone says that something is "only a theory," they usually mean that it doesn't have a lot of evidence in its favor. However, according to the scientific definition, the term "only a theory" is an oxymoron, since a theory must be backed by extensive evidence. Nevertheless, even scientists often use the word in both senses, so you have to analyze the context to decide which sense is meant.

Section 3.5

- (p. 83) This question asks students to think about the type of prediction made by a newspaper horoscope, as opposed to the more specific prediction of a weather forecast. This can lead to an interesting discussion about what constitutes a testable prediction. In class, you may wish to bring examples of more detailed horoscopes or do an experiment to test astrology.

Solutions to End-of-Chapter Problems (Chapter 3)

We do not include answers to the Review Questions, because these simply require students to review what is written in the text.

1. We all use the trial-and-error methods used in science frequently in our lives. Science is more systematic in its approach than we tend to be in more ordinary situations.
2. Ancient cultures studied astronomy to track the changes of the seasons. They needed this information to help them plant, grow, and harvest crops each year.

   Egyptians—Used Sun and stars to tell time, giving us our 12-hour day and 12-hour night.

   Anasazi—Created the Sun Dagger, which marks the solstices and equinoxes with special illuminations on those days. Understood lunar cycles.

   Babylonians—Were able to predict eclipses accurately.

   Chinese—Kept detailed records of the skies for thousands of years.

   Polynesians—Experts at navigation, including celestial navigation.
3. The days of the week are named for the seven wandering objects in the sky that the ancients knew: the Sun and Moon and the planets Mercury, Venus, Mars, Jupiter, and Saturn.
4. The ancient Egyptians used the shadows of the Sun, perhaps cast by obelisks, to measure the time of day. At night, they used the positions of the stars at different times of year to do the same thing. The Anasazi used the way spirals in the "Sun Dagger" were illuminated to determine the time of year. In particular, on the solstices and equinoxes the illumination was special to mark those days. The Aztecs built the Templo Mayor so that on the equinoxes the Sun would rise from exactly between two structures at the top when viewed from the opposite side of the plaza. Stonehenge may have been constructed for a similar purpose, with the Sun rising near certain stones on special days.
5. A lunar calendar is a calendar in which the months are tied to the Moon's 29-day cycle. As a result, a lunar calendar has 11 fewer days per year than a calendar that is based on the Earth's orbital period.

   The Metonic cycle is the 19-year cycle in which lunar phases recur on the same solar dates.

   Ramadan shifts through the year because the Muslim calendar is a lunar calendar with 11 fewer days per year than a solar year. So the dates of Ramadan shift through the seasons (on which our calendar is based) from one year to the next. The Jewish holidays stay close to a single date on our calendar, however, because the calendar that produces them compensates for this 11-day shortage by adding a month in 7 years of the 19-year Metonic cycle.

6. A scientific model is conceptual rather than physical and is used to explain and predict natural phenomena.
7. The Greek geocentric model goes back a long way into the past. Early developments include the idea of the celestial sphere (fifth century B.C., due to Anaximander), discovery that the Earth is round (Eratosthenes was able to actually measure its radius around 240 B.C.), and the notion (due to Plato) that all of the celestial objects traveled in perfect circles. Eudoxus added separate spheres for each planet, as well as for the Sun and the Moon. Ptolemy incorporated all of this, as well as other ideas, into his model.
8. Claudius Ptolemy was a Greek astronomer who lived in the second century B.C. He gathered together the works of many earlier Greek astronomers into his very successful model of the solar system. His model was able to explain retrograde motion by having the planets move on smaller circles attached to the larger circles on which they went around Earth.
9. The Copernican revolution was the overthrowing of the Ptolemaic model of the solar system, essentially changing the human view of the universe from one in which Earth was imagined to be central to one in which Earth is just one of many similar planets.
10. The Copernican model was not immediately accepted because it didn't do any better at predicting the motions of the planets than the Ptolemaic model did. It was also about equally complex, so there were few advantages to changing models.

    Tycho collected new, more precise data on the positions of the planets. When Johannes Kepler was able to look at these data, he realized that he could improve on Copernicus's basic model by making the orbits elliptical rather than circular with the Sun at one focus rather than the center of the orbit. These improvements, coupled with his other two laws, led to a substantially simpler model of the solar system that was also more accurate than the Ptolemaic model. This new model faced considerable resistance from some people, but it was strongly supported by influential scientists of the day. Most notably, Galileo not only advocated the Copernican model (with Kepler's improvements), but he also used the newly invented telescope to study the sky. In doing so, he made discoveries, such as the phases of Venus and the moons of Jupiter, that supported Copernicus's model and ruled out the basic tenets of the geocentric model.
11. An ellipse is an oval-like figure. We can draw an ellipse by putting two tacks down into a piece of paper and then running a loop of string around both of them. If we hook a pencil inside the string, pull the loop tight, and then drag the pencil around, keeping the string taut, we get our ellipse. The foci of the ellipse are the locations of the tacks. The eccentricity is a measure of how noncircular the ellipse is: 0 eccentricity is a circle, while higher values of the eccentricity make more stretched-out ellipses. (The maximum value of eccentricity for an ellipse is 1.)
12. i) Planets move in elliptical orbits around the Sun with the Sun at one focus. This describes the shape of the orbits (ellipses rather than the circles used by most previous models) and where the Sun is located relative to the orbits (at a focus rather than in the center).

    ii) A line from the planet to the Sun sweeps out equal areas in equal amounts of time. This law describes how fast the planets move in their orbits. When they are close to the Sun, they move faster and when they are far away they move slower.

iii) The law $p^2 = a^3$ relates the period, p, with the semimajor axis, a. This law says that the more distant planets orbit more slowly than the ones that are closer to the Sun. It also says that the only thing that affects the orbital period of the planets is the semimajor axis. So things like the mass of the planet and the orbital eccentricity do not matter.

13. A hypothesis in science is essentially an educated guess about why or how some phenomenon happens. If the hypothesis survives repeated tests and explains a broad enough range of phenomena, it may be elevated to the status of theory.
14. The hallmarks of science are that it seeks explanations for phenomenon using natural causes, relies on the creation and testing of models (and that the models should be as simple as possible), and uses testable predictions to determine if a model should be kept or discarded. In the Copernican revolution, the first hallmark shows in the way Tycho's data led Kepler to look for a natural explanation for the observations. The second shows in the way the Copernican model, with Kepler's improvements, proved better than any competing model, such as that of Ptolemy. The third shows in the way the models were carefully tested by looking for observations that each model predicted. The Ptolemaic model was then rejected and we do not use it today.

    Occam's razor is the idea that when faced with more than one model that seems to match the data, we should use the simplest one.

    Personal testimony does not count as evidence in science because it is impossible for other people to verify the testimony independently.
15. A pseudoscience is something that looks like a science but is not. For example, astrology is a pseudoscience because it makes predictions, but it does not reject models that make repeated inaccurate predictions. On the other hand, many fields do not make predictions and do not act likes sciences, so they are nonsciences but not pseudoscience. (For example, religions are ways of understanding the world around us, but because they do not generally make testable predictions, they are nonscience but not pseudoscience.)
16. The basic idea of astrology is that the positions of objects in the sky (the Sun, Moon, planets, and so forth) can affect our lives or predict our futures. For the ancients, astrology probably seemed to make sense because the Sun and Moon really do have effects on our lives: The changing of the seasons and the coming and going of the tides are tied to the Sun and Moon, after all. So it might seem natural to guess that other bodies in the sky might affect us as well. Today, however, we understand that that the Sun and Moon influence our lives through gravity (and light in the case of the Sun), and that the gravity of the planets is too weak to influence us at the distances at which they are located. Moreover, repeated tests have always shown astrological predictions to be no more accurate than would be expected by pure chance.
17. *Ancient astronomers failed to realize that Earth goes around the Sun because they just weren't as smart as people today.* This statement does not make sense; the human brain has not evolved significantly in historical times.
18. *In ancient Egypt, children whose parents gave them "1 hour" to play got to play longer in the summer than in the winter.* This statement makes sense, because the Egyptians divided the daylight hours into 12 equal parts, regardless of the time of year. Thus, the longer daylight in summer meant the Egyptian hour was longer in summer.

19. *If the planet Uranus had been identified as a planet in ancient times, we'd probably have 8 days in a week.* It's impossible to know whether this would be true or false, but it makes sense based on the origin of the names of the days of our week. With an eighth "planet" visible in the sky, we might well have had eight days in a week.
20. *The date of Christmas (December 25) is set each year according to a lunar calendar.* This statement does not make sense, because December 25 is a date on a solar calendar.
21. *When navigating in the South Pacific, the Polynesians found their latitude with the aid of the pointer stars of the Big Dipper.* This statement does not make sense, because the Big Dipper and north celestial pole are not visible from deep in the Southern Hemisphere.
22. *The Ptolemaic model reproduced apparent retrograde motion by having planets move sometimes counterclockwise and sometimes clockwise in their circles.* This statement is false. The Ptolemaic model did not vary the directions in which planets moved; it reproduced apparent retrograde motion by having the planets turn on small circles upon larger circles.
23. *According to Kepler's laws, Earth would take longer to orbit the Sun if it had a larger mass.* This statement is false; Kepler's third law tells us that a planet's orbital period depends only on its average orbital distance.
24. *In science, saying that something is a theory means that it is really just a guess.* This statement is false. In science, a theory must be well tested.
25. *A scientific theory should never gain acceptance until it has been proved true beyond all doubt.* This statement is false, as a theory can never be proved true beyond doubt.
26. *Ancient astronomers were convinced of the validity of astrology as a tool for predicting the future.* This statement is false; as evidenced by the quotation from Ptolemy in Section 3.5, ancient astronomers were well aware of the shortcomings of astrology as a means of prediction.
27. b; 28. a; 29. b; 30. a; 31. c;
32. b; 33. b; 34. c; 35. c; 36. b.
37. More than one answer is possible for each part of this question, but here are some samples: (a) Observing changes in the sky with latitude would show that Earth is not flat. (b) Showing that the Sun is in different positions (that is, different times of day) for different longitudes would show that Earth is curved east-west in addition to north-south. (c) Showing that changes in the sky with latitude and longitude are independent of where you start would show that Earth has spherical symmetry rather than some other shape.
38. The dates will vary depending on the year. The key points are: (a) Chanukah stays within a few-week range, because its date is chosen on a calendar that follows the Metonic cycle. (b) Ramadan moves through the year, because it is tied to a "pure" lunar calendar (one that does not use the Metonic cycle).
39. This question asks students to make a bulleted "executive summary" of the Copernican revolution. Answers will vary, so grades should be based on the clarity, conciseness, and completeness of the list.
40. Answers will vary depending on the idea chosen. The key in grading is for students to explain themselves clearly and to defend their opinions well.

41. This essay question can generate interesting responses. Of course, the impacts of the Copernican revolution involve opinion, so grade essays based on how well they are written and defended.
42. This question involves independent research. Answers will vary.
43. This problem requires students to devise their own scientific test of astrology. One example of a simple test is to cut up a newspaper horoscope and see whether others can correctly identify the one that applies to them.

44.–46. These questions involve independent research. Answers will vary.

47. First, we calculate the number of days in 19 years, given that there are 365.2422 days in a year:

$$19 \cancel{\text{yr}} \times \frac{365.2422 \text{ days}}{1 \cancel{\text{yr}}} = 6{,}939.60 \text{ days}$$

We can compare this to the number of days in 235 months, given that there are 29.5306 days in a month:

$$235 \cancel{\text{months}} \times \frac{29.5306 \text{ days}}{1 \cancel{\text{month}}} = 6{,}939.69 \text{ days}$$

Clearly, these two numbers are very close, less than a tenth of a day different. So 19 years is almost exactly 235 months.

We can use this fact to keep lunar calendars roughly synchronized with the seasons by making sure to add an extra month in several years out of every 19. By doing this, we ensure that there are 235 months in each 19-year cycle so that the lunar calendar is well synchronized with the seasons.

48. First, let us calculate how many months are in 60 years in the Chinese calendar. We are told that the typical year is 12 months long and that 22 years have an extra month. Using this information, we see that there are 22 years with 13 months and the remaining 38 years (60 years – 22 years) have 12 months. So the total number of months is:

$$38 \cancel{\text{yr}} \times \frac{12 \text{ months}}{1 \cancel{\text{yr}}} + 22 \cancel{\text{yr}} \times \frac{13 \text{ months}}{1 \cancel{\text{yr}}} = 742 \text{ months}$$

In Problem 47 we are told that the average month has 29.5306 days. So we can calculate the number of days in those 742 months:

$$742 \cancel{\text{months}} \times \frac{29.5306 \text{ days}}{1 \cancel{\text{month}}} = 21{,}912 \text{ days}$$

In Problem 47 we also see that there are 365.2422 days in 1 solar year, so we can also calculate how many days would be in 60 solar years:

$$60 \cancel{\text{yr}} \times \frac{365.2422 \text{ days}}{1 \cancel{\text{yr}}} = 21{,}915 \text{ days}$$

The Chinese calendar differs from a precise solar calendar by 3 days in every 60 years. This scheme is similar to the method discussed in Problem 47 and in the text (where there are 7 years with 13 months in every 19 years) because both add extra months to some years to compensate for the shift of the lunar calendars relative to the seasons.

49. We will follow Eratosthenes's method, from the Special Topic box. In the case of Nearth, we learn that the Sun is straight overhead at Nyene at the same time that it is 10° from the zenith at Alectown and that the two cities are 1,000 kilometers apart. So we can set up the same type of relationship as Eratosthenes did:

$$\frac{10\cancel{^\circ}}{360\cancel{^\circ}} \times (\text{circumference of Nearth}) = 1{,}000 \text{ km}$$

We solve for the circumference of Nearth:

$$\begin{aligned}\text{circumference of Nearth} &= 1{,}000 \text{ km} \times \frac{360\cancel{^\circ}}{10\cancel{^\circ}} \\ &= 36{,}000 \text{ km}\end{aligned}$$

The circumference of Nearth is 36,000 kilometers.

50. We will follow Eratosthenes's logic, from the Special Topic box. In the case of Tirth, we learn that the Sun is straight overhead at Tyene at the same time that it is 4° from the zenith at Alectown and that the two cities are 1,000 kilometers apart. So we can set up the same sort of relationship as Eratosthenes did:

$$\frac{4\cancel{^\circ}}{360\cancel{^\circ}} \times (\text{circumference of Tirth}) = 400 \text{ km}$$

We solve for the circumference of Tirth:

$$\begin{aligned}\text{circumference of Tirth} &= 400 \text{ km} \times \frac{360\cancel{^\circ}}{4\cancel{^\circ}} \\ &= 36{,}000 \text{ km}\end{aligned}$$

The circumference of Tirth is 36,000 kilometers.

51. We are told in Mathematical Insight 3.2 that the perihelion and aphelion distances between the Sun and a planet are, respectively:

$$\text{perihelion distance} = a(1 - e)$$
$$\text{aphelion distance} = a(1 + e)$$

where a is the semimajor axis and e is the eccentricity.
From Appendix E, Mars's orbital eccentricity is 0.093 and its semimajor axis is 1.524 AU. So we get:

$$\begin{aligned}\text{perihelion distance} &= (1.524 \text{ AU}) \times (1 - 0.93) \\ &= 1.382 \text{ AU}\end{aligned}$$

and

$$\begin{aligned}\text{aphelion distance} &= (1.524 \text{ AU}) \times (1 + 0.93) \\ &= 1.666 \text{ AU}\end{aligned}$$

Mars's minimum distance from the Sun is 1.382 AU and its maximum distance is 1.666 AU.

52. From Appendix E, Pluto has the largest eccentricity of the planets, at 0.248. (Note that we are ignoring the recently discovered Planet X, which is apparently larger than Pluto and has a greater eccentricity.) Given its semimajor axis, 39.54 AU, we can use the formulas from Mathematical Insight 3.2 to calculate its perihelion and aphelion distances from the Sun:

$$\begin{aligned}\text{perihelion distance} &= a(1-e)\\ &= 39.54\text{ AU} \times (1-0.248)\\ &= 29.73\text{ AU}\end{aligned}$$

and

$$\begin{aligned}\text{aphelion distance} &= a(1+e)\\ &= 39.54\text{ AU} \times (1+0.248)\\ &= 49.35\text{ AU}\end{aligned}$$

Pluto's perihelion distance from the Sun is 29.73 AU, and its aphelion distance from the Sun is 49.35 AU.

53. From Appendix E, Venus has the smallest eccentricity of the planets, at 0.007. Given its semimajor axis, 0.723 AU, we can use the formulas from Mathematical Insight 3.2 to calculate its perihelion and aphelion distances from the Sun:

$$\begin{aligned}\text{perihelion distance} &= a(1-e)\\ &= 0.723\text{ AU} \times (1-0.007)\\ &= 0.718\text{ AU}\end{aligned}$$

and

$$\begin{aligned}\text{aphelion distance} &= a(1+e)\\ &= 0.723\text{ AU} \times (1+0.007)\\ &= 0.728\text{ AU}\end{aligned}$$

Venus's perihelion distance from the Sun is 0.718 AU, and its aphelion distance from the Sun is 0.728 AU.

54. We can use Kepler's third law to find the period of Sedna. Kepler's third law states that

$$a^3 = p^2$$

where a is the semimajor axis in AU and p is the period in years. We are asked to find the period, so we can solve for p by taking the square root of both sides:

$$p = \sqrt{a^3}$$

Using the given value of a = 509 AU, we calculate:

$$\begin{aligned}p &= \sqrt{509^3}\\ &= 11{,}500\text{ yr}\end{aligned}$$

Sedna's orbital period is 11,500 years.

55. Since the mass of this new star is the same as the mass of the Sun, we can use Kepler's third law. According to Kepler's third law:

$$a^3 = p^2$$

where a is the semimajor axis in AU and p is the period in years. We are asked to find the period, so we can solve for p by taking the square root of both sides:

$$p = \sqrt{a^3}$$

We do not need the eccentricity at all (notice that it does not appear in Kepler's third law), but we do need to convert the new planet's semimajor axis into AU to use this formula. So converting 112,000,000 kilometers into AU:

$$112{,}000{,}000 \not{\text{km}} \times \frac{1 \text{ AU}}{150{,}000{,}000 \not{\text{km}}} = 0.747 \text{ AU}$$

Using this in Kepler's third law;

$$\begin{aligned} p &= \sqrt{(0.747)^3} \\ &= 0.645 \text{ yr} \end{aligned}$$

The new planet has an orbital period of 0.645 year, or about $7\frac{3}{4}$ months.

56. a. We can use Kepler's third law to find the semimajor axis of Halley's comet. Kepler's third law states that:

$$a^3 = p^2$$

where a is the semimajor axis in AU and p is the period in years. We are asked to find the semimajor axis, so we can solve for a by taking the cube root of both sides:

$$a = \sqrt[3]{p^2}$$

Since Halley has an orbital period of 76 years, we can calculate the semimajor axis:

$$\begin{aligned} a &= \sqrt[3]{76^2} \\ &= 17.9 \text{ AU} \end{aligned}$$

Comet Halley has a semimajor axis of 17.9 AU.

b. We use the formulas from Mathematical Insight 3.2. We are told that Halley's comet has an eccentricity of 0.97 and we calculated its semimajor axis to be 17.9 AU in part (a):

$$\begin{aligned} \text{perihelion distance} &= 17.9\ (1 - 0.97) \\ &= 0.537 \text{ AU} \end{aligned}$$

and

$$\begin{aligned} \text{aphelion distance} &= 17.9\ (1 + 0.97) \\ &= 35.3 \text{ AU} \end{aligned}$$

Halley's comet comes as close as 0.537 AU to the Sun and travels as far away as 35.3 AU from the Sun.

c. Halley's comet spends most of its time far from the Sun near aphelion. We know this from Kepler's second law, which tells us that bodies move faster when they are closer to the Sun in their orbits than when they are farther away. So Halley's comet moves most slowly at aphelion. Since it is moving most slowly there, Halley's comet also spends more time in that part of its orbit.

Chapter S1. Celestial Timekeeping and Navigation

This is a supplementary chapter, meaning that coverage is optional—nothing in this chapter is prerequisite for the rest of the book. This final chapter of Part 1 fills in details about apparent motions of the sky that have not already been covered in the first three chapters. Note that, although it covers fairly "basic" ideas about the sky, this material is often quite difficult for students. Understanding motions of the sky requires visualizing three-dimensional geometry, which some students simply cannot grasp in just a week or so. It certainly helps if you have access to a planetarium when covering the motions of the sky.

As always, when you prepare to teach this chapter, be sure you are familiar with the relevant media resources (see the complete, section-by-section resource grid in Appendix 3 of this Instructor's Guide) and the online quizzes and other study resources available on the Astronomy Place Web site.

What's New in the Fourth Edition That Will Affect My Lecture Notes?

As everywhere in the book, we have added learning goals that we use as subheadings, rewritten to improve the text flow, improved art pieces, and added new illustrations. The art changes, in particular, will affect what you wish to show in lecture. We have not made any substantial content or organizational changes to this chapter.

Teaching Notes (By Section)

Section S1.1 Astronomical Time Periods

This section introduces basic astronomical periods: the solar versus the sidereal day; the synodic versus the sidereal month; the tropical versus the sidereal year; and synodic versus sidereal periods of the planets.

- You may wish to do the demonstration described in the text to show the difference between a sidereal and a solar day; it's easy for students to watch you rotating as you walk around an object that represents the Sun.
- Technical note: The true rotation period of the Earth differs from the sidereal day by a few thousandths of a second because of the precession of Earth's axis.
- If your campus has a sundial, take a class field trip or ask your students to investigate it on their own.
- Many students will find the issues of calendar reform quite interesting from a historical point of view, especially the fact that, for centuries, not all countries agreed on the date, even if they were ostensibly using the same "Christian" calendar.

Section S1.2 Celestial Coordinates and Motion in the Sky

This section covers the coordinates of right ascension and declination and the variation in sky motions with latitude.

- In class, you may want to emphasize that these coordinates are easier to understand if we think of them as "celestial latitude" and "celestial longitude."
- If possible, it really helps if you can visit a planetarium to demonstrate the motions described in this section. In that case, you may wish to begin by pointing out how its dome distorts what we see in the real sky. In particular, the point in the planetarium representing the zenith generally is directly over the projector, rather than over any audience member's head. As a result, the planetarium sky looks most distorted above wherever you are sitting. The other major distortion is the smaller angular size of everything viewed in the planetarium compared to the real sky.
- We've found that the order in which sky motions at different latitudes are described is very important in helping students understand what is going on. The order we've chosen begins with the simplest latitudes—the North Pole and then the equator. In class, you should next do your own latitude, then generalize.
- Keep in mind that some otherwise bright students will have difficulty with this three-dimensional geometry regardless of what you do. Such students may simply have to memorize the rules rather than truly understand the geometry.

Section S1.3 Principles of Celestial Navigation

This section gives a brief overview of techniques of celestial navigation.

- The idea of determining your longitude by calling a friend in Greenwich may seem a bit far-fetched, especially in historical context, but we've found it effective at getting students to realize that it's really pretty easy to determine longitude—the only trick is that you need a clock to tell you the time someplace else.
- *Technical note:* Common Misconceptions: Compass Directions. In this box, we say "magnetic north" to mean the magnetic pole located in the Northern Hemisphere. In terms of response to magnetism, the magnetic pole in the Northern Hemisphere is a south magnetic pole; that is why the north end of a magnet is attracted in this direction. Similarly, by "south magnetic pole" we really mean the magnetic pole located in the Southern Hemisphere—which is a north magnetic pole, according to its magnetic properties.

Answers/Discussion Points for Think About It Questions

The Think About It questions are not numbered in the book, so we list them in the order in which they appear, keyed by section number.

Section S1.1

- (p. 93) At midnight, looking toward the meridian means looking in a direction 180° away from the Sun. Neither Mercury nor Venus ever ventures anywhere close to this far from the Sun as viewed from Earth; neither, therefore, ever appears on the meridian at midnight.
- (p. 96) 12:01 A.M. is 1 minute after midnight, while 12:01 P.M. is 1 minute after noon.

Section S1.2

- (p. 99) This exercise asks students to continue work with their own model of the celestial sphere, which we explain how to make in the main text (a few paragraphs earlier). On April 21, the Sun is about 1/12 of the way around the ecliptic from the spring equinox. On November 21, it is about 1/12 of the way short of being at the winter solstice.
- (p. 101) This exercise again asks students to continue building their own model of the celestial sphere.
- (p. 103) Again, students will mark their models of the celestial sphere. They should now be able to locate the Sun for any day, including their birthdays.
- (p. 104) No stars are circumpolar at the equator, and all portions of the celestial sphere become visible over the course of the day.
- (p. 106) At 30°S latitude, the celestial equator crosses the meridian at altitude $90° - 30° = 60°$ in the northern half of the sky. Stars with positive declinations follow short tracks across the northern sky. Stars with negative declinations follow long tracks across the southern sky (crossing into the north if they have a declination between 0° and –30°). Stars with declinations more negative than –60° are circumpolar.

Solutions to End-of-Chapter Problems (Chapter S1)

1. A sidereal day is shorter than a solar day because as Earth spins, it also moves around the Sun. So for each complete rotation, Earth must rotate a little extra before the Sun returns to the same place in the local sky.
2. A sidereal month is the time it takes for the Moon to go around the Earth relative to the background stars. This is shorter than a synodic month, which is the time it takes the Moon to go from new phase to the next new phase. The synodic month is longer than the sidereal month due to Earth's orbital motion, which means that the Moon must travel more than one complete orbit from new moon to new moon.

   A sidereal year is the time it takes Earth to complete an orbit relative to the positions of distant stars. This is about 20 minutes longer than a tropical year, which is the time between two successive vernal (spring) equinoxes. The difference is due to the precession of Earth's spin axis.

   A planet's sidereal period is the time it takes the planet to go around the Sun. (From the Sun's point of view, the planet would come back to the same stars. But don't try to stand on the Sun and check.) The planet's synodic period is the time it takes the planet to go from a particular alignment with the Earth back to that same relative alignment—for example, the time it takes Mars to go from being opposition to opposition again.
3. Opposition occurs when a planet is on the opposite side of Earth from the Sun. This can only happen for planets located farther than Earth from the Sun. (So Venus and Mercury are *never* in opposition.)

   A conjunction occurs when a planet is aligned with the Sun in our sky, either between the Sun and Earth or on the opposite side of the Sun from Earth. For planets beyond Earth, only a superior conjunction (when the planet is on the opposite side of the Sun from Earth) is possible. For Mercury and Venus, which orbit closer to the Sun than we do, an inferior conjunction (when the planet is between the Sun and Earth) also occurs.

Greatest elongation only occurs for planets closer to the Sun than the Earth—that is, for Mercury and Venus. This is the arrangement where the planet will appear to be as far from the Sun in our sky as it will get during an orbit.

4. We will see a transit only for the planets Mercury and Venus (at least for planets within our solar system). A transit occurs when a planet appears to pass in front of the Sun's disk as seen from Earth, so it can occur only when the planet is in inferior conjunction. Moreover, because Mercury and Venus's orbits are tilted relative to Earth's orbit, the planets usually are slightly north or south of the Sun at inferior conjunction. However, when the planet's orbits line up just right, we get a transit. Transits of Mercury occur about a dozen times a century, while Venus transits occur in pairs separated by more than 100 years, with the second transit of a pair occurring about 8 years after the first.

5. Apparent solar time is the time based on the actual location of the Sun in the sky, while mean solar time is based on making all days 24 hours long. The two differ because of the tilt of Earth's spin access and because Earth moves at different speeds in its orbit as it gets slightly closer to or farther from the Sun, so that the actual length of a day based on the Sun's position in the sky actually varies over the course of the year, with 24 hours as the average. Apparent solar time can be read directly from a sundial, while mean solar time is easier to obtain with mechanical or electronic clocks because all days are the same length.

   Standard time is the mean solar time over an east-west swath of Earth known as a time zone. Daylight saving time is simply standard time advanced by 1 hour. Universal time is the mean solar time for Greenwich, England.

6. The Julian calendar was developed at Julius Caesar's orders to correct problems with the older Roman calendar. It added an extra day every 4 years to keep the calendar more closely synchronized with the seasons. However, over many years this was still not quite accurate enough, and the Julian calendar drifted relative to the seasons. Pope Gregory XIII ordered the calendar to be fixed. His Gregorian calendar modified the leap-year pattern so that while it still normally occurs every 4 years, we do not have leap year in century years, except for years that are divisible by 400. Today we use the Gregorian calendar.

7. When we describe equinoxes and solstices as points on the celestial sphere, we refer to the positions among the stars that the Sun actually occupies on those days. For example, the spring equinox in the sky is the point in the constellation Pisces at which the Sun is located on the day of the spring equinox.

8. Declination and right ascension are the coordinates we use to pinpoint positions on the celestial sphere. These are quite similar to latitude and longitude: Like latitude, declination is zero at the equator and is measured in degrees. But rather than say "north" or "south," we assign declinations positive or negative values. Right ascension is like longitude, except that it is usually measured in hours, minutes, and seconds rather than in degrees.

9. The Sun's celestial coordinates change over the year because, as we orbit the Sun, the Sun appears to drift eastward around the celestial sphere at a rate of about 2 hours of right ascension per month. It also moves somewhat north and south with the seasons (due to Earth's tilt), so that the Sun has its largest positive (northmost) declination on the summer solstice and its largest negative (southmost) declination on the winter solstice, and is at 0° declination on the equinoxes.

10. At the North Pole, the north celestial pole is directly overhead and the celestial equator circles the horizon.

    At the equator, the north celestial pole is on the northern horizon, and the celestial equator runs from due east to due west and through the zenith.

    At 40° north latitude, the north celestial pole is due north, 40° above the horizon. The celestial equator runs from due east to due west, passing through an altitude of 50° due south.

11. At 40°N latitude, the Sun will rise due east and set due west on the equinoxes. On these days it will cross the meridian at altitude 50° due south. On the summer solstice, the Sun will rise north of due east, cross the meridian at altitude 73.5° due south, and set north of due west. On the winter solstice, the Sun rises south of due east, crosses the meridian at altitude 26.5° due south, and sets south of due west.

    At the equator on the equinoxes, the Sun will rise due east, pass directly overhead, and set due west. On the summer solstice, the Sun will rise north of due east, cross the meridian at 66.5° to the *north* and set north of due west. On the winter solstice, the mirror image will occur: The Sun will rise south of due east, cross the meridian at 66.5° altitude to the south, and set south of due west.

    At the North Pole on the equinoxes, the Sun will skim the horizon, making a circle all around the horizon each day. On the summer solstice the Sun will circle 23.5° above the horizon all day. On the winter solstice, we would not see the Sun at all.

    At the South Pole on the equinoxes, the Sun will skim the horizon, making a circle all around the horizon each day. On the summer solstice, the Sun would not appear at all. On the winter solstice, the Sun will circle 23.5° above the horizon all day.

12. The tropics of Cancer and Capricorn are the most extreme latitudes (farthest north and south) where the Sun can ever be seen directly overhead. On the solstices (summer for Cancer, winter for Capricorn), the Sun will rise north of due east (south, for winter at Capricorn), pass directly overhead, and set north of due west (south, for winter at Capricorn).

    The Arctic and Antarctic Circles are the northmost and southmost points where the Sun will be seen at least briefly on every day of the year. Alternatively, these are the southmost and northmost latitudes where the Sun will appear above the horizon for an entire 24-hour day at least once a year. So on the summer solstice at the Arctic Circle, the Sun will be due north at midnight, just at the horizon. It will work its way east, growing higher in the sky, until it comes around to the meridian to the south, when it will be at 47° above the horizon. It will then continue off to the west, then north, sinking back toward the horizon. On the winter solstice, the Sun will just barely appear above the horizon due south at noon. For the Antarctic circle, this situation is exactly reversed.

13. If we know the time and date, we can use the Sun to determine our latitude and longitude. The Sun's declination on a given date is well known, so based on the Sun's altitude when it's on the meridian, we can work out our latitude. We could alternately use a star at night to do the same thing. Longitude is trickier. To do this, we need a clock that keeps Universal Time accurately. However, if we have this, we can use when a given star crosses the meridian (or when the Sun does, although we have to account for the Sun's variations with the day of the year) to tell our longitude.

14. The Global Positioning System is a set of satellites designed to allow people on Earth to determine their locations with extreme accuracy.

15. *Last night I saw Venus shining brightly on the meridian at midnight.* This statement does not make sense, because it would require that Venus be at opposition to the Sun in the sky—and, because Venus is closer to the Sun than is Earth, it is never at opposition.
16. *The apparent solar time was noon, but the Sun was just setting.* This statement does not make sense, because apparent solar noon is defined as the time when the Sun is at its highest point on the meridian. If the Sun is at its highest point, it cannot be setting.
17. *My mean solar clock said it was 2:00 P.M., but my friend who lives east of here had a mean solar clock that said it was 2:11 P.M.* Mean solar time is different for every different longitude, so this statement makes sense if the friend lives the equivalent of 11 minutes of longitude east of you.
18. *When the standard time is 3:00 P.M. in Baltimore, it is 3:15 P.M. in Washington, D.C.* This statement does not make sense, because standard time zones must be 1 hour apart, not 15minutes apart. (Also, Baltimore and Washington, D.C., are in the same time zone.)
19. *Last night around 8 P.M., I saw Jupiter at an altitude of 45° in the south.* This statement makes sense, because it describes the position of Jupiter in your local sky.
20. *The latitude of the stars in Orion's belt is about 5°N.* This statement does not make sense; Orion's belt is not on the Earth, and hence does not have a latitude.
21. *Today the Sun is at an altitude of 10° on the celestial sphere.* This statement does not make sense, because altitude is a coordinate of the local sky, not of the celestial sphere.
22. *Los Angeles is west of New York by about 3 hours of right ascension.* This statement does not make sense, because right ascension is a coordinate of the celestial sphere, not of the Earth.
23. *The summer solstice is east of the vernal equinox by 6 hours of right ascension.* This statement is true at all times.
24. *Even though my UT clock had stopped, I was able to find my longitude by measuring the altitudes of 14 different stars in my local sky.* This statement does not make sense, because longitude determination requires comparing the positions of stars (or the Sun) in your location with their positions in a known location, and the latter requires knowing the time in that location. Thus, without a clock, you can't determine your longitude.
25. b; 26 a; 27. c; 28. b; 29. a;
30. a; 31. c; 32. c; 33. a; 34. c.
35. If Earth rotated in the opposite direction, our orbit around the Sun would mean we'd need to turn through *less* than one full rotation from noon one day to noon the next, so the solar day would be *shorter* than the sidereal day.
36. If Earth's axis did not precess, there would be no difference between the sidereal and tropical years.
37. Answers will vary with latitude (except for part (b)); the following is a sample answer for 40°N latitude:
    a. The north celestial pole appears in your sky at an altitude of 40°, in the direction due north.

b. The meridian is a half-circle that stretches from the point due south on the horizon, through the zenith, to the point due north on the horizon.
c. The celestial equator is a half-circle that stretches from the point due east on the horizon, through an altitude of 50° due south, to the point due west on the horizon.
d. The Sun can appear at the zenith only in the tropics. At latitude 40°N, the Sun is never at the zenith.
e. Because the north celestial pole appears due north at an altitude of 40°, a star is circumpolar if it is within 40° of the north celestial pole. The north celestial pole has a declination of +90°, so within 40° means declinations greater than +50°.
f. The celestial equator reaches a maximum altitude of 50° in the southern sky. Thus, any star that is more than 50° south of the celestial equator is never visible above the horizon. More than 50° south of the celestial equator means declinations more negative than –50°.

38. For Sydney, 34°S latitude:
a. The south celestial pole appears in your sky at an altitude of 34°, in the direction due south.
b. The meridian is a half-circle that stretches from the point due south on the horizon, through the zenith, to the point due north on the horizon.
c. The celestial equator is a half-circle that stretches from the point due east on the horizon, through an altitude of 56° due north, to the point due west on the horizon.
d. The Sun can appear at the zenith only in the tropics. At latitude 34°S, the Sun is never at the zenith.
e. Because the south celestial pole appears due south at an altitude of 34°, a star is circumpolar if it is within 34° of the south celestial pole. The south celestial pole has a declination of –90°, so within 34° means declinations more negative than –56°.
f. The celestial equator reaches a maximum altitude of 56° in the northern sky. Thus, any star that is more than 56° north of the celestial equator is never visible above the horizon. More than 56° north of the celestial equator means declinations greater than +56°.

39. Answers will vary with latitude; the following is a sample answer for 40°N latitude:
a. On the spring or fall equinox, the Sun rises due east, reaches an altitude of 50°S on the meridian, and sets due west.
b. On the summer solstice, the Sun rises more than *23.5°* north of due east, reaches an altitude of 50° + 23.5° = 73.5° on the meridian in the south, and sets more than *23.5°* north of due west.
c. On the winter solstice, the Sun rises more than *23.5°* south of due east, reaches an altitude of 50° – 23.5° = 26.5° on the meridian in the south, and sets more than *23.5°* south of due west.
d. Answers depend on the date.

40. In Sydney (lat. 34°S): The celestial equator goes due east, through 90° – 34° = 56° in the north, to due west.
a. On the spring or fall equinox, the Sun rises due east, reaches an altitude of 56°N on the meridian, and sets due west.

b. On the summer solstice, the Sun rises more than 23.5° north of due east, reaches an altitude of 56° – 23.5° = 32.5° on the meridian in the north, and sets more than 23.5° north of due west.

c. On the winter solstice, the Sun rises more than 23.5° south of due east, reaches an altitude of 56° + 23.5° = 79.5° on the meridian in the north, and sets more than 23.5° south of due west.

d. Answers depend on the date.

41. a. Your latitude is 15°N. Because it is the vernal equinox, the Sun follows the path of the celestial equator through the sky. Thus, the Sun's meridian altitude of 75°S tells you that this also is the altitude at which the celestial equator crosses the meridian. Because we know that the celestial equator crosses the meridian at 90°– [your latitude], your latitude is 90° – 75° = 15°; it is north latitude because the celestial equator is in your southern sky.

b. Your longitude is 150°W. The Sun is on your meridian, so it is noon for you. The UT clock reads 22:00, or 10 P.M., so Greenwich is 10 hours ahead of you. Each hour represents 15° of longitude, so 10 hours means 150°; you are west of Greenwich because your time is behind.

c. You are very close to Hawaii.

42. a. You are on the equator. Because it is the summer solstice, the Sun crosses the meridian 23.5° north of the celestial equator. Thus, the Sun's meridian altitude of 67.5°N tells you that the celestial equator is passing through your zenith and hence that you are on the Earth's equator.

b. Your longitude is 90°E. The Sun is on your meridian, so it is noon for you. The UT clock reads 06:00, or 6 A.M., so Greenwich is 6 hours behind you. Each hour represents 15° of longitude, so 6 hours means 90°; you are east of Greenwich because your time is ahead.

c. You are in the Indian Ocean, not too far west of the island of Sumatra.

43. a. Your latitude is within 1° of 67°N, which you know because that is the altitude of Polaris in your sky.

b. Your longitude is 15°W. Your local time is midnight and the UT clock reads 01:00, or 1 A.M., so Greenwich is 1 hour ahead of you. Thus, you are 15° west of Greenwich.

c. You are just off the northern coast of Iceland.

44. a. Your latitude is 33°S, which you know because that is the altitude of the south celestial pole in your sky.

b. Your longitude is 75°W. Your local time is 6 A.M. and the UT clock reads 11:00, or 11 A.M., so Greenwich is 5 hours ahead of you. Thus, you are $5 \times 15° = 75°$ west of Greenwich.

c. You are off the coast of Chile, nearly due west of Santiago.

45. Because Mars has nearly the same axis tilt as Earth, the path of the Sun in the summer is very much like the path of the Sun on Earth in the summer for the same latitude. Because the latitude is only 15°N, the Sun is crossing through the northern sky. The operators should tilt the panels a little toward the north.

46. The range of latitudes for which the Sun can reach the zenith is slightly larger, as is the range of latitudes for which the Sun can be circumpolar. Both ranges increase with axis tilt, so the slightly larger axis tilt of Mars means slightly greater ranges. (Sketch not shown.)
47. If Earth went around the Sun in 6 months rather than a year, we would travel twice as great of an angle around the Sun in 1 day as we do now. Right now, we travel about 1° around the Sun in our orbit, so we'd travel 2° per day in the 6-month year. The solar day would have to be longer than the sidereal day by enough for Earth to spin an extra 2°. How long would this take? Well, Earth spins 360° in about 24 hours (actually, in 23 hours and 56 minutes, but this should be good enough). So converting from degrees to hours, we get:

$$2^{\circ} \times \frac{24 \text{ hr}}{360^{\circ}} = 0.13 \text{ hr}$$

We had better convert this to minutes, though:

$$0.13 \text{ hr} \times \frac{60 \text{ min}}{1 \text{ hr}} = 8 \text{ min}$$

So Earth's solar day would be 8 minutes longer than the sidereal day. Since the sidereal day is 23 hours and 56 minutes, that comes out to 24 hours and 4 minutes. Note that the extra 8 minutes is exactly twice the 4-minute difference we experience with the 12-month year. Since we'd be traveling twice as far per day (in angle) in the 6-month year, this should make sense.
48. In Mathematical Insight S1.1, we learn that the orbital period for planets farther from the Sun than Earth is related to the synodic period by the relationship:

$$P_{orb} = P_{syn} \times \frac{1 \text{ yr}}{(P_{syn} - 1 \text{ yr})}$$

where P_{orb} is the sidereal orbital period and P_{syn} is the synodic orbital period. So we just need to get Saturn's synodic period in years so that the units match up. So we convert from days to years:

$$378.1 \text{ days} \times \frac{1 \text{ yr}}{365.25 \text{ days}} = 1.0352 \text{ yr}$$

Plugging this value for P_{syn}, we get:

$$P_{orb} = 1.0352 \text{ yr} \times \frac{1 \text{ yr}}{(1.0352 - 1 \text{ yr})}$$

$$= 29.41 \text{ yr}$$

Saturn's orbital period must be 29.41 years. Note that our answer agrees quite well with the value in Appendix E of 29.42 years.
49. In Mathematical Insight S1.1, we learn that the orbital period for planets closer to the Sun than Earth is related to the synodic period by the relationship:

$$P_{orb} = P_{syn} \times \frac{1 \text{ yr}}{(P_{syn} + 1 \text{ yr})}$$

where P_{orb} is the sidereal orbital period and P_{syn} is the synodic orbital period.

We need to convert Mercury's synodic period into years so that the units match up:

$$115.9 \cancel{\text{days}} \times \frac{1 \text{ yr}}{365.25 \cancel{\text{days}}} = 0.3173 \text{ yr}$$

Plugging in to the formula, we get:

$$P_{orb} = 0.3173 \cancel{\text{yr}} \times \frac{1 \text{ yr}}{(0.3173 + 1 \cancel{\text{yr}})}$$

$$= 0.2409 \text{ yr}$$

Mercury's orbital period is 0.2409 year, which is in excellent agreement with the value in Appendix E.

50. Because the synodic period of the asteroid is longer than Earth's orbital period, it must orbit farther from the Sun than the Earth does. In this case, we will use the relationship for synodic and sidereal orbital periods for those planets:

$$P_{orb} = P_{syn} \times \frac{1 \text{ yr}}{(P_{syn} - 1 \text{ yr})}$$

where P_{orb} is the sidereal orbital period and P_{syn} is the synodic orbital period. So we just need to get the asteroid's synodic period in years so that the units match up. So we convert from days to years:

$$429 \cancel{\text{days}} \times \frac{1 \text{ yr}}{365.25 \cancel{\text{days}}} = 1.17 \text{ yr}$$

Plugging this value for P_{syn}, we get

$$P_{orb} = 1.17 \cancel{\text{yr}} \times \frac{1 \text{ yr}}{(1.17 - 1 \cancel{\text{yr}})}$$

$$= 6.88 \text{ yr}$$

The asteroid's orbital period is 6.88 years.

51. To answer this, we will need to use the equation of time plot (Figure 2 in Special Topic: Solar Days and the Analemma). Reading off of the graph for February 15, we see that mean solar time is ahead of the apparent solar time by about 14 minutes. So if the sundial says that the time is 18 minutes until noon, we have to add 14 minutes to this to get a mean solar time of 4 minutes to noon, or 11:56 A.M.

52. To answer this, we will need to use the equation of time plot (Figure 2 in Special Topic: Solar Days and the Analemma). Reading off of the graph for July 1, we see that the mean solar time is about 5 minutes ahead of apparent solar time. So if the sundial says 3:30 P.M., the mean solar time would be 5 minutes later, or 3:35 P.M.

53. On the day of the spring equinox, the Sun is located at the position of the spring equinox in the sky. Thus, at 4 P.M., both the Sun and the spring equinox are 4 hours past the meridian, so the local sidereal time is 4 hours.

54. Vega has RA = $18^h\,35^m$. Thus, at LST = 19:30, Vega's hour angle is:

$$HA_{Vega} = LST - RA_{Vega} = 19{:}30 - 18{:}35 = 00{:}55$$

Vega crossed your meridian about 55 minutes ago, which means it will cross again in about 23 hours 5 minutes.

55. We are given HA_{star} = 3 hr and LST = 8:15. Thus, we have the equation:

$$HA_{star} = LST - RA_{star}$$

Solving for the RA of the star, we find:

$$RA_{star} = LST - HA_{star} = 08{:}15 - 03{:}00 = 05{:}15$$

The star has a right ascension of 5 hours 15 minutes.

56. We are told in Mathematical Insight S1.2 that

$$HA_{object} = LST - RA_{object}$$

where HA_{object} is the hour angle, LST is the local sidereal time, and RA_{object} is the object's right ascension. We are told that the local sidereal time is 7:00 and that the Orion Nebula has a right ascension of $5^h\ 25^m$, so the hour angle of the Orion Nebula must be $1^h\ 35^m$. This puts the Orion Nebula west of the meridian by about 25°.

Since the Orion Nebula has a declination of –5.5°, it is near the celestial equator. At our latitude, the celestial equator has an altitude of 50° (90°-latitude) where it crosses the meridian in the south. The Orion Nebula will have sunk a bit lower since then, so we should look for it at about 40° altitude, somewhere in the southeast.

57. We will compute this in a way similar to the way in which we found the difference between the mean solar and sidereal days. The Moon's orbital period is about 27.5 days. In this time, it travels 360° around the celestial sphere. Thus, each day, it travels relative to the stars through an angle of:

$$1\ \cancel{\text{day}} \times \frac{360°}{27.3\ \cancel{\text{days}}} = 13.2°$$

We now need to figure out how long it takes the Earth to cover this angle it rotates. Since the Earth spins 360° in 24 hours, we can use this to convert:

$$13.2\cancel{°} \times \frac{24\text{ hr}}{360\cancel{°}} = 0.88\text{ hr}$$

This number would probably be better if it were in minutes, so we convert:

$$0.88\ \cancel{\text{hr}} \times \frac{60\text{ min}}{1\ \cancel{\text{hr}}} \approx 53\text{ min}$$

The synodic period of the Moon is about 24 hours and 53 minutes.

Doing the same for Phobos, we see from Appendix E that its orbital period is 0.319 Earth day. First, we should convert this to Martian days. Appendix E also tells us that Mars spins on its axis every 1.026 Earth days. So Phobos's orbital period in Martian days is:

$$0.319\ \cancel{\text{Earth day}} \times \frac{1\text{ Martian day}}{1.026\ \cancel{\text{Earth days}}} = 0.327\text{ Martian day}$$

In this time, it travels 360° around the Martian sky. However, this time Mars has spun a around a bit so now Phobos has to catch up to be back on the meridian.

How far has Mars spun? Well, Mars spins 360° in 1 Martian day, so we convert 0.327 Martian day to degrees:

$$0.327\ \cancel{\text{Martian day}} \times \frac{360°}{1\ \cancel{\text{Martian day}}} = 118°$$

How long does it take Phobos to cover that extra distance? Well, if it travels 360° in 0.327 day, then we can convert 118° into the time it takes for Phobos to cover that distance:

$$118\cancel{°} \times \frac{0.327\ \text{Martian day}}{360\cancel{°}} = 0.107\ \text{Martian day}$$

So adding this to the orbital period, we discover that Phobos takes 0.434 Martian day to go from the meridian back around to the meridian again—which means that Phobos circles the Martian sky more than twice with each Martian day.

58. Let's see how far Mercury's local meridian has to move to catch the Sun after one orbit. The Sun moves at 360° per 88 days. So we can find out how far it has moved in on a 58.6-day spin period:

$$58.6\ \cancel{\text{days}} \times \frac{360°}{88\ \cancel{\text{days}}} = 240°$$

So Mercury's spin has to catch up 240°. But wait! The Sun will still be moving at its 360°/88-day rate across the sky while Mercury's meridian tries to catch up at its 360°/58.6-day rate. (For Earth, we didn't include this because the Earth spins so much faster than it orbits. But for Mercury, the two speeds are too close to ignore this effect.) So we can set up a relationship between how fast Mercury's meridian moves in time t and how far it has to go to catch up:

$$\frac{360°}{58.6\ \text{days}}t = \frac{360°}{88\ \text{days}}t + 240°$$

We can group the terms with t:

$$\frac{360°}{58.6\ \text{days}}t - \frac{360°}{88\ \text{days}}t = 240°$$

$$\left(\frac{360°}{58.6\ \text{days}} - \frac{360°}{88\ \text{days}}\right)t = 240°$$

and then solve for t:

$$t = \frac{240\cancel{°}}{\left(\frac{360\cancel{°}}{58.6\ \text{days}} - \frac{360\cancel{°}}{88\ \text{days}}\right)}$$

$$= 117\ \text{days}$$

We have to add this to our 58.6-day spin period, which already happened, so we get 176 days for the solar day. That's exactly twice the orbital period. The solar day on Mercury is twice the length of the year!

Part II: Key Concepts for Astronomy

Chapter 4. Making Sense of the Universe: Understanding Motion, Energy, and Gravity

This chapter focuses on three major ideas and their astronomical applications: (1) Newton's laws of motion; (2) the laws of conservation of energy and angular momentum; and (3) the law of gravity.

As always, when you prepare to teach this chapter, be sure you are familiar with the relevant media resources (see the complete, section-by-section resource grid in Appendix 3 of this Instructor's Guide) and the online quizzes and other study resources available on the Mastering Astronomy Web site.

What's New in the Fourth Edition That Will Affect My Lecture Notes?

Those who have taught from previous editions of *The Cosmic Perspective* should be aware of the following organizational or pedagogical changes to this chapter (i.e., changes that will influence the way you teach) from the third edition:

- Note that we have eliminated the old Chapter 4, integrating its content on energy and energy conservation into the new Section 4.3. Other topics from the old Chapter 4 are now integrated into the new Chapter 5.
- Section 4.3 is new to this edition, providing a consolidated look at conservation laws in astronomy.
- We have consolidated the old Sections 5.4 through 5.6 into the new Section 4.5.

Teaching Notes (By Section)

Section 4.1 Describing Motion: Examples from Daily Life

Most nonscience majors are unfamiliar with the basic terminology of motion. For example, few students enter our astronomy classes with an understanding of why acceleration is measured in units of length over time squared, of the definitions of force and momentum, or of how mass and weight differ. This section introduces all these ideas in the context of very concrete examples that should be familiar from everyday life.

- Classroom demonstrations can be particularly helpful in this and the next section; for example, demonstrate that all objects accelerate the same under gravity, or use an air track to show conservation of momentum.
- Note that, aside from a footnote, we neglect the distinction between weight (or "true weight") and apparent weight. The former is often defined in physics texts as mg, whereas the latter also includes the effects of other accelerations (such as

the acceleration due to Earth's rotation or the acceleration in an elevator). While this distinction is sometimes useful in setting up physics problems, it can become very confusing in astronomy, where, for example, it is difficult to decide how to define "true weight" for objects located between the Earth and the Moon. Moreover, the distinction is unimportant from the point of view of general relativity, so our discussion works well to set the stage for the general relativity discussion in Chapter S3.

- Note also that, in stating that astronauts in orbit are weightless, we are neglecting the tiny accelerations, including those due to tidal forces, that affect objects in orbiting spacecraft. Because of these small accelerations, NASA and many space scientists have taken to referring to the conditions in orbiting spacecraft as "microgravity," rather than "weightlessness." In our opinion, the term *microgravity* is a poor one for students and tends to feed the common misconception that gravity is absent in space—when, in fact, the acceleration of gravity is only a few percent smaller in low-Earth orbit than on the ground. Perhaps a better term for the conditions in orbit would be *microacceleration*, but we feel it is pedagogically more useful to simply neglect the small accelerations and refer to the conditions as "weightlessness due to free-fall." If you want to be truly accurate, you might refer to the conditions as "near-weightlessness" and explain why small accelerations still are present.

Section 4.2 Newton's Laws of Motion

Having described the terminology of motion, we next discuss Newton's laws of motion. This discussion should solidify students' grasp of how their everyday experiences reflect Newtonian physics.

Section 4.3 Conservation Laws in Astronomy

This section covers conservation of angular momentum and conservation of energy, along with a discussion of the various forms of energy.

- When introducing angular momentum, you may wish to do a demonstration of conservation of angular momentum using a bicycle wheel and a rotating platform.
- Note that we discuss conservation of energy in a modern sense, with mass-energy included as a form of potential energy.
- Note that we do not introduce a formula for gravitational potential energy, because the general formula would look too complex at this point (coming before the law of gravity), and the formula mgh (which will be familiar to some of your students) is a special case that applies only on the surface of the Earth. However, you may wish to mention the formula mgh in class, particularly if your students are already familiar with it.

Section 4.4 The Universal Law of Gravitation

The pieces now are all in place to introduce Newton's universal law of gravitation. This section discusses gravity generally, and describes how Newton explained and expanded on Kepler's laws.

- Note that, as in Chapter 1, we are using average distance to mean a semimajor axis distance.

- Note also that, while we mention parabolas and hyperbolas as allowed orbital paths, the bold term introduced to include both these cases is *unbound orbits*. Similarly, we refer to elliptical orbits as *bound orbits*. We feel that the terms *bound* and *unbound* are far more intuitive for students than precise mathematical shapes.

Section 4.5 Orbits, Tides, and the Acceleration of Gravity

We now apply Newton's universal law of gravitation to explain fundamental ideas in astronomy, including orbital energy and changes, escape velocity, tides, and the acceleration of gravity.

- Note our emphasis on the idea that orbits cannot change spontaneously—they can change only if there is an exchange of orbital energy. We have found that this is a very important point that students often fail to grasp unless it is made very explicitly. We encourage you to keep reminding them of this point throughout your course whenever you are explaining gravitational capture of any kind—from an asteroid being captured by a planet, to the gravitational collapse of a cloud of gas into a star, to the infall of material into an accretion disk.
- Note that our discussion of tides includes an explanation of the cause of the Moon's synchronous rotation, which was first introduced in Chapter 2, as well as other examples of synchronous rotation, including the case of Pluto and Charon and the 3-to-2 ratio of orbital period to rotation rate for Mercury.
- The final, short subsection brings closure to the historical aspects of this chapter by explaining why, at least in the context of Newton's law of gravity, all falling objects fall with the same acceleration of gravity. It also mentions the fact that Newton still saw this as an extraordinary coincidence, thus setting the stage for our discussion of general relativity in Chapter S3. This subsection should be considered optional.

Answers/Discussion Points for Think About It Questions

The Think About It questions are not numbered in the book, so we list them in the order in which they appear, keyed by section number.

Section 4.1

- (p. 118) Students should realize that by crumpling the paper, they make it less subject to air resistance and hence can see the effects of gravity more easily.
- (p. 119) This question asks students to try a small demonstration that will help them understand the difference between mass and weight. Weight changes with acceleration, but is not affected at constant speed in the elevator.
- (p. 120) The "throw yourself at the ground and miss" idea is similar to the idea that an object in orbit is constantly falling to the ground, but moving forward so fast that it always "misses" the ground as it falls.

Section 4.3

- (p. 126) As the water gets closer to the drain, it moves in a smaller circle and thus must circle the drain faster to conserve its angular momentum.
- (p. 128) Just as a pot of hot water transfers thermal energy to you much more rapidly than hot air of the same temperature, your body will lose some of its thermal energy (meaning you get colder) much more quickly in cold water than in cold air. Thus, falling into a cold lake can cause you to lose heat rapidly, making it very dangerous.

Section 4.4

- (p. 130) If the distance increases to $3d$, the gravitational attraction decreases by a factor of $3^2 = 9$. If the distance decreases to $0.5d$, the gravitational attraction increases by a factor of $2^2 = 4$.

Section 4.5

- (p. 137) Because the tidal force declines rapidly with distance (in fact, as the cube of distance), the other planets would have to be extremely large in mass (e.g., like the Sun) to have any noticeable tidal effect. Because other planets are very low in mass compared to the Sun, their effects are negligible.

Solutions to End-of-Chapter Problems (Chapter 4)

1. The term "speed" is used to describe how fast something is moving. "Velocity" carries that same information, but it also tells us in which direction the object is going.
   An example of traveling at a constant speed but not at a constant velocity is traveling in a circular orbit around the Earth. Such an object always travels at the same speed. But the direction in which the object is traveling is constantly changing as the object moves around the circle, so the velocity is constantly changing.
2. Acceleration is the rate of change of velocity in time. The acceleration of gravity, g, is the acceleration downward due to gravity. We give acceleration in units of m/s^2 to mean that the velocity is changing by so many m/s for every second during which the acceleration continues.
3. Momentum is the product of mass and velocity (mass $\times$ velocity). By Newton's second law of motion, the only way to change the momentum of an object is to apply a force to it. However, the object's momentum will respond only to a net force, the force that is left when we add all of the forces together. Even if the individual forces are large, if they cancel out and leave no net force, then there is no change in momentum.
4. Free-fall is the state of falling without any resistance to the fall. Objects in free-fall are weightless because they are not pushing against anything to give them weight. Astronauts in the Space Station are in constant free-fall as they fall around the Earth (always missing it), so they are weightless.
5. i) An object moves at a constant velocity if there is no net force on it. This is why objects that are at rest do not start moving spontaneously. (They remain at constant—zero—velocity.)
   ii) Force = mass $\times$ acceleration. This law tells us that it takes a lot more force to push a car forward than a bicycle at the same acceleration.
   iii) For every force, there is an equal and opposite force. This explains the recoil someone feels when she fires a gun: The gun is applying a force to the bullet, but the bullet applies a force back on the gun.
6. The conservation laws say that momentum, angular momentum, and energy are all conserved. That is, in a particular system, the total amount of each of these quantities does not change.

An example of conservation of momentum in astronomy is a rocket. If we start with a rocket at rest and fire the engines, we expel a lot of hot gas out the back. To conserve momentum, the rocket must move forward.

Conservation of angular momentum is what gives us Kepler's second law: When a planet gets closer to the Sun, it has to move faster to conserve its angular momentum.

Conservation of energy tells us that as an object falls to the Earth's surface, it loses gravitational potential energy. To conserve energy, the object has to move faster as it falls. Eventually, it hits the ground and stops. At this point, its kinetic energy is converted to heat and sound.

7. Kinetic energy is the energy an object has due to its motion. Two examples are a car driving down the highway or a cup of hot coffee. (In the latter case, the motion is in the random movement of the molecules, not the overall motion of the liquid.)

   Potential energy is energy that is stored. It could be, for example, chemical, mechanical (e.g., by springs), gravitational, or nuclear. Two examples of potential energy are breakfast cereal, which has chemical potential energy, and a heavy book on the top shelf at the library, which has gravitational potential energy.

   Radiative energy is energy in light. Sunlight carries this form of energy.

8. Thermal energy is the amount of energy stored in the random motions of the molecules of some object. Temperature is a measurement of the average kinetic energy of these random motions. The two concepts are related since both deal with the energy in the random motions of the particles. However, thermal energy measures the total energy in all of the particles, while temperature measures the average energy per particle. Thus, two objects with the same temperature can have different amounts of thermal energy if they differ in size or density.

9. The rock I hold out the window of this 10-story building has more gravitational potential energy than a rock on the ground. This is because the gravitational potential energy depends on how far the object can fall as well as on the mass of the object.

10. Mass-energy is the potential energy that is stored in the form of matter. All matter can be converted to energy. The amount of energy available is given by Einstein's famous equation, $E = mc^2$: The energy stored is equal to the mass times the speed of light squared.

11. Newton's law of universal gravitation states that every object in the universe attracts every other object. The force of the attraction depends on the product of the masses and decreases as 1 over the distance between the objects squared. In mathematical form:

$$F_g = G\frac{M_1 M_2}{d^2}$$

   where F_g is the force of gravity, G is the gravitational constant, M_1 and M_2 are the masses of the objects, and d is the distance separating the centers of the objects.

12. A bound orbit is one where the orbiting object goes around again and again, while in an unbound orbit the object just makes one orbit before leaving. Orbits can be any conic section: circle, ellipse, parabola, or hyperbola.

13. If we want to measure the mass of an object with Newton's version of Kepler's third law, we need to observe a small object orbiting a more massive one. We can then determine the mass of the massive object from the orbital period and orbital distance of the less massive object. If both objects are comparable in mass, we can determine the sum of their masses from the their mutual orbital period and distance.

14. Because of conservation of energy, objects do not change orbits spontaneously. As long as nothing changes the energy of the object, it must remain in the same orbit because different orbits have different energies associated with them. However, if something like atmospheric drag takes away energy from the object, it will have to change to a smaller, lower-energy orbit.

    Similarly, if a third body gets close and gravitationally attracts an orbiting object, the orbiting body can gain or lose energy and change its orbit. If enough energy is added to the object, it can achieve escape velocity and leave orbit entirely and fly away. Human-made objects can also achieve escape velocity by firing their engines and adding energy to their orbits until they are moving fast enough to leave orbit.

15. The Moon creates tides on the Earth by pulling on the different parts of Earth with different forces. The nearest parts of the Earth get stronger tugs, according to Newton's law of universal gravitation, so they try to move more toward the Moon than the farther-away parts. This actually creates two bulges: The nearest point to the Moon bulges toward the Moon because it is more attracted to the Moon than the average part of the planet, while the farthest point bulges away because it is less attracted to the Moon than the average part of the planet. This leads to two high and two low tides each day as we spin around under the Moon.

16. Tides vary with the phases of the Moon because the Sun also creates tides on the Earth. When the Moon is either new or full (in the same direction as the Sun from the Earth or on the opposite side of the Earth from the Sun), the two tidal bulges add together to create higher, or spring, tides. When the Moon is in first- or third-quarter, the two tidal bulges tend to cancel each other out somewhat so that we get lower, or neap, tides.

17. Tidal friction is the loss of spin energy due to the movement of the tidal bulges. This friction causes Earth's rotation to slow gradually with time. To conserve angular momentum in this process, the Moon gradually moves farther from Earth. In the case of Earth, tidal friction hasn't slowed us down to the point where we keep the same face pointing toward the Moon. However, tidal friction long ago made the Moon slow down to that point. This is the cause of its synchronous rotation.

18. You would not fall at the same rate on the Moon as on the Earth. Because the Moon has a different mass and a different radius, Newton's law of universal gravitation predicts different accelerations due to gravity for the two bodies. In the case of the Moon, the radius is $\frac{1}{4}$ that of Earth, but it is about 80 times less massive. When we combine these two competing effects, we find that the acceleration due to gravity on the Moon is about $\frac{1}{6}$ that of the acceleration on the Earth. This is why astronauts on the Moon moved and fell differently than they would have on the Earth.

19. *If you could go shopping on the Moon to buy a pound of chocolate, you'd get a lot more chocolate than if you bought a pound on Earth.* This statement makes sense, because pounds are a unit of weight and objects weigh less on the Moon than on Earth.

20. *Suppose you could enter a vacuum chamber (on Earth)—that is, a chamber with no air in it. Inside this chamber, if you dropped a hammer and a feather from the same height at the same time, both would hit the bottom at the same time.* This statement is true. Without air resistance, all objects will fall under gravity at the same rate.

21. *When an astronaut goes on a space walk outside the Space Station, she will quickly float away from the station unless she has a tether holding her to the station or she constantly fires thrusters on her space suit.* This statement is false. She and the Space Station share the same orbit and will stay together unless they are pushed apart (which could happen, for example, if she pushed off the side).
22. *Newton's version of Kepler's third law allows us to calculate the mass of Saturn from orbital characteristics of its moon Titan.* This statement makes sense, because we can calculate the mass of Saturn by knowing the period and average distance for Titan.
23. *If we could magically replace the Sun by a giant rock with precisely the same mass, the Earth's orbit would not change.* This statement is true, because the rock would have the same gravitational effect on Earth as does the Sun.
24. *The fact that the Moon rotates once in precisely the time it takes to orbit the Earth once is such an astonishing coincidence that scientists probably never will be able to explain it.* This statement is false, because the synchronous rotation is not a coincidence at all and its cause has been well explained.
25. *Venus has no oceans, so it could not have tides even if it had a moon (which it doesn't).* This statement does not make sense, because tides affect an entire planet, not just the oceans. Thus, if Venus had a moon, it could have "land tides."
26. *If an asteroid passed by Earth at just the right distance, it would be captured by the Earth's gravity and become our second moon.* This statement does not make sense, because objects cannot spontaneously change their orbits without having some exchange of energy with anotherobject.
27. *When I drive my car at 30 miles per hour, it has more kinetic energy than it does at 10 miles per hour.* This statement makes sense, because kinetic energy depends on the square of the speed. Thus, tripling the speed means a factor of $3^2 = 9$ increase in kinetic energy.
28. *Someday soon, scientists are likely to build an engine that produces more energy than it consumes.* This statement does not make sense, because such an engine would violate the law of conservation of energy.
29. c; 30. b; 31. a; 32. a; 33. b;
34. c; 35. b; 36. c. 37. c; 38. a.
39. a. The rock falls with the acceleration of gravity, which is approximately 10 m/s^2, meaning that its speed increases by about 10 m/s with each passing second. Thus, after falling for 4 seconds, the rock will be falling at a speed of 40 m/s.
    b. If you sled down a street with an acceleration of 4 m/s^2, after 5 seconds you would be going at a speed of $5 \times 4 = 20$ m/s.
    c. The acceleration of –20 mi/hr/s means that your speed slows by 20 mi/hr with each passing second. Thus, it will take you 3 seconds to slow from 60 mi/hr to a stop.
40. Astronauts are *not* weightless during either launch or return to Earth, because they are not in free-fall at those times. Astronauts feel forces due to acceleration as they launch into space and forces due to air resistance when the spacecraft slows as it returns to Earth.
41. a. At the same height, a more massive object has more gravitational potential energy than a less massive object. The reason is that both objects will fall at the same rate due to gravity, but the more massive object will have more kinetic energy at any particular speed. Thus, the more massive object must also have more gravitational potential energy, since its total (kinetic plus gravitational potential) energy is conserved as it falls.

b. The diver has more gravitational potential energy on the higher board because she has a greater distance that she can fall.

c. The satellite in Jupiter orbit has more gravitational potential energy because Jupiter's stronger gravity means that the satellite, if it fell, would fall at a greater speed and hence have more kinetic energy. Since energy must be conserved, the satellite must also have more gravitational potential energy.

42. a. In the equation $E = mc^2$, E is energy, m is mass, and c is the speed of light. (In international units, we measure the energy in joules, the mass in kilograms, and the speed of light in meters per second.) The equation states that mass and energy are equivalent; under certain circumstances it is possible to convert mass into energy, and vice versa.

b. The Sun, which is necessary for life on Earth, produces energy by nuclear fusion. In nuclear fusion, mass is converted to energy as described by Einstein's formula. (The efficiency of fusion is 0.7%. For every 1,000 grams of hydrogen, fusion results in 993 grams of helium; the remaining 7 grams is converted to energy.)

c. The formula also explains the operation of nuclear bombs, in which mass is converted to energy during nuclear fission (uranium and plutonium bombs and "triggers") or nuclear fusion ("H-bombs" or "thermonuclear bombs"). Thus, the equivalence of mass and energy is intimately tied to both our ability to live and our ability to self-destruct.

43. a. Quadrupling the distance between two objects decreases the gravitational attraction between them by a factor of $4^2 = 16$.

b. If the Sun were magically replaced by a star with twice as much mass, the gravitational attraction between the Earth and the Sun would double.

c. If Earth were moved to $\frac{1}{3}$ of its current distance from the Sun, the gravitational attraction between Earth and Sun would *increase* by a factor of $3^2 = 9$.

44. a. Newton's version of Kepler's third law tells us that a planet's orbit around a star depends on the sum of the masses of the star and planet. We can generally neglect the planet's mass since a star is usually so much more massive than any planet, which means that the planet orbit depends on the star's mass but not its own mass. Thus, Earth's orbit could not stay the same if the Sun were replaced by a more massive star, because the orbit DOES depend on the star's mass.

b. Changing Earth's mass would not affect its orbit, because as discussed above, the planet mass is not important to orbital properties.

45. The tidal force acting on you depends on the difference between the gravitational force acting on your head and on your toes. But the gravitational force on any part of your body depends on the distance of the body part from the center of the Earth. Because the length of your body is negligible compared to the radius of the Earth, there's no noticeable difference in gravitational force between your head and your toes. (As discussed in Chapter 18, this would no longer be the case if you could stand on a very compact object such as a neutron star, or if you were falling into a black hole.)

46. Yes. If the Moon had rotated more slowly when it formed, the tidal forces would have acted to speed its rotation up until the rotation became synchronized with the orbit.

47. a. A geostationary satellite must remain above the same point on Earth and hence must orbit with the Earth's actual rotation period. The actual rotation period of the Earth is a sidereal day, not a solar day.

b. Only an equatorial orbit can keep a satellite at geosynchronous altitude over the same point on Earth. A satellite in a polar orbit, for example, would still be moving over different latitudes even if it took exactly 1 day to complete each orbit.

c. Essay question; answers will vary, but the key points should be that a geostationary orbit keeps a satellite above the horizon at all times and eliminates the need for any tracking by ground receivers.

48. a. Because it is at the altitude of geosynchronous orbit, the top of the elevator would be moving with precisely the necessary orbital speed for this altitude. If you placed an object outside the top of the elevator, the object would also be moving with the orbital speed needed for this altitude (assuming that you do not give it any kind of push that gives it a speed relative to the top of the elevator). Thus, the object would simply remain in orbit next to the elevator top, rather than falling.

b. Placing a satellite into geosynchronous orbit would require nothing more than taking it up the elevator. That is, rather than needing a huge rocket to reach orbit, satellites could simply be taken up the elevator shaft, which requires far less energy than a rocket. (In fact, the energy cost of getting a satellite to geosynchronous altitude in the elevator would be only a few dollars, as opposed to millions for a rocket launch.) Once at the top, a small rocket could be used to launch a satellite to a higher orbit, to the Moon, or into deep space.

49. a. If 2.5×10^{16} joules represents the energy of a major earthquake, the energy of a 1-megaton bomb is smaller by a factor of:

$$\frac{2.5 \times 10^{6} \ \cancel{\text{joule}}}{5 \times 10^{15} \ \cancel{\text{joule}}} = \frac{25 \times 10^{15}}{5 \times 10^{15}} = 5$$

A major earthquake releases as much energy as five 1-megaton bombs.

b. The annual U.S. energy consumption is about 10^{20} joules, and a liter of oil yields about 1.2×10^{7} joules. Thus, the amount of oil needed to supply all the U.S. energy for a year would be:

$$\frac{10^{20} \ \cancel{\text{joule}}}{1.2 \times 10^{7} \ \cancel{\text{joule}}/\text{liter}} = 8 \times 10^{12} \text{ liter}$$

or about 8 trillion liters of oil (roughly 2 trillion gallons).

c. We can compare the Sun's annual energy output to that of the supernova by dividing; to be conservative, we use the lower number from the 10^{44}–10^{46} range for supernova energies:

$$\frac{\text{supernova energy}}{\text{Sun's annual energy output}} = \frac{10^{44} \ \cancel{\text{joule}}}{10^{34} \ \cancel{\text{joule}}} = 10^{10}$$

The supernova puts out about 10 billion times as much energy as the Sun does in an entire year. That is why a supernova can shine nearly as brightly as an entire galaxy, though only for a few weeks.

50. We are seeking the speed at which a 0.2 kilogram candy bar must move to have a kinetic energy equal to its chemical potential energy of 10^{6} joules released by metabolism. We

set the formula for the kinetic energy of the candy bar equal to 10^6 joules, then solve for the speed:

$$\frac{1}{2}(0.2\text{ kg})\times v^2 = 10^6\frac{\text{kg}\times\text{m}^2}{\text{s}^2} \Rightarrow v = \sqrt{\frac{10^6\frac{\cancel{\text{kg}}\times\text{m}^2}{\text{s}^2}}{\frac{1}{2}\times(0.2\ \cancel{\text{kg}})}} = \sqrt{10^7\frac{\text{m}^2}{\text{s}^2}} = 3{,}162\frac{\text{m}}{\text{s}}$$

We now convert this speed from m/s to km/hr:

$$3{,}162\frac{\cancel{\text{m}}}{\cancel{\text{s}}}\times\frac{1\text{ km}}{1000\ \cancel{\text{m}}}\times\frac{60\ \cancel{\text{s}}}{1\ \cancel{\text{min}}}\times\frac{60\ \cancel{\text{min}}}{1\text{ hr}} = 11{,}383\frac{\text{km}}{\text{hr}}$$

The candy bar would have to be traveling at a speed of more than 11,000 km/hr to have as much kinetic energy as the energy it will release through metabolism.

51. Let's assume that your mass is about 50 kilograms. Then the potential energy contained in your massis:

$$E = mc^2 = (50\text{ kg})\times\left(3\times10^8\frac{\text{m}}{\text{s}}\right)^2 = 4.5\times10^{18}\text{ joules}$$

This is nearly 1,000 times greater than the 5×10^{15} joules released by a 1-megaton bomb.

52. First, we find the U.S. energy consumption per minute by converting the annual energy consumption into units of joules per minute:

$$\frac{10^{20}\text{ joules}}{\cancel{\text{yr}}}\times\frac{1\ \cancel{\text{yr}}}{365\ \cancel{\text{days}}}\times\frac{1\ \cancel{\text{day}}}{24\ \cancel{\text{hr}}}\times\frac{1\ \cancel{\text{hr}}}{60\text{ min}} \approx \frac{1.9\times 10^{14}\text{ joules}}{60\text{ min}}$$

Next we divide this energy consumption per minute by the amount of energy available through fusion of 1 liter of water (from Table 4.1):

$$\frac{1.9\times10^{14}\frac{\cancel{\text{joules}}}{\text{min}}}{7\times10^{13}\frac{\cancel{\text{joules}}}{\text{liter}}} \approx 2.7\frac{\text{liters}}{\text{min}}$$

In other words, it would take less than 3 liters of water per minute—which is less than 1gallon per minute—to meet all U.S. energy needs through nuclear fusion. This is somewhat less than the rate at which water flows from a typical kitchen faucet. So if we could simply attach a nuclear fusion reactor to your kitchen faucet, we could stop producing and importing oil, remove all the hydroelectric dams, shut down all the coal-burning power plants, and still have energy to spare.

53. As long as a planet's mass is small compared to the Sun, the planet's orbital period is independent of its mass, because only the sum of the planet's mass and the Sun's mass appears in the equation for Newton's version of Kepler's third law. Thus, the orbital period of a planet at Earth's distance but with twice the mass of Earth would still be 1 year.

54. Newton's version of Kepler's third law has the form:

$$p^2 = \frac{4\pi^2}{G(M_1+M_2)}a^3$$

Because the square of the period varies inversely with the sum of the masses, the orbital period itself depends on the inverse square root of the object masses:

$$p = \sqrt{\frac{4\pi^2}{G(M_1 + M_2)} a^3}$$

Thus, if we have a star four times as massive as the Sun, the period of a planet orbiting at 1 AU will be $1/\sqrt{4} = 1/2$ that of the Earth, or 6 months.

55. a. Using the Moon's orbital period and distance and following the method given in Mathematical Insight 4.3, we find the mass of the Earth to be about:

$$M_{\text{Earth}} \approx \frac{4\pi^2}{G} \frac{(a_{\text{Moon}})^3}{(p_{\text{Moon}})^2}$$

Making sure that we use appropriate units, we find:

$$M_{\text{Earth}} \approx \frac{4\pi^2 \left(384{,}000 \cancel{\text{km}} \times 1000 \frac{\text{m}}{\cancel{\text{km}}}\right)^3}{\left(6.67 \times 10^{-11} \frac{\text{m}^3}{\text{kg} \times \text{s}^2}\right)\left(27.3 \cancel{\text{days}} \times 24 \frac{\cancel{\text{hr}}}{\cancel{\text{day}}} \times 3600 \frac{\text{s}}{\cancel{\text{hr}}}\right)^2} = 6.0 \times 10^{24} \text{ kg}$$

b. Using Io's orbital period and distance and following the method given in Mathematical Insight 4.3, we find the mass of Jupiter to be about:

$$M_{\text{Jupiter}} \approx \frac{4\pi^2 \left(422{,}000 \cancel{\text{km}} \times 1000 \frac{\text{m}}{\cancel{\text{km}}}\right)^3}{\left(6.67 \times 10^{-11} \frac{\text{m}^3}{\text{kg} \times \text{s}^2}\right)\left(42.5 \cancel{\text{hr}} \times 3600 \frac{\text{s}}{\cancel{\text{hr}}}\right)^2} = 1.9 \times 10^{27} \text{kg}$$

We find the same answer using Europa's orbital properties; Kepler's third law does not depend on the mass of either moon because neither moon has a significant mass in comparison to the mass of Jupiter.

c. We again start with Newton's version of Kepler's third law:

$$p^2 = \frac{4\pi^2}{G(M_1 + M_2)} a^3$$

We solve for the semimajor axis of the planet with a little algebra:

$$a = \sqrt[3]{\frac{G(M_1 + M_2)}{4\pi^2} p^2}$$

We convert the planet's orbital period of 63 days into seconds:

$$63 \cancel{\text{days}} \times \frac{24 \cancel{\text{hr}}}{1 \cancel{\text{day}}} \times \frac{60 \cancel{\text{min}}}{1 \cancel{\text{hr}}} \times \frac{60 \text{ s}}{1 \cancel{\text{min}}} = 5.44 \times 10^6 \text{ s}$$

Planets are much less massive than their stars are, so we can approximate $M_1 + M_2 \approx M_{\text{star}}$. From Appendix A, the mass of the Sun is about 2×10^{30} kg. So we can calculate the semimajor axis:

$$a = \sqrt[3]{\frac{\left(6.67 \times 10^{-11} \frac{\text{m}^3}{\text{kgs}^2}\right)(2 \times 10^{30}\ \text{kg})}{4\pi^2}(5.44 \times 10^6\text{s})^2}$$

$$= 4.64 \times 10^{10}\ \text{m}$$

Of course, this number would probably be more useful in astronomical units, so we should convert:

$$4.64 \times 10^{10}\ \cancel{\text{m}} \times \frac{1\ \text{AU}}{1.5 \times 10^{11}\ \cancel{\text{m}}} = 0.31\ \text{AU}$$

The new planet is only 0.31 AU from its star, which is closer than Mercury is to the Sun.

56. a. If we are working with Charon's orbit around Pluto, Newton's version of Kepler's third law takes the form:

$$(p_{\text{Charon}})^2 = \frac{4\pi^2}{G(M_{\text{Pluto}} + M_{\text{Charon}})}(a_{\text{Charon}})^3$$

In this case we are looking for the combined mass of the two worlds, so we do not simplify the equation further. We simply solve for the masses, then plug in the numbersgiven:

$$M_{\text{Pluto}} + M_{\text{Charon}} = \frac{4\pi^2 \times (a_{\text{Charon}})^3}{G \times (p_{\text{Charon}})^2}$$

$$= \frac{4\pi^2 \times \left(19{,}700\ \cancel{\text{km}} \times 1000 \frac{\cancel{\text{m}}}{\cancel{\text{km}}}\right)^3}{\left(6.67 \times 10^{-11} \frac{\cancel{\text{m}}^3}{\text{kg} \times \cancel{\text{s}^2}}\right)\left(6.4\ \cancel{\text{days}} \times 24 \frac{\cancel{\text{hr}}}{\cancel{\text{day}}} \times 3600 \frac{\cancel{\text{s}}}{\cancel{\text{hr}}}\right)^2}$$

$$= 1.5 \times 10^{22}\ \text{kg}$$

Now we compare this mass for Pluto and Charon to the Earth's mass of 6.0×10^{24} kg:

$$\frac{M_{\text{Earth}}}{M_{\text{Pluto}} + M_{\text{Charon}}} = \frac{6.0 + 10^{24}\ \cancel{\text{kg}}}{1.5 \times 10^{22}\ \cancel{\text{kg}}} = 400$$

The Earth's mass is about 400 times greater than the combined mass of Pluto and Charon.

b. The Space Shuttle is much less massive than the Earth, so we can follow the method of Mathematical Insight 4.3 to write:

$$(P_{\text{Shuttle}})^2 = \frac{4\pi^2}{G(M_{\text{Earth}} + M_{\text{Shuttle}})}(a_{\text{Shuttle}})^3 \approx \frac{4\pi^2}{GM_{\text{Earth}}}(a_{\text{Shuttle}})^3$$

Remember that a_{Shuttle} represents the Shuttle's average distance from the center of the Earth. The radius of the Earth is about 6,400 kilometers, so:

$$a_{\text{Shuttle}} = 6{,}400 \text{ km} + 300 \text{ km} = 6{,}700 \text{ km}$$

or 6.7 × 106 m. The mass of the Earth is about $M_{\text{Earth}} \approx 6.0 \times 1024$ kg Substituting these values and solving the equation for p_{Shuttle} by taking the square root of both sides yields:

$$p_{\text{Shuttle}} \approx \sqrt{\frac{4\pi^2}{\left(6.67 \times 10^{-11} \frac{\cancel{\text{m}^3}}{\cancel{\text{kg}} \times \text{s}^2}\right)(6.0 \times 10^{24} \ \cancel{\text{kg}})}(6.7 \times 10^6 \ \cancel{\text{m}})^3} \approx 5{,}400 \text{ s}$$

The Shuttle orbits the Earth in about 5,400 seconds, or 90 minutes.

c. We again begin with Newton's version of Kepler's third law:

$$p^2 = \frac{4\pi^2}{G(M_1 + M_2)} a^3$$

Since we want the mass of the galaxy, we need to solve for mass. Luckily, the mass of the galaxy is much larger than the mass of the Sun, so $M_1 + M_2 \approx M_{\text{galaxy}}$. Solving for this mass in Newton's version of Kepler's third law:

$$M_{\text{galaxy}} = \frac{4\pi^2}{Gp^2} a^3$$

We are given that the Sun orbits every 230,000,000 years and its orbital semimajor axis is 28,000 light-years. We need to convert these to seconds and meters respectively:

$$230{,}000{,}000 \text{ yr} \times \frac{365 \ \cancel{\text{days}}}{1 \text{ year}} \times \frac{24 \ \cancel{\text{hr}}}{1 \ \cancel{\text{day}}} \times \frac{60 \ \cancel{\text{min}}}{1 \ \cancel{\text{hr}}} \times \frac{60 \text{ s}}{1 \ \cancel{\text{min}}} = 7.25 \times 10^{15} \text{s}$$

Using the fact that there are 9.46×10^{12} km in a light-year, found in Appendix A:

$$28{,}000 \ \cancel{\text{light-years}} \times \frac{9.46 \times 10^{12} \ \cancel{\text{km}}}{1 \ \cancel{\text{light-year}}} \times \frac{1000 \text{ m}}{1 \ \cancel{\text{km}}} = 2.65 \times 10^{20} \text{ m}$$

Now we can calculate the mass of the galaxy:

$$M_{\text{galaxy}} = \frac{4\pi^2}{\left(6.67 \times 10^{-11} \frac{\cancel{\text{m}^3}}{\text{kg} \times \cancel{\text{s}^2}}\right)(7.25 \times 10^{15} \ \cancel{\text{s}})^2}(2.65 \times 10^{20} \text{ m})^3$$

$$= 2.10 \times 10^{41} \text{ kg}$$

This number is probably easier to understand as multiples of the mass of the Sun, so let's convert:

$$2.10 \times 10^{41} \ \cancel{\text{kg}} \times \frac{1 \text{ solar mass}}{2 \times 10^{30} \ \cancel{\text{kg}}} = 1.05 \times 10^{11} \text{ solar masses}$$

Based on the given data, the mass of the galaxy is about 105 billion times the mass of the Sun.

57. In parts (a) and (c), the easiest way to find the escape velocities with the given data is by comparison to the escape velocity from Earth:

$$\frac{v_{\text{escape planet}}}{v_{\text{escape Earth}}} = \frac{\sqrt{\dfrac{2\cancel{G}M_{\text{planet}}}{R_{\text{planet}}}}}{\sqrt{\dfrac{2\cancel{G}M_{\text{Earth}}}{R_{\text{Earth}}}}} = \sqrt{\frac{M_{\text{planet}}}{M_{\text{Earth}}} \times \frac{R_{\text{Earth}}}{R_{\text{planet}}}}$$

Given that the escape velocity from Earth's surface is about 11 km/s, this formula becomes:

$$v_{\text{escape planet}} = 11\frac{\text{km}}{\text{s}} \times \sqrt{\frac{M_{\text{planet}}}{M_{\text{Earth}}} \times \frac{R_{\text{Earth}}}{R_{\text{Planet}}}}$$

a. From the surface of Mars, the escape velocity is:

$$v_{\text{escape}} = 11\frac{\text{km}}{\text{s}} \times \sqrt{\frac{M_{\text{Mars}}}{M_{\text{Earth}}} \times \frac{R_{\text{Earth}}}{R_{\text{Mars}}}} = 11\frac{\text{km}}{\text{s}} \times \sqrt{0.11 \times \frac{1}{0.53}} = 5.0\frac{\text{km}}{\text{s}}$$

b. From the surface of Phobos, the escape velocity is:

$$v_{\text{escape}} = \sqrt{\frac{2 \times \left(6.67 \times 10^{-11} \dfrac{\text{m}^3}{\cancel{\text{kg}} \times \text{s}^2}\right) \times (1.1 \times 10^{16}\ \cancel{\text{kg}})}{12{,}000\ \text{m}}} = 11\frac{\text{m}}{\text{s}} = 0.11\frac{\text{km}}{\text{s}}$$

c. From the surface of Jupiter, the escape velocity is:

$$v_{\text{escape}} = 11\frac{\text{km}}{\text{s}} \times \sqrt{\frac{M_{\text{Jupiter}}}{M_{\text{Earth}}} \times \frac{R_{\text{Earth}}}{R_{\text{Jupiter}}}} = 11\frac{\text{km}}{\text{s}} \times \sqrt{317.8 \times \frac{1}{11.2}} = 58.6\frac{\text{km}}{\text{s}}$$

d. To find the escape velocity from the solar system, starting from the Earth's orbit, we use the mass of the Sun (since that is the mass we are trying to escape) and the Earth's distance from the Sun of 1 AU, or 1.5×10^{11} m:

$$v_{\text{escape}} = \sqrt{\frac{2 \times \left(6.67 \times 10^{-11} \dfrac{\text{m}^3}{\cancel{\text{kg}} \times \text{s}^2}\right) \times (2.0 \times 10^{30}\ \cancel{\text{kg}})}{1.5 \times 10^{11}\ \text{m}}} = 42{,}200\frac{\text{m}}{\text{s}} = 42.2\frac{\text{km}}{\text{s}}$$

e. To find the escape velocity from the solar system starting from Saturn's orbit, we use the mass of the Sun (since that is the mass we are trying to escape) and Saturn's distance from the Sun of 1.4×10^{12} m:

$$v_{\text{escape}} = \sqrt{\frac{2 \times \left(6.67 \times 10^{-11} \dfrac{\text{m}^3}{\cancel{\text{kg}} \times \text{s}^2}\right) \times (2.0 \times 10^{30}\ \cancel{\text{kg}})}{1.4 \times 10^{12}\ \text{m}}} 13{,}800\frac{\text{m}}{\text{s}} = 13.8\frac{\text{km}}{\text{s}}$$

58. In parts (a) through (d), the easiest way to find the accelerations with the given data is by comparison to the acceleration of gravity on Earth:

$$\frac{a_{\text{planet}}}{a_{\text{Earth}}} = \frac{G\dfrac{M_{\text{planet}}}{R^2_{\text{planet}}}}{G\dfrac{M_{\text{Earth}}}{R^2_{\text{Earth}}}} = \frac{M_{\text{planet}}}{M_{\text{Earth}}} \times \left(\frac{R_{\text{Earth}}}{R_{\text{planet}}}\right)^2$$

a. On the surface of Mars, the acceleration of gravity is:

$$a_{\text{Mars}} = a_{\text{Earth}} \times \frac{M_{\text{Mars}}}{M_{\text{Earth}}} \times \left(\frac{R_{\text{Earth}}}{R_{\text{Mars}}}\right)^2 = 9.8\frac{\text{m}}{\text{s}^2} \times 0.11 \times \left(\frac{1}{0.53}\right)^2 = 3.8\frac{\text{m}}{\text{s}^2}$$

b. On the surface of Venus, the acceleration of gravity is:

$$a_{\text{Mars}} = a_{\text{Venus}} \times \frac{M_{\text{Venus}}}{M_{\text{Earth}}} \times \left(\frac{R_{\text{Earth}}}{R_{\text{Venus}}}\right)^2 = 9.8\frac{\text{m}}{\text{s}^2} \times 0.82 \times \left(\frac{1}{0.95}\right)^2 = 8.9\frac{\text{m}}{\text{s}^2}$$

c. At Jupiter's cloud tops, the acceleration of gravity is:

$$a_{\text{Jupiter}} = a_{\text{Earth}} \times \frac{M_{\text{Jupiter}}}{M_{\text{Earth}}} \times \left(\frac{R_{\text{Earth}}}{R_{\text{Jupiter}}}\right)^2 = 9.8\frac{\text{m}}{\text{s}^2} \times 317.8 \times \left(\frac{1}{11.2}\right)^2 = 25\frac{\text{m}}{\text{s}^2}$$

Because Jupiter has no solid surface, you could weigh yourself only if you were standing on a surface held at a steady altitude in Jupiter's atmosphere, such as on the floor of an airplane or a balloon flying in Jupiter's atmosphere.

d. On the surface of Europa, the acceleration of gravity is:

$$a_{\text{Europa}} = a_{\text{Earth}} \times \frac{M_{\text{Europa}}}{M_{\text{Earth}}} \times \left(\frac{R_{\text{Earth}}}{R_{\text{Europa}}}\right)^2 = 9.8\frac{\text{m}}{\text{s}^2} \times 0.008 \times \left(\frac{1}{0.25}\right)^2 = 1.3\frac{\text{m}}{\text{s}^2}$$

e. On the surface of Phobos, the acceleration of gravity is:

$$a_{\text{Phobos}} = G\frac{M_{\text{Phobos}}}{R^2_{\text{Phobos}}} = \left(6.67 \times 10^{-11}\frac{\text{m}^3}{\cancel{\text{kg}} \times \text{s}^2}\right) \times \frac{1.1 \times 10^{16}\ \cancel{\text{kg}}}{(12{,}000\ \text{m})^2} = 0.0051\frac{\text{m}}{\text{s}^2}$$

59. a. An acceleration of 6 gees means $6 \times 9.8\ \text{m/s}^2$, or $58.8\ \text{m/s}^2$.

b. The force you will feel from this acceleration of 6 gees will be six times your normal weight.

c. It is unlikely that you could survive this acceleration of 6 gees for very long. It would be rather like lying on a table with six times (actually five to six times, since you have your own weight as well) your normal weight in bricks stacked up over you. You could survive fine for a while, but eventually this compression would probably cause serious damage.

60. a. Swisscheese's orbital period is longer than that of the Moon because, by Newton's version of Kepler's third law, more distant objects have longer orbital periods.

b. We start with Newton's version of Kepler's third law:

$$p^2 = \frac{4\pi^2}{G(M_1 + M_2)} a^3$$

We take the square root of both sides to solve for the period:

$$p = \sqrt{\frac{4\pi^2}{G(M_1 + M_2)} a^3}$$

Taking the hint in the problem, we'll form a ratio to the period of the Moon to get the period of Swisscheese:

$$\frac{p_{\text{Swisscheese}}}{p_{\text{Moon}}} = \frac{\sqrt{\frac{\cancel{4\pi^2}}{G(\cancel{M}_{\text{Earth}} + \cancel{M}_{\text{Swisscheese}})} a^3_{\text{Swisscheese}}}}{\sqrt{\frac{\cancel{4\pi^2}}{G(\cancel{M}_{\text{Earth}} + \cancel{M}_{\text{Moon}})} a^3_{\text{Moon}}}}$$

We can cancel a lot of common terms in this, especially since $M_{\text{Swisscheese}} = M_{\text{Moon}}$ (although even if this were not the case, the mass of the Earth is so much larger than the mass of any likely moon that the masses of the two moons would have been unimportant anyway). We simplify the equation to find:

$$\frac{p_{\text{Swisscheese}}}{p_{\text{Moon}}} = \frac{\sqrt{a^3_{\text{Swisscheese}}}}{\sqrt{a^3_{\text{Moon}}}}$$

We can also use a little bit of algebra and put both numerator and denominator under the same exponent:

$$\frac{p_{\text{Swisscheese}}}{p_{\text{Moon}}} = \sqrt{\left(\frac{a_{\text{Swisscheese}}}{a_{\text{Moon}}}\right)^3}$$

Now, we know that $a_{\text{Swisscheese}} = 2a_{\text{Moon}}$. This allows us to compute the ratio of the periods:

$$\frac{p_{\text{Swisscheese}}}{p_{\text{Moon}}} = \sqrt{\left(\frac{2\cancel{a}_{\text{Moon}}}{\cancel{a}_{\text{Moon}}}\right)^3}$$

$$= \sqrt{2^3}$$

$$= 2.83$$

The period of Swisscheese is 2.83 times that of the Moon, or about 2.83 months.

c. Swisscheese would create a third tidal bulge on the Earth. When the moons were on the same side of the Earth or on opposite sides, these bulges would add together to create extra large tides. When they were 90° apart, the bulges would partially cancel, creating smaller tides. When the moons were lined up in their new or full phases, the bulges they raised would also add with the tidal bulge due to the Sun, creating extra large tides.

Chapter 5. Light and Matter: Reading Messages from the Cosmos

This chapter focuses on the nature of light and matter, so that students can understand how we interpret the messages carried by light.

As always, when you prepare to teach this chapter, be sure you are familiar with the relevant media resources (see the complete, section-by-section resource grid in Appendix 3 of this Instructor's Guide) and the online quizzes and other study resources available on the Mastering Astronomy Web site.

What's New in the Fourth Edition That Will Affect My Lecture Notes?

Those who have taught from previous editions of *The Cosmic Perspective* should be aware of the following organizational or pedagogical changes to this chapter (i.e., changes that will influence the way you teach) from the third edition:

- Section 5.3, Properties of Matter, contains ideas of atoms that formerly appeared in Chapter 4 of the third edition.

Teaching Notes (By Section)

Section 5.1 Light in Everyday Life

Continuing the pattern of the chapters in Part II, we begin with a section on light in everyday life. We use this section to define basic terminology such as power, spectrum, emission, absorption, transmission, and reflection.

Section 5.2 Properties of Light

This section introduces several important concepts: wave properties of wavelength, frequency, and speed; wave-particle duality; the idea that light comes in the form of photons; the concept of a field; and the idea of light as an electromagnetic wave. All of these concepts are probably unfamiliar to your students and therefore warrant some discussion in class.

- Note that throughout the book we use the term light as a synonym for electromagnetic radiation in general, as opposed to meaning only visible light. Thus, we are explicit in saying visible light when that is what we mean.

Section 5.3 Properties of Matter

There are three key points in this section: (1) We want students to understand the basic concepts and terminology of atoms, including the size scale of atoms, the constituents of atoms (i.e., protons, neutrons, electrons), and the electrical charge of atoms and their constituent particles; (2) we want students to understand how different substances can differ by element (i.e., atomic number), isotope, or ionization; and (3) we want students to understand phase changes.

- Note that we never give the Bohr picture of the atom, instead discussing only the modern picture, albeit in rather vague terms (e.g., stating "electrons in atoms are 'smeared out,' forming a kind of cloud . . ."). This reflects our belief that the Bohr model, while useful for purposes of calculation, tends only to reinforce misconceptions about atoms that most students bring with them to our courses—namely, the belief that electrons look and act like miniature planets orbiting a miniature Sun. If you have the chance to cover it, our discussion of atomic structure becomes a little less vague in Chapter S4, where we discuss the uncertainty principle and quantum mechanics.
- Note that our discussion of phases is framed in terms of bond breaking, which makes molecular dissociation and ionization part of the phase-change chain. Describing phase changes in these terms should make it easier for students to understand the processes that take place in planets and stars.
- A note on atomic terminology: Astronomers usually refer to the number of protons + neutrons in an atom as its "atomic mass." However, chemists use this term for the actual mass as a weighted average of isotopes found on Earth (i.e., the mass shown on the periodic table). Thus, the formal name for the number of protons + neutrons is "atomic mass number." We use this term so that students will not be confused if they have had chemistry and have used the term "atomic mass" in its chemistry sense.

Section 5.4 Learning from Light

This section is probably the most important in the chapter, because it covers the interpretation of astronomical spectra. Note that it begins with Figure 5.13 showing a schematic spectrum and then later shows this spectrum analyzed in Figure 5.20.

- A classroom demonstration of spectroscopy can be very useful if available. We like to hand out inexpensive plastic diffraction gratings, which students can use to see spectra of various discharge tubes with different gases such as hydrogen, helium, sodium, and neon, along with an incandescent light bulb to serve as a white light source.
- We have found that this material, while somewhat complex, is not difficult for most students to grasp. However, the jargon often used by astronomers tends to confuse students. Therefore we have tried to eliminate jargon. Note in particular:
  - Aside from a brief note, we do not give a name to Kirchhoff's laws; we do, of course, describe them, both in the text and in Figure 5.14.
  - We use the term *thermal radiation* rather than *blackbody radiation*.
  - We describe the Stefan-Boltzmann law and Wien's law simply as two "rules" that describe the temperature dependence of a thermally emitting object.
  - When discussing atomic transitions in hydrogen, we are explicit in stating the energy levels between which the transitions occur, rather than introducing the jargon of Lyman α, etc.

Section 5.5 The Doppler Shift

The final section of this chapter describes the Doppler effect and how we can measure it using spectral lines as reference points. We also include a brief discussion of how Doppler line broadening allows us to determine stellar rotation rates.

Answers/Discussion Points for Think About It Questions

The Think About It questions are not numbered in the book, so we list them in the order in which they appear, keyed by section number.

Section 5.1

- (p. 147) This question asks students to try to see the individual red, green, and blue colors from which TV images are constructed on the screen.

Section 5.2

- (p. 151) The third one, with the shortest wavelength, has the most energy. The top one, with the longest wavelength, has the least energy.

Section 5.3

- (p. 155) ^{3}He represents helium containing two protons and one neutron.
- (p. 156) Global warming should increase evaporation, and hence increase cloud cover.
- (p. 159) Yes, for a jump from level 2 to level 4.

Section 5.4

- (p. 161) In a cold cloud of hydrogen gas, nearly all the electrons will be in the ground state and hence cannot fall to a lower energy level. Thus, we do not see emission lines from cold clouds of hydrogen gas.

Section 5.5

- (p. 167) A line shifted from a rest wavelength of 121.6 nanometers to 120.5 nanometers has actually shifted to a position farther from the blue end of the visible spectrum, which begins around 400 nanometers. Nevertheless, we call this a *blueshift* because it is a shift to a shorter (as opposed to a longer) wavelength.

Solutions to End-of-Chapter Problems (Chapter 5)

1. Power is the rate of energy transfer. Power is measured in Watts, which are Joules per second.
2. A spectrum is light that has been spread out into its different colors, like in a rainbow. We see a spectrum by passing light through a prism or reflecting it off of a diffraction grating.
3. Emission: A light bulb emits light.
   Absorption: A person lying in the summer sun absorbs light.
   Transmission: A window transmits light.
   Reflection/Scattering: A mirror reflects light.
4. Particles are individual things, while waves are spread out and carry energy without necessarily carrying any matter along with them.
   Wavelength—The distance between two successive peaks in a wave.
   Frequency—The number of wave peaks to pass by a point in a second.
   Speed of a wave—How fast the wave itself travels, equal to $\text{wavelength} \times \text{frequency}$.

5. Because light is a vibration of electric and magnetic fields, we say that light is an electromagnetic wave. Because all electromagnetic waves travel at the speed of light, the product of their frequencies and wavelengths is always the speed of light. (This is because the product of frequency and wavelength is always the speed of the wave. For light, that speed is constant.) Mathematically:

$$\text{frequency} = \frac{\text{speed of light}}{\text{wavelength}}$$

6. A photon is a particle of light. Unlike an ordinary wave, light has a smallest unit that cannot be subdivided. However, light also has wavelike properties, such as having characteristic wavelengths and frequencies.
7. From lowest to highest energy, the electromagnetic spectrum runs: radio, infrared, visible, ultraviolet, X rays, gamma rays. This is the same order we would get if we listed them in order of frequency (lowest to highest) since energy is directly proportional to frequency. However, it is the reverse of what we get if we list them in order of wavelength, because frequency and wavelength are inversely related.
8. An atom has a tiny nucleus in the center with the protons and neutrons. Around the nucleus are clouds of electrons. Atoms are tiny, less than one-millionth the size of the period at the end of this sentence. However, the nuclei are much smaller still.
9. An atom's atomic number is the number of protons it has in its nucleus. Its atomic mass number is the number of protons plus the number of neutrons. Two atoms can have the same number of protons (have the same atomic number) and have different numbers of neutrons. In this case, we say that these atoms are different isotopes of the same element. A molecule is a group of two or more atoms bound together.
10. Electrical charge is a measure of how strongly something will interact with electromagnetic fields. We define a proton as having +1 unit of charge and electrons as having −1 unit when talking about atoms and particles. Two particles with different sign charges will attract each other, so protons and electrons will attract. Particles with charges of the same sign will repel each other, so two electrons will repel each other.
11. We begin with ice, water's solid form. In this state, the molecules of water are bound to their neighbors and cannot move significantly. As we heat it up (apply more energy), the molecules vibrate faster until they have so much energy that they break the bonds between themselves and their neighbors. At this point, molecules can slip past each other and we have a liquid. In this state, the molecules are not directly bound to their neighbors, but they are attracted to each other enough to stay grouped together. As the liquid is heated further, the molecules move faster and faster. Eventually, even the attraction felt between molecules in the liquid state is not enough to overcome the molecules' random motions and the water becomes a gas. In the gas state, the individual molecules fly about freely without feeling attractions to each other (although they do collide with each other). If we keep heating the water vapor, eventually we will give it so much energy that the molecules will break apart, and we no longer have water, just hydrogen and oxygen (or, initially, H and OH ions). If we apply even more energy, the electrons will be removed from their atoms, and we will have a hot plasma, an electrically charged gas.
12. When we say that energy levels in atoms are "quantized," we mean that the electrons can have only particular amounts of electrical potential energy in atoms. Electrons can make a transition from one level to another by taking in or emitting a

specific amount of energy. If too much or too little energy is offered, the electron cannot make the transition.

13. We convert the rainbow-like spectrum into a graph by plotting the intensity of the light at each wavelength as a function of the wavelength. This sort of graph is easier to interpret than the rainbow-like spectrum is.
14. A continuous spectrum is seen when we have a hot object emitting light across a broad range of wavelengths. We see an emission line spectrum when we look at a cloud of thin gas because we see only the wavelengths of light that correspond to the atomic transitions allowed in that gas. We see an absorption line spectrum when we look at a hot object (like a star) through a thin cloud of gas. The continuous spectrum loses photons of the specific frequencies that the atoms in the thin gas like to absorb.

    The spectrum at the start of the chapter shows a continuous spectrum from the Sun's hot "surface." However, before the light reaches us, it passes through the Sun's thin atmosphere and photons of certain specific wavelengths are absorbed by the gas. This creates the dark lines in the spectrum.
15. Each atom tends to absorb and emit different wavelengths of light. Similarly, every molecule absorbs or emits different bands of wavelengths. So when we look at an absorption or emission spectrum, we can see the "fingerprints" of the different atoms (or ions) or molecules. In this way, we can learn what an object is made of without ever sampling the object.
16. We know that a hotter object emits more light at every wavelength. So the intensity of light from the 8,000 K star will be higher at every wavelength. We also know that the spectrum of the hotter object will peak at shorter wavelengths. So the 8,000 K star will show a peak at shorter wavelength.
17. Figure 5.20 has two overall humps. The one on the left corresponds to the Sun's thermal emission and tells us the temperature of the Sun. The one on the right is the thermal emission from the planet and tells us the planet's surface temperature. We see the planet's peak farther to the right than the star's because the planet is cooler and so peaks at longer wavelengths (as Wein's law tells us it should).

    We also see UV emission lines from the upper atmosphere, telling us that that part of the atmosphere is hot enough for atoms to have enough energy to emit in those short wavelengths. There are also broad absorption bands in the spectrum, telling us that the object has molecules in its atmosphere. (In this case, the bands tell us that there is carbon dioxide in particular.)

    Finally, we see that the solar spectrum is missing much of the bluer light, telling us that the object absorbs the blue light and reflects the red. This makes the object reddish in color.
18. The Doppler effect is the change in frequency in light due to the source's motion toward or away from the observer. When the source is coming toward us, the light we see has a shorter wavelength (higher frequency), and we say that it is blueshifted. If the object is moving away from us, the light has a longer frequency than we would expect, and we say that it is redshifted.

    For rotating objects, we see part of the object coming toward us and part of it moving away from us. The light from the part coming toward us is blueshifted, and the light from the part going away is redshifted. If we can't see the different parts of the object clearly, we will see the blueshifted, the redshifted, and the unshifted light all at once, making it look like the lines have been broadened.

19. *The walls of my room are transparent to radio waves.* This statement makes sense; radio waves do indeed pass through most walls, which is why you can receive radio broadcasts and cell phone calls inside a house.
20. *Because of their higher frequency, X rays must travel through space faster than radio waves.* This statement does not make sense; all light travels through space at the same speed of light.
21. *If you could see infrared light, you would see a glow from the backs of your eyelids when you closed your eyes.* This statement makes sense, because your eyelids are warm and emit infrared radiation.
22. *If you had X-ray vision, then you could read this entire book without turning any pages.* This statement does not make sense. The book does not emit X rays, so X-ray vision wouldn't do you any good at all.
23. *Two isotopes of the element rubidium differ not only in their numbers of neutrons, but also in their numbers of protons.* This statement does not make sense, because an element is defined by its number of protons (atomic number); if two atoms have different numbers of protons, then they cannot be the same element.
24. *A "white-hot" object is hotter than a "red-hot" object.* This statement makes sense; white is a mix of all visible colors, so an object glowing white-hot must be emitting some blue light, making it hotter than a red-hot object.
25. *If the Sun's surface became much hotter (while the Sun's size remained the same), the Sun would emit more ultraviolet light but less visible light than it currently emits.* This statement does not make sense; if the Sun's surface were hotter, it would emit more thermal radiation at all wavelengths of light.
26. *If you could view a spectrum of light reflecting off a blue sweatshirt, you'd find the entire rainbow of color.* This statement does not make sense; the blue sweatshirt reflects only blue visible light, so the spectrum of this reflected light would not contain other colors such as red.
27. *Nearly all galaxies show redshifts, so they must all be red in color.* This statement does not make sense; a redshift does not mean red color. For example, a blue galaxy with a small redshift will still appear blue.
28. *If a distant galaxy has a substantial redshift (as viewed from our galaxy), then anyone living in that galaxy would see a substantial redshift in a spectrum of the Milky Way Galaxy.* This statement makes sense; the redshift means that we see the galaxy moving away from us, so observers in that galaxy must also see us moving away from them—which means they see us redshifted as well.
29. c; 30. a; 31. b; 32. a; 33. c;
34. a; 35. a; 36. b; 37. a; 38. a.
39.
    a. The iron has atomic number 26, atomic mass 26 + 30 = 56, and, if it is neutral, 26 electrons to balance the charge of its 26 protons.
    b. Atoms 2 and 3 are isotopes of each other, because they have the same number of protons but different numbers of neutrons.
    c. An O^{+5} ion is five times ionized and is missing five of its eight electrons; thus, the ion has three electrons.
40.
    a. Fluorine with 9 protons and 10 neutrons has atomic number 9 and atomic weight 19. If we added a proton to this nucleus, the result would have a different atomic number and therefore would no longer be fluorine. If we added a neutron to the fluorine nucleus, the atomic number would be unchanged, so the result

would still be fluorine. However, because the atomic weight would change, we would have a different isotope of fluorine.

b. Gold with atomic number 79 and atomic weight 197 has nuclei containing 79 protons and 197 – 79 = 118 neutrons. If the gold is electrically neutral, its atoms have 79 electrons to offset the charge of the 79 protons. If the gold is triply ionized, it is missing 3 of its electrons and thus has 76 electrons.

c. Uranium has atomic number 92 and hence has 92 protons. Thus, ^{238}U contains 238 – 92 = 146 neutrons and ^{235}U contains 235 – 92 = 143 neutrons.

41. a. Nearly all the matter in the Sun is in the plasma phase because the high temperatures inside the Sun keep nearly all the atoms fully ionized.

b. Most of the ordinary matter (i.e., not dark matter) in the universe is found in stars or in very hot intergalactic gas. Because this matter is in the plasma phase, plasma is the most common phase of matter in the universe.

c. Plasma is rare on Earth because the relatively low temperatures on the Earth's surface are too low for ionization of atoms.

42. a. Transition B could represent an electron that gains 10.2 eV of energy because it jumps up from 0 eV to 10.2 eV.

b. Transition C represents the electron that loses 10.2 eV of energy because the electron is jumping down.

c. Transition E represents an electron that is breaking free of the atom because the electron has enough energy to be ionized.

d. Transition D is not possible because electrons can jump only from one allowed energy level to another, not to energies in between.

e. Transition A represents an electron falling from level 3 to level 1, emitting 12.1 eV of energy in the process. This is more energy than emitted in transition C.

43. a. We can determine the chemical composition of an object by identifying the specific spectral lines due to various elements.

b. If the spectrum is nearly a thermal radiation spectrum, we can determine the object's surface temperature from the peak wavelength of emission. Otherwise, we can determine the surface temperature by studying the ionization states present among the chemicals in the object.

c. A thin cloud of gas will have a nearly "pure" emission or absorption line spectrum. A more substantial object will have a thermal radiation spectrum with emission or absorption lines superimposed.

d. An object such as a planet with a hot upper atmosphere will have emission lines in the ultraviolet because atoms in this atmosphere will be in excited states that allow emission.

e. An object that is reflecting blue light from its star will have a spectrum with a general shape like that of a thermal radiation spectrum but with red light otherwise missing.

f. We can determine the speed at which an object is moving toward or away from us by measuring the Doppler shift of lines in its spectrum.

g. We can determine the rotation rate of an object by the degree of broadening of its spectral lines, which occurs because different parts of the object have different Doppler shifts due to the rotation.

44. A glowing cloud of gas will produce an emission line spectrum.

45. a. The sketch should have red light missing instead of blue light from the reflected thermal radiation in the visible part of the spectrum.
    b. With a warmer temperature, the infrared hump due to the planet's own thermal radiation should grow larger, and its peak should shift to a slightly shorter wavelength.
    c. With faster rotation, its spectral lines would be wider.
46. We are considering a spectral line with a rest wavelength of 121.6 nanometers that appears at 120.5 nanometers in Star A, 121.2 nanometers in Star B, 121.9 nanometers in Star C, and 122.9 nanometers in Star D. Because the line is shifted to a wavelength shorter than its rest wavelength in Stars A and B, these two stars are moving toward us; Stars C and D are moving away from us. The star showing the greatest shift is Star D, in which the line is redshifted by 1.3 nanometers; thus, of these four stars, Star D is moving the fastest relative to us.
47. Converting from calories to joules, we find that the daily energy usage of a typical adult is:

$$2{,}500\ \cancel{\text{cal}} \times 4{,}000 \frac{\text{joules}}{\cancel{\text{cal}}} = 10^7 \text{ joules}$$

To find the average power requirement of an adult in watts, we divide the energy usage from part (a) by the number of seconds in a day:

$$\frac{10^7 \text{ joules}}{24\ \cancel{\text{hr}} \times 3600 \frac{\text{s}}{\cancel{\text{hr}}}} = 116 \text{ watts}$$

The average adult runs on about 120 watts of power. Note that this is quite similar to the power needed for a typical light bulb.
48. We are asked to convert 1 kilowatt-hour into joules. First, we will convert from kilowatt-hours to kilowatt-seconds:

$$1 \text{ kilowatt-hour} \times \frac{60\ \cancel{\text{min}}}{1 \text{ hr}} \times \frac{60 \text{ s}}{1\ \cancel{\text{min}}} = 3{,}600 \text{ kilowatt-seconds}$$

Next, we convert from kilowatts to watts:

$$3{,}600 \text{ kilowatt-seconds} \times \frac{1000 \text{ watts}}{1 \text{ kilowatt}} = 3{,}600{,}000 \text{ watt-seconds}$$

Finally, note that a watt is a joule per second. So 1 watt × 1 s = 1 joule, leaving us with:

$$1 \text{ kilowatt-hour} = 3{,}600{,}000 \text{ joules}$$

If your house used 900 kilowatt-hours in a month (according to your utility bill), the energy in joules was:

$$900\ \cancel{\text{kilowatt-hours}} \times \frac{3{,}600{,}000 \text{ joules}}{1\ \cancel{\text{kilowatt-hour}}} = 3.24 \times 10^9 \text{ joules}$$

You used 3.24 billion joules of energy.

49. For a frequency of 1,120 kilohertz, the wavelength (λ) and energy (E) are:

$$\lambda = \frac{c}{f} = \frac{3 \times 10^8 \frac{\text{m}}{\cancel{\text{s}}}}{1120 \times 10^3 \frac{1}{\cancel{\text{s}}}} = 268\text{m}$$

$$E = h \times f = (6.626 \times 10^{-34} \text{ joule} \times \text{s}) \times 1{,}120 \times 10^3 \frac{1}{\text{s}} = 7.42 \times 10^{-28} \text{ joule}$$

50. For a photon with wavelength 120 nanometers, the frequency (f) and energy (E) are:

$$f = \frac{c}{\lambda} = \frac{3 \times 10^8 \frac{\text{m}}{\text{s}}}{120 \times 10^{-9} \text{ m}} = 2.5 \times 10^{15} \frac{1}{\text{s}} = 2.5 \times 10^{15} \text{ Hz}$$

$$E = h \times f = (6.626 \times 10^{-34} \text{ joule} \times \text{s}) \times 2.5 \times 10^{15} \frac{1}{\text{s}} = 1.7 \times 10^{-18} \text{ joule}$$

51. A photon energy of 10 keV is equivalent to an energy of:

$$10{,}000 \text{ eV} \times 1.60 \times 10^{-19} \frac{\text{joule}}{\text{eV}} = 1.60 \times 10^{-15} \text{joule}$$

The frequency and wavelength of this photon are:

$$f = \frac{E}{h} = \frac{160 \times 10^{-15} \cancel{\text{joule}}}{6.626 \times 10^{-34} \cancel{\text{joule}} \times \text{s}} = 2.41 \times 10^{18} \frac{1}{\text{s}} = 2.41 \times 10^{18} \text{ Hz}$$

$$\lambda = \frac{c}{f} = \frac{3 \times 10^8 \frac{\text{m}}{\cancel{\text{s}}}}{2.41 \times 10^{18} \frac{1}{\cancel{\text{s}}}} = 1.25 \times 10^{-10} \text{m} = 0.125 \text{ nm}$$

52. a. The energy of a single photon with wavelength 600 nanometers is:

$$E = \frac{hc}{\lambda} = \frac{(6.626 \times 10^{-34} \text{joule} \times \cancel{\text{s}}) \times 3 \times 10^8 \frac{\cancel{\text{m}}}{\cancel{\text{s}}}}{600 \times 10^{-19} \cancel{\text{m}}} = 3.31 \times 10^{-19} \text{joule}$$

b. A 100-watt light bulb emits 100 joules of energy each second. Thus, the number of 600-nanometers photons it emits each second is:

$$\text{number of photons} = \frac{\text{total energy emitted}}{\text{energy emitted per photon}} = \frac{1000 \text{ joule}}{3.31 \times 10^{-19} \frac{\text{joule}}{\text{photon}}} = 3 \times 10^{20} \text{ photons}$$

c. Because the light bulb emits more than 10^{20} photons each second, there is no chance that we will notice these individual photons acting like particles in our everyday life. Instead, we notice only their collective wave effects.

53. We can use the Stefan-Boltzmann law to compute the energy emitted by the object from every square meter:

$$\text{emitted power} = \sigma T^4$$

where T is the temperature and $\sigma = 5.7 \times 10^{-8} \dfrac{\text{watt}}{\text{K}^4\ \text{m}^2}$.

For our 3,000 K object:

$$\text{emitted power} = \left(5.7 \times 10^{-8} \frac{\text{watt}}{\cancel{\text{K}^4}\ \cancel{\text{m}^2}}\right)(3{,}000\ \cancel{\text{K}})^4$$

$$= 4.6 \times 10^{6} \frac{\text{watt}}{\cancel{\text{m}^2}}$$

We find the wavelength of the peak emission with Wein's law;

$$\lambda_{\text{peak}} = \frac{2{,}900{,}000}{T}\text{nm}$$

where T is again the temperature, measured in Kelvins. For a 3,000 K object, we get:

$$\lambda_{\text{peak}} = \frac{2{,}900{,}000}{3000}\text{nm}$$
$$= 967\ \text{nm}$$

The object emits 4.6 million watts per square meter and has a spectrum that peaks at 967 nanometers.

54. We can use the Stefan-Boltzmann law to compute the energy emitted by the object from every square meter:

$$\text{emitted power} = \sigma\ T^4$$

where T is the temperature and $\sigma = 5.7 \times 10^{-8} \dfrac{\text{Watt}}{\text{K}^4\text{m}^2}$.

For a 50,000 K object;

$$\text{emitted power} = \left(5.7 \times 10^{-8} \frac{\text{watt}}{\cancel{\text{K}^4}\ \text{m}^2}\right)\left(50{,}000\ \cancel{\text{K}}\right)^4$$

$$= 3.6 \times 10^{11} \frac{\text{watt}}{\text{m}^2}$$

We find the wavelength of the peak emission with Wein's law:

$$\lambda_{\text{peak}} = \frac{2{,}900{,}000}{T}\text{nm}$$

where T is again the temperature, measured in Kelvins. For a 50,000 K object, we get:

$$\lambda_{\text{peak}} = \frac{2{,}900{,}000}{50{,}000}\text{nm}$$
$$= 58 \text{ nm}$$

The 50,000 K object emits 360 billion watts per square meter and has a spectrum that peaks at 58 nanometers.

55. a. To solve this, we will set up a ratio using the Stefan-Boltzmann law:

$$\frac{\text{energy emitted by hotter Sun}}{\text{energy emitted by the Sun}} = \frac{\sigma\, T_{\text{hotter}}^{4}}{\sigma T_{\text{Sun}}^{4}}$$
$$= \left(\frac{T_{\text{hotter}}}{T_{\text{Sun}}}\right)^{4}$$

So if $T_{\text{hotter}} = 2T_{\text{Sun}}$, we get:

$$\frac{\text{energy emitted by hotter Sun}}{\text{energy emitted by the Sun}} = \left(\frac{2\cancel{T_{\text{Sun}}}}{\cancel{T_{\text{Sun}}}}\right)^{4}$$
$$= (2)^{4}$$
$$= 16$$

If the Sun were twice as hot, it would emit 16 times as much power per square meter.

b. We will again use a ratio to work this problem, using Wein's law this time:

$$\frac{\text{peak wavelength of hotter Sun}}{\text{peak wavelength of Sun}} = \frac{\dfrac{\cancel{2{,}900{,}000}}{T_{\text{hotter}}}\cancel{\text{nm}}}{\dfrac{\cancel{2{,}900{,}000}}{T_{\text{Sun}}}\cancel{\text{nm}}}$$
$$= \frac{T_{\text{Sun}}}{T_{\text{hotter}}}$$

where we have canceled the 2,900,000 in the numerator and denominator and we have used some of the rules for fractions to simplify the expression. Using the fact that $T_{\text{hotter}} = 2T_{\text{Sun}}$, we get:

$$\frac{\text{peak wavelength of hotter Sun}}{\text{peak wavelength of Sun}} = \frac{T_{\text{Sun}}}{2T_{\text{Sun}}}$$
$$= \frac{1}{2}$$

If the Sun were twice as hot, its peak wavelength would be at $\frac{1}{2}$ the peak wavelength of the real Sun.

c. It is doubtful that life could exist on Earth around this star. There would be so much energy coming from the star that our planet would probably be much too hot.

56. a. We find the average power radiated per square meter on the Sun's surface by dividing the Sun's total power output by its surface area:

$$\text{power per m}^2 = \frac{\text{total power}}{\text{surface area}} = \frac{4 \times 10^{26}\,\text{watt}}{4 \times \pi \times (7 \times 10^8\,\text{m})^2} = 6 \times 10^7\,\frac{\text{watt}}{\text{m}^2}$$

b. According to the Stefan-Boltzmann law, the emitted power per unit area is equal to the quantity σT^4. Solving for the surface temperature, we find:

$$\text{power per unit area } \sigma T^4 \Rightarrow T = \sqrt[4]{\frac{\text{power per unit area}}{\sigma}}$$

We can now calculate the Sun's surface temperature from the power per unit area found in part (a):

$$T = \sqrt[4]{\frac{\text{power per unit area}}{\sigma}} = \sqrt[4]{\frac{6 \times 10^7\,\frac{\cancel{\text{watt}}}{\cancel{\text{m}^2}}}{5.7 \times 10^{-8}\,\frac{\cancel{\text{watt}}}{\cancel{\text{m}^2} \times \text{K}^4}}} = 6 \times 10^3\ \text{K}$$

We have found that the Sun's surface temperature is about 6,000 K.

57. We use the Doppler shift formula to find the speeds of the stars:

Star A: $$\upsilon = \frac{\Delta\lambda}{\lambda_0} \times c = \frac{120.5\ \cancel{\text{nm}} - 1216\ \cancel{\text{nm}}}{121.6\ \cancel{\text{nm}}} \times 300{,}000\frac{\text{km}}{\text{s}} = -2{,}714\frac{\text{km}}{\text{s}}$$

The negative value indicates that Star A is moving toward us.

Star B: $$\upsilon = \frac{\Delta\lambda}{\lambda_0} \times c = \frac{121.2\ \cancel{\text{nm}} - 1216\ \cancel{\text{nm}}}{121.6\ \cancel{\text{nm}}} \times 300{,}000\frac{\text{km}}{\text{s}} = -987\frac{\text{km}}{\text{s}}$$

The negative value indicates that Star B is moving toward us.

58. We use the Doppler shift formula to find the speeds of the stars:

Star C: $$\upsilon = \frac{\Delta\lambda}{\lambda_0} \times c = \frac{121.9\ \cancel{\text{nm}} - 1216\ \cancel{\text{nm}}}{121.6\ \cancel{\text{nm}}} \times 300{,}000\frac{\text{km}}{\text{s}} = 740\frac{\text{km}}{\text{s}}$$

The positive value indicates that Star C is moving away from us.

Star D: $$\upsilon = \frac{\Delta\lambda}{\lambda_0} \times c = \frac{122.9\ \cancel{\text{nm}} - 1216\ \cancel{\text{nm}}}{121.6\ \cancel{\text{nm}}} \times 300{,}000\frac{\text{km}}{\text{s}} = 3207\frac{\text{km}}{\text{s}}$$

The positive value indicates that Star D is moving away from us.

59. a. For a coil with a temperature of 3,000 K, the wavelength of maximum intensity is:

$$\lambda_{\text{max}} = \frac{2{,}900{,}000}{3000\ (\text{Kelvin})}\text{nm} = 966\ \text{nm}$$

Note that this wavelength is considerably longer than the 500-nanometers wavelength of maximum emission from the Sun and lies in the infrared portion of the spectrum. Thus, light bulbs with coils at 3,000 K emit much of their energy in the infrared, rather than as visible light.

b. Because light from standard light bulbs has a spectrum that peaks in the infrared, it is generally redder in color than sunlight. Thus, to record "true" colors, film for indoor photography must compensate for the fact that indoor light bulbs emit more red light by having enhanced sensitivity to the less abundant blue light.
c. Standard light bulbs emit thermal radiation, so they must emit over a wide range of wavelengths. In fact, because their thermal emission peaks in the infrared, they actually emit most of their energy as infrared light, rather than visible light.
d. Because a standard light bulb emits much of its light in the infrared, this light is "wasted" as far as electrical lighting is concerned. In contrast, a fluorescent light bulb that produces emission lines only in the visible portion of the spectrum would have no "wasted" light. Thus, a fluorescent bulb of the same wattage as a standard bulb can actually produce much more visible light.
e. Despite the higher cost of a compact fluorescent light bulb compared to a standard light bulb, the former can save money over the long run for two principal reasons: (1) lower cost for operation because of its lower energy usage, and (2) longer life, so it needs replacement less often. The latter can particularly save money for businesses, since businesses must pay for the labor involved in changing light bulbs; this labor can be substantial if the light bulb is in a hard-to-reach area.

Chapter 6. Telescopes: Portals of Discovery

This chapter focuses on telescopes and their uses. Note that, although these instruments are fundamental to modern astronomy, most of the material in this chapter is not prerequisite to later chapters. Thus, you can consider this chapter to be optional.

As always, when you prepare to teach this chapter, be sure you are familiar with the relevant media resources (see the complete, section-by-section resource grid in Appendix 3 of this Instructor's Guide) and the online quizzes and other study resources available on the Astronomy Place Web site.

What's New in the Fourth Edition That Will Affect My Lecture Notes?

As everywhere in the book, we have added learning goals that we use as subheadings, rewritten to improve the text flow, improved art pieces, andadded new illustrations. The art changes, in particular, will affect what you wish to show in lecture. We have not made any substantial content or organizational changes to this chapter.

Teaching Notes (By Section)

Section 6.1 Eyes and Cameras: Everyday Light Sensors

As with all the chapters in Part II, we begin this chapter with a section on "everyday" light collection, discussing the human eye and cameras.

Section 6.2 Telescopes: Giant Eyes

This section describes the general design of optical telescopes.

- Note our emphasis on two principal properties of telescopes: light-collecting area and angular resolution.
- Note that while we show Cassegrain, Newtonian, and Nasmyth/Coudé foci in a figure, we do not expect students to learn the names, and we give the names only for reference.
- Although different observers tend to categorize observations differently, we have chosen to categorize observations as either imaging, spectroscopy, or timing. We believe that this categorization is pedagogically useful, because it most closely corresponds to the figures that students see in the book and in news reports: photographs (imaging), spectra (spectroscopy), and light curves (timing).

Section 6.3 Telescopes and the Atmosphere

In this section, we turn to the atmospheric effects due to light pollution and turbulence (twinkling), leading to a discussion of how observing sites are chosen and of adaptive optics.

- This is where we point out that most wavelengths of light do not penetrate the atmosphere, and introduce the rationale for space telescopes.

Section 6.4 Telescopes and Technology

This section covers telescopes designed to collect light of different wavelengths, and interferometry.

Answers/Discussion Points for Think About It Questions

The Think About It questions are not numbered in the book, so we list them in the order in which they appear, keyed by section number.

Section 6.1

- (p. 176) The pupil will be wider in the eye exposed to light. Doctors dilate your pupils so that they can see through them into your eye.
- (p. 177) This question is intended simply to get students to see the correspondence between the ideas they are learning in this chapter and the cameras that they use at home.

Section 6.2

- (p. 178) This question can make for a short discussion, as students state how the total light-collecting area of the eyes of people in their homes compares to that of a 10-meter telescope.
- (p. 184) CAT scans and MRIs use false-color displays to show different types of tissue or different elements within tissue, allowing doctors effectively to "see" inside your body.

Section 6.3

- (p. 187) The coin appears to move because the water is moving and changing the path of the light as it comes through the water toward our eyes. It's basically the same reason that air turbulence causes twinkling.

Section 6.4

- (p. 193) If you've never tried this trick for seeing grazing incidence reflections of visible light, you and your students may all be surprised by how it works. Note that this demonstration of visible-light grazing incidence works only with fairly smooth and shiny surfaces. It is impressive if you've never noticed it before.

Solutions to End-of-Chapter Problems (Chapter 6)

1. The eye focuses light by bending (refracting) it so that parallel rays of incoming light all meet up on the same point in the back of our eye. A glass lens does the same thing, but unlike the eye it cannot adjust its shape to change its focus. The focal plane of a lens is the place where an image appears in focus.
2. Cameras are better than eyes for astronomy because they are more reliable than people sketching what they see. In addition, they can use long exposure times to make images of faint objects.

   CCDs are better than film because they are more sensitive. Thus, we can get the same image with a shorter exposure time. They are also sensitive to a much broader range of brightness at one time, making it possible to take a picture of brighter and fainter objects at the same time. And finally, they automatically save their images digitally, which means we can import them into computers to manipulate them easily.
3. Telescopes have two key properties: light-collecting area and angular resolution. One is light-collecting area. A telescope can collect a lot more light than the human eye can because it is larger. This is important because the objects astronomers want to look at are usually very faint. The other important property of telescopes is angular resolution: Telescopes can make out finer details than our eyes can. This is important because the objects we want to study appear small in our sky.
4. The diffraction limit is a natural, physical limit on the finest resolution a given telescope can achieve at a particular wavelength. As the telescope gets larger, the diffraction limit decreases (leading to the possibility of finer resolution). However, as the wavelength increases, the diffraction limit increases so that longer wavelengths require larger telescopes to achieve the same resolution.
5. The difference between a reflecting telescope and a refracting telescope lies in how the light is focused. In a refracting telescope, a lens is used to focus the light, whereas in a reflecting telescope we use mirrors. Today, we mainly use reflectors because they allow for larger mirrors and less loss of light.
6. The three basic types of observing are imaging, spectroscopy, and timing. For imaging, the telescope functions like a camera, taking a picture of the object of interest. For spectroscopy, astronomers use a diffraction grating (or some other means) to spread the light into its component wavelengths and then measure the brightness of each wavelength. Finally, for timing observations, astronomers use multiple images or otherwise measure the incoming light at known times to measure how the brightness varies.
7. An image in a nonvisible wavelength is just like an image in a visible wavelength, except that our eyes would never be able to see it without help. However, the colors in the image are not the actual colors of the object. They are simply used to aid us in studying the object.

8. Spectral resolution measures how fine the details are that we can see in a spectrum. With better spectral resolution, we can make out features closer together in wavelength in the spectrum. High spectral resolution requires longer exposures (or brighter sources) than we need for images, because the light is spread out when we make a spectrum.
9. The atmosphere has three negative effects on observations. First, it blocks light of most wavelengths from ever reaching the ground. Second, the atmosphere scatters human-generated light and makes it more difficult to make observations. And finally, the constant movement of air in the atmosphere (the source of turbulence when flying) causes stars to appear to twinkle, which effectively blurs telescopic images. Putting a telescope in space overcomes these problems because the telescope no longer has to look through the atmosphere. Adaptive optics is a technology that can essentially undo the blurring caused by twinkling.
10. Figure 6.22 shows that the only wavelengths that make it all the way to the ground are the narrow range of visible light and the radio wavelengths. A few other wavelengths can be detected from high mountains or from aircraft, but most wavelengths require us to put telescopes in space, which is why space-based astronomy is so important.
11. Radio astronomy uses dishes like satellite dishes to observe the objects. In some cases, the dishes work together to form images at a much higher resolution than any one of them could achieve alone. An example of a radio telescope is the Very Large Array in New Mexico.

    Infrared telescopes are very similar to optical telescopes, except that they use different detectors to sense the light. An example of an infrared telescope is the Spitzer Infrared Space Telescope.

    Most of the ultraviolet wavelengths act enough like the visible wavelengths that we can use the same sorts of telescopes, with different detectors. (At the shortest wavelengths, the ultraviolet waves actually act more like X rays so we use telescopes more like X-ray telescopes.) An example of an ultraviolet telescope is the Hubble Space Telescope.

    X-ray telescopes have to gently redirect the light because it tends to go straight through things in its path. X-ray mirrors are arranged so that the light just grazes them. An example of an X-ray telescope is the Chandra space telescope.

    Gamma-ray telescopes must be massive in order to intercept this high-energy light. An example of a gamma-ray telescope is Swift.
12. Interferometry is the process of linking up multiple telescopes so that they combine their observations to make much higher resolution images than the individual telescopes can. This is a massive bonus since it is often impossible to make single telescopes as big as we would like. However, for a much smaller sum of money, a few cheaper telescopes can be used to get the same resolution (although not the same light-gathering power).
13. *The image was blurry, because the photographic film was not placed at the focal plane.* This statement makes sense, because light is generally in focus only at the focal plane.
14. *By using a CCD, I can photograph the Andromeda Galaxy with a shorter exposure time than I would need with photographic film.* This statement makes sense, because a CCD is more sensitive than photographic film and hence can record an image in less time.

15. *I have a reflecting telescope in which the secondary mirror is bigger than the primary mirror.* This statement does not make sense. Remember that the secondary mirror is placed in front of the primary mirror in a reflecting telescope. Thus, if the secondary mirror was bigger than the primary mirror, it would block all light from reaching the primary mirror, rendering the telescope useless.
16. *The photograph shows what appear to be just two distinct stars, but each of those stars is actually a binary star system.* This statement makes sense. If the angular resolution of the telescope isn't good enough to resolve the individual stars in each pair, then each pair will appear as a single star.
17. *I built my own 14-inch telescope that has a lower diffraction limit than most large professional telescopes.* This statement does not make sense, because the diffraction limit depends only on the telescope's size: A small telescope cannot have a lower diffraction limit than a large telescope.
18. *Now that I've bought a spectrograph, I can use my home telescope for spectroscopy as well as imaging.* This statement makes sense, as long as you can hook up the spectrograph to your telescope. However, unless your telescope is unusually large, you'll only be able to get spectra of relatively bright objects.
19. *If you lived on the Moon, you'd never see stars twinkle.* This statement makes sense, because twinkling is caused by the atmosphere and the Moon doesn't have an atmosphere.
20. *New technologies will soon allow astronomers to use X-ray telescopes on the Earth's surface.* This statement does not make sense, because X rays do not reach the ground and there is no technology we can use for a telescope that will change this basic fact.
21. *Thanks to adaptive optics, the telescope on Mount Wilson can now make ultraviolet images of the cosmos.* This statement does not make sense, because ultraviolet light does not reach the ground and there is no technology we can use for a telescope that will change this basic fact.
22. *Thanks to interferometry, a properly spaced set of 10-meter radio telescopes can achieve the angular resolution of a single, 100-kilometer radio telescope.* This statement makes sense, because interferometry allows multiple small telescopes to achieve the angular resolution of a larger telescope.
23. b; 24. b; 25. b; 26. c; 27. c;
28. c; 29. a; 30. a; 31. a; 32. b.
33. You cannot gain any detail by blowing up a magazine or newspaper photograph, beyond the detail that is already there. In much the same way, additional magnification with an astronomical telescope cannot provide any more detail than the telescope is capable of obtaining as a result of its size and optical quality.
34. a. Timing, since we are looking for variation.
    b. Spectroscopy, because we learn composition by identifying spectral lines.
    c. Spectroscopy, because we must compare the wavelengths of the object's spectral lines to the wavelengths of the same lines in the laboratory in order to determine the Doppler shift, from which we determine the speed.
35. Answers will vary; the key point is to be sure students have considered the major factors that affect astronomical observations from the ground.

36. The five telescopes will be much more valuable if linked together by interferometry, because it will improve their angular resolution. Adaptive optics will do nothing at all for telescopes in space, since this technology is designed only to counteract blurring caused by Earth's atmosphere.
37. Through a filter that transmits only red light, you would see the flag's red stripes, but the rest would be black. Through a filter that transmits only blue light, you would see the blue background but not the red stripes. (Depending on the filter, you might see white portions of the flag through either filter, since white light contains all the colors.)
38. The answer depends on the type of observation. Sometimes it is useful to see a wide field, and sometimes it is more useful to see more detail in a small field of view.
39. The advantages of an observatory on the Moon's surface over observatories on Earth are essentially the same as those of space telescopes. The advantage of the lunar surface over orbiting telescopes is ease of construction operation. Disadvantages include the higher cost of reaching the Moon in the first place, and the fact that the telescopes can observe only a portion of the sky (the portion above the horizon) at any one time.
40. This is a project that should help students realize that brighter stars tend to twinkle more than dimmer ones and that stars twinkle more when nearer to the horizon then when higher overhead.
41. This is a project requiring some research into how radio bands are assigned.
42. This telescope review project requires some personal research.
43. a. The 10-meter Keck telescope has twice the diameter of the 5-meter Hale telescope, so its light-collecting area is $2^2 = 4$ times as much.
    b. A 100-meter telescope would have 10 times the diameter of the 10-meter Keck telescope, so its light-collecting area would be $10^2 = 100$ times as much.
44. We seek the angular separation of two stars with real separation $s = 100$ million km $= 10^8$ km at a distance $d = 100$ light-years. We begin by converting the distance from light-years to kilometers:

$$100 \cancel{\text{ly}} \times 10^{13} \frac{\text{km}}{\cancel{\text{ly}}} = 10^{15} \text{ km}$$

Now we use the angular separation formula:

$$\alpha = \frac{s}{2\pi d} \times 360^\circ = \frac{10^8 \cancel{\text{km}}}{2\pi \times 10^{15} \cancel{\text{km}}} \times 360^\circ \approx (6 \times 10^{-6})^\circ$$

Finally, we convert from degrees to arcseconds so that we can compare it to the 0.05-arcsecond resolution of the Hubble Space Telescope:

$$(6 \times 10^{-6})\cancel{^\circ} \times \frac{60\cancel{'}}{1\cancel{^\circ}} \times \frac{60''}{1\cancel{'}} = 0.02''$$

The angular separation of the two stars in this binary system is below the angular resolution of the Hubble Space Telescope, so they will appear as a single point of light in Hubble Space Telescope images.

45. a. Mathematical Insight 6.1 tells us that we can calculate the angular separation of two objects via the formula:

$$\text{angular separation} = 206{,}265'' \times \frac{\text{physical separation}}{\text{distance}}$$

where the physical separation and distance have to be measured in the same units.

To use this formula for the Sun and Jupiter, we need the Sun-Jupiter distance. Appendix E says that this is 7.783×10^8 km. Before we can use the formula, we'll need to convert the distance, 10 light-years, to kilometers. Appendix A says that 1 light-year is 9.46×10^{12} km, so we convert:

$$10 \cancel{\text{light-years}} \times \frac{9.46 \times 10^{12}\ \text{km}}{1 \cancel{\text{light-year}}} = 9.46 \times 10^{13}\ \text{km}$$

Calculating the angular separation:

$$\text{angular separation} = 206{,}265'' \times \frac{7.783 \times 10^8\ \text{km}}{9.46 \times 10^{13}\ \text{km}}$$
$$= 1.70''$$

From 10 light-years away, Jupiter would appear to be 1.70 arcseconds from the Sun.

b. Using the same formula with the same distance, this time with the Earth-Sun separation (1.496×10^8 km), we get:

$$\text{angular separation} = 206{,}265'' \times \frac{1.496 \times 10^8\ \text{km}}{9.46 \times 10^{13}\ \text{km}}$$
$$= 0.326''$$

Earth would appear to be only 0.326 arcsecond from the Sun from 10 light-years away.

c. Hubble has an angular resolution of 0.05 arcsecond. So both observations should, in theory, be possible. However, the Sun is much, much brighter than either planet. So the planetary light will probably be lost in the glare of the Sun, even though we should be able to resolve the planets.

46. a. We use the formula from Mathematical Insight 6.2 to find the diffraction limit of the human eye, but instead of having a telescope diameter, we use the lens diameter of the eye, given as 0.8 centimeter:

$$\text{diffraction limit (arcseconds)} = 2.5 \times 10^5 \times \left(\frac{\text{wavelength}}{\text{lens diameter}}\right)$$
$$= 2.5 \times 10^5 \times \left(\frac{500 \times 10^{-9}\ \cancel{\text{m}}}{0.008\ \cancel{\text{m}}}\right) = 16''$$

(Note that the actual angular resolution of the human eye, about 1 arcminute, is not as good as the diffraction limit, which is a theoretical limit for a perfect optical system.)

b. For a 10-meter telescope, the diffraction limit resolution is:

$$\text{diffraction limit (arcseconds)} = 2.5 \times 10^5 \times \left(\frac{\text{wavelength}}{\text{telescope diameter}} \right)$$

$$= 2.5 \times 10^5 \times \left(\frac{500 \times 10^{-9}\ \cancel{\text{m}}}{10\ \cancel{\text{m}}} \right) = 0.0125''$$

The diffraction limit of the 10-meter telescope is smaller than the diffraction limit of the human eye by a factor of about 16/0.0125 = 1,280, or close to 1,300.

47. For a 100-meter (10^4-centimeter) radio telescope observing radio waves with a wavelength of 21 centimeters, the diffraction limit is:

$$\text{diffraction limit (arcseconds)} = 2.5 \times 10^5 \times \left(\frac{\text{wavelength}}{\text{telescope diameter}} \right)$$

$$= 2.5 \times 10^5 \times \left(\frac{21\ \cancel{\text{cm}}}{10^4\ \cancel{\text{cm}}} \right) = 525''$$

This angular resolution of over 500 arcseconds is about 10,000 times poorer than the Hubble Space Telescope's 0.05-arcsecond resolution for visible light. In order to achieve significantly better angular resolution when observing 21-centimeter radio waves, a radio telescope must have an effective diameter much larger than 100 meters. Because it would be impractical to build such huge telescopes, radio astronomers use the technique of interferometry to make many small radio telescopes achieve the angular resolution that a single very large one would achieve.

48. We are told in Mathematical Insight 6.2 that the diffraction limit is given by the equation:

$$\text{diffraction limit} = (2.5 \times 10^5 \text{ arcseconds}) \frac{\text{wavelength of light}}{\text{diameter of telescope}}$$

where the wavelength of light and the diameter of the telescope must be measured in the same units. In our case, the wavelength is 21 centimeters and the telescope is 0.5 meter across. Converting from meters to centimeters:

$$0.5\ \cancel{\text{m}} \times \frac{100 \text{ cm}}{1\ \cancel{\text{m}}} = 50 \text{ cm}$$

We are ready to use the formula for the diffraction limit:

$$\text{diffraction limit} = (2.5 \times 10^5 \text{ arcseconds}) \frac{21 \text{ cm}}{50 \text{ cm}}$$

$$= 1.05 \times 10^5$$

The telescope would have a resolution of 1.05×10^5 arcseconds. We can convert this to degrees, to make it clearer how large this is:

$$1.05 \times 10^5\ \cancel{\text{arcseconds}} \times \frac{1\ \cancel{\text{arcminute}}}{60\ \cancel{\text{arcseconds}}} \times \frac{1^\circ}{60\ \cancel{\text{arcminutes}}} = 29.3^\circ$$

The angular resolution of your satellite dish is only 29.3°—roughly 60 times the 0.5° angular size of the full moon! Thus, your dish would not be very useful for making images; it's also too small to gather much light, and the poor angular resolution means we couldn't pinpoint an object's location well enough for the dish to do useful spectroscopy either.

49. a. The angular area of the advanced camera's field of view is about $(0.06°)^2 = 0.0036$ square degree.

    b. Given that the angular area of the entire sky is about 41,250 square degrees, taking pictures of the entire sky would require:

$$\frac{41{,}250 \text{ square degrees}}{0.0036 \text{ square degree}} = 11.5 \text{ million}$$

    separate photographs by the advanced camera.

50. If each of 11.5 million photographs required 1 hour to take, the total time required would be 11.5 million hours, which is the same as:

$$11.5 \text{ million } \cancel{\text{hr}} \times \frac{1 \cancel{\text{day}}}{24 \cancel{\text{hr}}} \times \frac{1 \text{ yr}}{365 \cancel{\text{day}}} = 1{,}310 \text{ yr}$$

That is, it would take more than a thousand years for the Hubble Space Telescope to photograph the entire sky, assuming 1 hour per photograph. Clearly, we'll need many more telescopes in space if we hope to get Hubble Space Telescope-quality photos of the entire sky.

51. We are told in Mathematical Insight 6.2 that the diffraction limit is given by the equation:

$$\text{diffraction limit} = (2.5 \times 10^5 \text{ arcseconds}) \frac{\text{wavelength of light}}{\text{diameter of telescope}}$$

where the wavelength of light and the diameter of the telescope must be measured in the same units. We are told that the wavelength is 500 nanometers and that the diameter of the telescope is 300 meters. To make the units consistent, we will convert 500 nanometers into meters, recalling that a nanometer is one-billionth of a meter:

$$500 \cancel{\text{nm}} \times \frac{1 \text{ m}}{1 \times 10^9 \cancel{\text{nm}}} = 5.00 \times 10^{-7} \text{ m}$$

Now, using the formula for angular resolution:

$$\text{diffraction limit} = (2.5 \times 10^5 \text{ arcseconds}) \frac{5.00 \times 10^{-7} \cancel{\text{m}}}{300 \cancel{\text{m}}}$$

$$= 4.17 \times 10^{-4} \text{ arcsecond}$$

This telescope would have a diffraction limit of 4.17×10^{-4} arcsecond.

52. This is a geometry problem. If the mirrors are 4 meters in diameter, they are 2 meters in radius. We want to know how far the photons at the outside, 2 meters from the center, must travel after being deflected 2° to meet with a photon that comes right down the

center of the telescope. We can sketch the geometry as a right triangle. The photon that's coming down the telescope's central axis is the vertical leg of the triangle, while the photon that was deflected 2° out at the edge of the telescope is the hypotenuse:

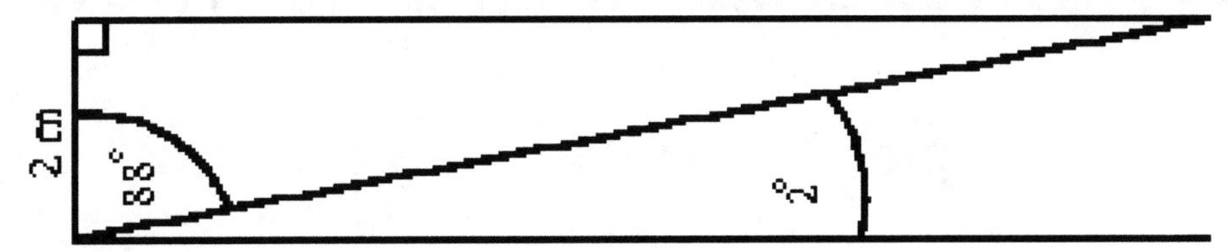

For right triangles, we know that:

$$\tan(\theta) = \frac{\text{opposite leg}}{\text{adjacent leg}}$$

Solving for the opposite leg, which will be the length of the telescope in this case:

$$\text{opposite leg} = (\text{adjacent leg})\tan(\theta)$$

Using $\theta = 88°$ and 2 meters for the adjacent leg (and remembering to set our calculators to degrees, not radians!), we get:

$$\begin{aligned}\text{opposite leg} &= 2\tan(88°)\\ &= 57\text{ m}\end{aligned}$$

So Chandra should be around 57 meters long! In reality, Chandra is about 13 meters long. This is possible because it uses more than one set of mirrors to focus the X rays.

Part III: Learning from Other Worlds

Chapter 7. Our Planetary System

This chapter offers an introduction to our solar system, with an emphasis on giving students a common background in the essential information about the planets that will allow a deeper look in coming chapters. We begin by imagining an alien spacecraft coming in from afar and mapping the broad features of the solar system. Then we focus in on individual worlds, taking a tour of the Sun and the nine planets. Finally, we discuss spacecraft exploration of the solar system.

As always, when you prepare to teach this chapter, be sure you are familiar with the relevant media resources (see the complete, section-by-section resource grid in Appendix 3 of this Instructor's Guide) and the online quizzes and other study resources available on the Mastering Astronomy Web site.

What's New in the Fourth Edition That Will Affect My Lecture Notes?

As everywhere in the book, we have added learning goals that we use as subheadings, rewritten to improve the text flow, improved art pieces, and added new illustrations. The art changes, in particular, will affect what you wish to show in lecture.

- We have switched the order of topics in the chapter, allowing students to first be exposed to the "planetary tour," and to then see how the planetary properties fit into important patterns. The presentation of a factual basis first allows students to proceed from the specific to the general, providing a strong base for the chapter and all of Part 3.
- Table 7.1, Planetary Facts, has been simplified to its essentials, allowing students to absorb key information on a single page rather than two. Of course, all the data are shown in full in Appendix E.
- The discussion of how spacecraft work now uses the Mars Exploration Rovers and *Cassini/Huygens* as examples.
- This section also has a new historical Special Topic on the determination of the astronomical unit.
- The long-anticipated discovery of an object larger than Pluto ("2003 UB313") is integrated into this chapter. Students' attention should be drawn back to the Special Topics box on page 9. Although the object is introduced in this chapter, the full discussion comes in Chapter 12, along with further discussion of how many planets our solar system has.

Teaching Notes (By Section)

Section 7.1 Studying the Solar System

- This section introduces the "big picture" layout of the solar system in figures followed by our "tour" of the major worlds in the solar system. The Sun and each planet get a full page, with sizes and distances referenced to the same scale (the 1-to-10-billion Voyage scale) introduced in Section 1.2. The section includes a useful one-page table of planetary facts essential for understanding the behavior of planets. This section also introduces and emphasizes the concept of comparative planetology, the idea that similarities and differences among the planets can be traced to common physical processes. This is the unifying theme behind all of Part III and should focus students on the importance of learning processes over facts.

Section 7.2 Patterns in the Solar System

This section builds upon the first section by summarizing all the features of our solar system that offer clues to its formation. We focus on four major features that must be explained by a scientific theory of solar system formation: (1) Large bodies have orderly motions; (2) planets fall into two main categories; (3) many small bodies populate the solar system; and (4) there are a few exceptions to the rules, such as the tilt of Uranus or our unusually large Moon (relative to the size of Earth).

- Note that we refer to residents of the Kuiper belt as comets in order to emphasize their icy compositions; not all planetary scientists refer to these objects in this way.

Section 7.3 Spacecraft Exploration of the Solar System

- This section is designed to give students a brief overview of the use of spacecraft in planetary exploration.
  - Note that we divide robotic spacecraft into four categories: Earth orbiters, flybys, orbiters (of other worlds), and probes/landers. This helps focus students on the key differences among the types of missions.
  - This section also has a new historical Special Topic on the determination of the astronomical unit, allowing students to see how Kepler's *relative* scale of the system could be converted to *absolute* units.

Answers/Discussion Points for Think About It Questions

The Think About It questions are not numbered in the book, so we list them in the order in which they appear, keyed by section number.

Section 7.1

- (p. 201) This question is designed to encourage students to discover patterns in the solar system instead of simply being told the answers. It is a good question to ask in class with Table 7.1 displayed—students will either remember the answers from reading the whole chapters, or will glean them from the table. The answers, of course, form the basis of this section and Chapter 8. The patterns of motion are listed on page 215.
- (p. 203) Another question designed to encourage students to discover patterns in the solar system. The properties of the two groups of planets (best gleaned from Table 7.1) are summarized in Table 7.2. Most of the patterns and distinctions are clear cut: No planets orbit the sun backward, and all jovian planets are much larger than the largest

terrestrial planet. All jovian planets have rings and moons. But some patterns are not "rules": Not all planets rotate the same way, and many of the smaller moons orbit their planets backward. Some terrestrial planets have moons, but not all.

- (p. 210) This question asks students to look for the latest on the *Cassini* mission to Saturn. Its latest encounters with moons and most recent rings discoveries should be of interest to students. Its nominal mission lasted until 2004, but an extended mission beyond is under consideration.

Section 7.3

- (p. 219) No. Because of their differing orbital periods, they no longer have a simple alignment like they did during the 1980s. In particular, Uranus has caught up with Neptune by now, so we won't be able to visit both with a single slingshot from Saturn.

Solutions to End-of-Chapter Problems (Chapter 7)

1. Comparative planetology is a way of studying the solar system that relies on comparing the objects in the solar system to each other. The objects do not need to be planets; moons, rings, asteroids, and comets are all part of the field, for example.
2. If we looked at the solar system with our naked eye from beyond the orbit of Pluto, we would see points of light for the various planets. The planets would not appear as much more than bright stars, although if we watched them for a while, we could see that they orbit the Sun.
3. From Figures 7.1 and 7.2 we can see several patterns to the solar system's layout:
    - The planets' orbits are almost all nearly circular.
    - The planets almost all orbit in the same plane.
    - Planets orbit in the same direction around the Sun.
    - Most planets spin in the same direction as they orbit.
    - The large moons almost all orbit their planets in the same direction as the planets spin and orbit.
    - The four inner planets are smaller and more closely spaced than the outer planets are.
4. Some possible answers are:

    Mercury
    - Rotates three times for every two orbits
    - Appears to have shrunk early in its life
    - The most metal-rich planet
    - Has 3-month-long days and nights
    - Temperatures varying from extremely hot to extremely cold over the course of a full solar day
    - Has no atmosphere

    Venus
    - Spins backward compared to its orbit
    - Hottest surface temperature of any planet due to its extreme greenhouse effect
    - Has a dense atmosphere
    - Roughly the same size and mass as Earth

Earth
- Has a surprisingly large moon
- Only planet with life (as far as we know)
- Only planet with oxygen in the atmosphere

Mars
- Has two tiny moons that might have once been asteroids
- Has volcanoes larger than any on Earth
- Has a canyon that runs one-fifth of the way around the planet
- Has polar ice caps
- Shows signs of once having flowing water
- May have once had life
- Has a thin atmosphere
- Most studied planet apart from Earth

Jupiter
- Largest planet, more than 300 times Earth's mass and 1,000 times Earth's volume
- Has no solid surface
- Has dozens of moons and thin rings
- Has three moons that may have subsurface oceans
- Has a moon, Io, that is the most volcanic body in the solar system

Saturn
- Has the most spectacular ring system
- Has many moons
- Has a moon, Titan, that has a dense nitrogen atmosphere

Uranus
- Spins on its side
- Has over a dozen moons
- Looks blue due to methane in its atmosphere
- Has a ring system

Neptune
- Slightly smaller in radius than Uranus, but more massive
- Has rings and many moons
- Has a moon, Triton, that is larger than Pluto and appears to have geysers
- Has a moon, Triton, that is the only large moon to orbit opposite the direction of its planet

Pluto
- Has a semimajor axis larger than Neptune's, but sometimes comes within Neptune's orbit
- The radius of its only moon, Charon, half that of Pluto's radius
- Never visited by a spacecraft
- Made of rock and ice, not gas

5. Four orderly motions in the solar system are
   - Planets all orbit in nearly the same plane and have nearly circular orbits.
   - Planets orbit the Sun in the same direction.
   - Most planets spin in the same direction as they orbit.
   - Most of the large moons in the solar system also orbit in the same sense as the planets—in nearly circular orbits, generally in nearly the same plane.
6. Terrestrial planets orbit close to the Sun and are tightly spaced together. They are made mostly of rock and metal and are smaller and denser than jovian planets. Terrestrial planets tend to have few, if any, moons and none has rings.

   In contrast, jovian planets are more distant from the Sun and are separated by much larger distances. They are made mostly of hydrogen, helium, and hydrogen compounds. They are much less dense than terrestrial planets. They are also larger than terrestrial planets. Jovian planets have many moons and have ring systems.

   The terrestrial planets are Mercury, Venus, Earth, and Mars. The jovian planets are Jupiter, Saturn, Uranus, and Neptune. Pluto does not fit into either class.
7. Hydrogen compounds are chemical compounds that use hydrogen and another common element. The most important hydrogen compounds are water (H_2O), ammonia (NH_3), and methane (CH_4). Hydrogen compounds are important ingredients in the jovian planets, Pluto, and comets.
8. Pluto is like a terrestrial planet in that it is small, relatively dense, and solid. It also includes rock as an important component, a terrestrial planet property. However, like the jovian planets, hydrogen compounds are also an important ingredient of Pluto. Furthermore, it has a distant orbit from the Sun like the jovian planets. Unlike either type of planet, Pluto has an eccentric and inclined orbit and is much smaller than even the terrestrial planets.
9. Asteroids are rocky or metallic bodies that orbit the Sun, but are much smaller than the planets. Most asteroids are found between the orbits of Mars and Jupiter in the "asteroid belt."
10. Comets are small, icy bodies that orbit the Sun. The difference between comets and asteroids is composition: While comets are made mostly of ices, asteroids are made mostly of rock and metals. Also, comets usually spend most or all of their time much farther from the Sun than asteroids.
11. The Kuiper belt is a disk of comets just past the orbit of Neptune. The Oort cloud is a spherical cloud of comets that surrounds our solar system much farther from the Sun. Comets in the Kuiper belt orbit the Sun more or less in the same plane as the planets and travel around the Sun in the same sense as the planets do. Oort cloud comets orbit the Sun with tilts of any value and often orbit "backward" relative to the planets.
12. Exceptions to the rule include:
   - Pluto has a high orbital eccentricity and inclination.
   - Venus spins backward.
   - Uranus and Pluto spin on their sides.
   - Triton and many small moons orbit their planets backward.
   - The Earth's Moon is huge considering it orbits a terrestrial planet.

13. Flybys are missions that pass by a planet or moon without going into orbit or landing. Most of the data are therefore taken in a burst of activity when the spacecraft is close to the planet. The advantage of flybys is that they get to their targets faster and are cheaper since they can use less fuel and are simpler to plan and build.

    An orbiter goes into orbit around the planet. The main advantage of this is that they take data for a much longer period of time, often years. Interesting moons or sites on the planet can be reimaged after they are discovered, unlike with flybys. Orbiters can also make maps of the planet and its moons with far better resolution than flybys because they are generally closer to the targets and are able to take their pictures over longer periods, therefore doing a more careful, detailed job.

    Landers or probes descend into the planet's atmosphere or down to the surface and land, taking data from up close. While landers are able to examine only much smaller parts of their planets, they are able to perform types of analyses that are not possible from a distance, such as determining chemical compositions. The pictures they return are also much more detailed, giving us a better idea of what the geology of the surface is like.

    Sample return missions gather up bits of their target (whether it's a planet, a comet, or something else) and return it to Earth for study by humans in laboratories. Because laboratories here on Earth are much better equipped to study the nature of the samples and because humans are more adaptable in our ability to study something than a robotic spacecraft, samples are considered very valuable.
14. Student answers to this will vary. Table 7.2 lists over a dozen missions of interest.
15. *Pluto orbits the Sun in the opposite direction of all the other planets.* False.
16. *If Pluto were as large as the planet Mercury, we would classify it as a terrestrial planet.* False. Terrestrial planets are rocky and Kuiper belt objects areicy.
17. *Comets in the Kuiper belt and Oort cloud have long, beautiful tails that we can see when we look through telescopes.* False. The vast majority of comets lie far from the Sun, too far to create tails, etc.
18. *Our Moon is about the same size as moons of the other terrestrial planets.* True, though there are more moons that are smaller than are larger.
19. *The mass of the Sun compared to the mass of all the planets combined is like the mass of an elephant compared to the mass of a cat.* True.
20. *On average, Venus is the hottest planet in the solar system—even hotter than Mercury.* True.
21. *The weather conditions on Mars today are much different than they were in the distant past.* True.
22. *Moons cannot have atmospheres, active volcanoes, or liquid water.* False.
23. *Saturn is the only planet in the solar system with rings.* False.
24. *We could probably learn more about Mars by sending a new spacecraft on a flyby than by any other method of studying the planet.* False.
25. b; 26. c; 27. c; 28. b; 29. c;
30. b; 31. a; 32. b; 33. c; 34. c.
35. Many possible answers, based on student interest.

36. Students should coherently summarize the patterns of motion listed as bullets on page 215, and make the basic observation that these patterns would not have arisen if every planet had formed from a different cloud at a different time. There would be no reason for the motions of the planets to be similar. Students will gain a deeper understanding of the reasons in Chapter 8.
37. This question asks students to coherently rephrase Table 7.2 in their own words. Students should include a discussion of density, composition, and distance. Students should note Pluto as an exception: It has the size and mass of a very small terrestrial planet, as well as having a solid surface. But it lies at a distance (and hence has a temperature) more in line with the jovian planets. Its composition and density do not match either type of planet, though they do resemble jovian planet moons.
38. a. Planets closest to the Sun are the warmest, as they absorb more solar energy (per unit area). Venus violates the trend, due to its greenhouse effect.
    b. Density: Jovian planets are all less dense than 2 grams per cubic centimeter, while terrestrial planets are all more than 3 grams per cubic centimeter. Composition: Jovian planets are all dominated by H, He, and hydrogen compounds, while terrestrial planets are mostly rocks and metals. Distance: Jovian planets are all 5 AU and beyond, while terrestrial planets are all inside 2 AU. Pluto's misfit status is also the subject of Problem 37.
    c. The orbital periods increase with the planets' semimajor axes, according to Kepler's third law ($p^2 = a^3$).
39. a. The column "Rotation Period" gives the time for a day—though note that it is a day relative to the stars (that is, a sidereal day—see Section S1.1), not the Sun. For all planets other than Mercury or Venus, the difference is small. Jovian planets have shorter days than terrestrial planets.
    b. Planets with significant axis tilts will have seasons: Earth, Mars, Saturn, Uranus, Neptune, Pluto.
    c. In Appendix E, students should notice that planets with different min/max distances (i.e., significant eccentricities) will have temperature variations due to changing distance from the Sun: Mercury, Mars, and Pluto.
40. Since we are asked "how many," this problem seems best approached with ratios. We will take the ratio of the volume of Jupiter to the volume of Earth, which will give us the number of Earths that could fit inside Jupiter. Our ratio looks like:

$$\frac{\text{volume of Jupiter}}{\text{volume of Earth}} = \frac{\frac{4}{3}\pi r_{\text{Jupiter}}^3}{\frac{4}{3}\pi r_{\text{Earth}}^3}$$

We can save ourselves a lot of mistakes in calculating these if we use some algebra and cancel terms in the numerator and the denominator:

$$\frac{\text{volume of Jupiter}}{\text{volume of Earth}} = \frac{r_{\text{Jupiter}}^3}{r_{\text{Earth}}^3}$$

We can also help ourselves by recalling that $a^n/b^n = (a/b)^n$, making our final expression:

$$\frac{\text{volume of Jupiter}}{\text{volume of Earth}} = \left(\frac{r_{\text{Jupiter}}}{r_{\text{Earth}}}\right)^3$$

This final expression is particularly nice because Appendix E tells us that the radius of Jupiter is 11.19 Earth radii. In other words, $r_{\text{Jupiter}}/r_{\text{Earth}} = 11.19$. So we can compute the ratio quite easily now:

$$\frac{\text{volume of Jupiter}}{\text{volume of Earth}} = (11.19)^3$$
$$= 1{,}400$$

This tells us that around 1,400 Earths could fit inside of Jupiter.

41. Kepler's third law tells us:

$$p^2 = a^3$$

where p is the planet's period in years and a is its semimajor axis in astronomical units. So we can solve for the semimajor axis by taking the cube root of both sides of the equation:

$$a = p^{2/3}$$

We are told that Ceres has a period of 2.77 years. So we can calculate the semimajor axis:

$$a = (2.77)^{2/3}$$
$$= 1.97 \text{ AU}$$

So Ceres orbits with a semimajor axis of 1.97 AU. Looking at Table 7.1, we see that that puts it between the orbits of Mars and Jupiter.

42. We know that density is defined as mass/volume. So we need the mass, which we are given, and the volume, which we will have to calculate. Luckily, we recall that the volume of a sphere is given by the expression:

$$\text{volume} = \frac{4}{3}\pi r^3$$

But before we calculate anything, we should convert our radius into centimeters because we want our density to be in grams per cubic centimeter. So we convert, first to meters then to centimeters:

$$12{,}800 \cancel{\text{km}} \times \frac{1000 \cancel{\text{m}}}{1 \cancel{\text{km}}} \times \frac{100 \text{ cm}}{1 \cancel{\text{m}}} = 1.28 \times 10^9 \text{ cm}$$

Now we calculate the volume:

$$\text{volume} = \frac{4}{3}\pi(1.28 \times 10^9 \text{ cm})^3$$
$$= 8.78 \times 10^{27} \text{ cm}^3$$

As mentioned earlier, we want our final number to be in grams per cubic centimeter. So we had better convert our mass into grams before finding the density. This is easily done:

$$5.97 \times 10^{25} \cancel{\text{kg}} \times \frac{1000 \text{ g}}{1 \cancel{\text{kg}}} = 5.97 \times 10^{28} \text{ g}$$

So finally, we find the density using the definition:

$$\begin{aligned} \text{density} &= \frac{\text{mass}}{\text{volume}} \\ &= \frac{5.97 \times 10^{28}\ \text{g}}{8.78 \times 10^{27}\ \text{cm}^3} \\ &= 6.80\ \text{g/cm}^3 \end{aligned}$$

The density of the planet is therefore $6.80\ \text{g/cm}^3$. Such a large density means that this is almost certainly a terrestrial planet.

43. The first and most obvious trend is that the more massive planets all have higher escape speeds than the less massive planets. This is not unexpected since escape speed increases with mass. (It decreases with rising radius, but the masses seem to jump up faster than the radii do, so the mass wins.) Particularly interesting are Jupiter and Saturn, which have almost the same radius but are quite different in mass. Jupiter's escape speed is much higher, almost twice as large as Saturn's. This makes sense for the more massive planet.

    Also interesting are Uranus and Neptune. Neptune is somewhat more massive than Uranus *and* has a slightly smaller radius. Both of these factors tend to increase the escape speed. Neptune's escape speed is higher than Uranus's, so that makes sense.

44. In Chapter 5 we learned that the expression for weight is:

$$\text{weight} = \text{mass} \times (\text{surface gravity})$$

We're told that our weight here on Earth is 100 pounds. So we can solve for our mass:

$$\text{mass} = \frac{100\ \text{lb}}{\text{Earth's surface gravity}}$$

(Usually, we'd convert into metric units, but this time we won't need to.) So we can plug this into the expression for weight to find that:

$$\text{weight} = (100\ \text{lb}) \times \frac{\text{planet's } \cancel{\text{surface gravity}}}{\text{Earth's } \cancel{\text{surface gravity}}}$$

(Note that the units of surface gravity cancel out. So we are OK leaving the weight in pounds this time.)

Now, this is convenient. In Appendix E, we are given the planet's surface gravity in Earth units. So all we have to do for each planet is multiply 100 pounds by the surface gravity that we are given, and we have the weight. So computing:

Planet	Surface Gravity	Weight
Mercury	0.38	38
Venus	0.91	91
Earth	1	100
Mars	0.38	38
Jupiter	2.53	253
Saturn	1.07	107
Uranus	0.91	91
Neptune	1.14	114
Pluto	0.07	7

45. We will assume that the spacecraft travels in a straight line and that it goes from Earth's average distance from the Sun to Pluto's average distance on the shortest path possible. So the distance it travels is:

$$a_{\text{Pluto}} - a_{\text{Earth}}$$

where the a's are the semimajor axes of the two orbits. We know that Earth's semimajor axis is 1 AU and Appendix E tells us that Pluto's is 39.5 AU. So the distance traveled will be 39.5 AU – 1 AU = 38.5 AU.

We are told that the spacecraft will take 9 years to get to Pluto. So, recalling that speed is distance divided by time, we can find the speed:

$$\begin{aligned} \text{speed} &= \frac{\text{distance}}{\text{time}} \\ &= \frac{38.5 \text{ AU}}{9 \text{ yr}} \\ &= 4.3 \text{ AU/yr} \end{aligned}$$

We are also asked to find this speed in km/hr, so we need to convert:

$$4.3 \cancel{\text{AU}}/\cancel{\text{yr}} \times \frac{1.496 \times 10^8 \text{ km}}{1 \cancel{\text{AU}}} \times \frac{1 \cancel{\text{yr}}}{365.25 \cancel{\text{days}}} \times \frac{1 \cancel{\text{day}}}{24 \text{ hr}} = 73{,}000 \text{ km/hr}$$

The spacecraft will travel an average speed of 4.3 AU/yr or 73,000 km/hr.

46. To add up the size of the Sun and all of the planets, we need to add the diameters:

$$d_{\text{Sun}} + d_{\text{Mercury}} + d_{\text{Venus}} + d_{\text{Earth}} + d_{\text{Mars}} + d_{\text{Jupiter}} + d_{\text{Saturn}} + d_{\text{Uranus}} + d_{\text{Neptune}} + d_{\text{Pluto}}$$

where d is the diameter of each body. However, we're given the radii, not the diameters. We *could* multiply each radius by 2 to get each diameter and then add those up. But that's annoying, especially because we can use a bit of algebra to save us that effort. Because $d = 2r$, we can rewrite the expression above:

$$2r_{\text{Sun}} + 2r_{\text{Mercury}} + 2r_{\text{Venus}} + 2r_{\text{Earth}} + 2r_{\text{Mars}} + 2r_{\text{Jupiter}} + 2r_{\text{Saturn}} + 2r_{\text{Uranus}} + 2r_{\text{Neptune}} + 2r_{\text{Pluto}}$$

which can be factored to get:

$$2(r_{\text{Sun}} + r_{\text{Mercury}} + r_{\text{Venus}} + r_{\text{Earth}} + r_{\text{Mars}} + r_{\text{Jupiter}} + r_{\text{Saturn}} + r_{\text{Uranus}} + r_{\text{Neptune}} + r_{\text{Pluto}})$$

Now we just have to add the radii up and multiply by 2. So looking up the radii in Appendix E and adding them up (and sparing the reader the details of the addition), we find that the sum of the radii is 2.02×10^5 km, so the sum of the diameters (and the distance that the planets and the Sun would stretch end to end) is 4.03×10^5 km.

Now the distance to Pluto makes a decent measurement of the size of our solar system. Pluto's semimajor axis is 39.5 AU. Converting to kilometers:

$$39.5 \cancel{\text{AU}} \times \frac{1.496 \times 10^8 \text{ km}}{1 \cancel{\text{AU}}} = 5.91 \times 10^9 \text{ km}$$

So Pluto's orbit is more than 10,000 times larger than the distance the planets and the Sun would stretch if they were set end to end. Clearly, the solar system is mostly empty space and the occasional planet or star is the exception rather than the rule.

47. To find the distance to Venus, we use the figure in "Special Topic: How Did We Learn the Distance to the Planets?" This figure suggests that the parallax we measure on Earth is the angular size of Earth as seen from Venus. Recall that in Mathematical Insight 2.1 we learned that distance, angular size, and physical size are related by the expression:

$$\frac{\text{angular size}}{360°} = \frac{\text{physical size}}{2\pi \times \text{distance}}$$

We can solve this for distance:

$$\text{distance} = \frac{360° \times (\text{physical size})}{2\pi \times (\text{angular size})}$$

We know the physical size (diameter) of the Earth, since the radius is 6,378 km: It is just twice the radius, or 12,756 km. We know the angular size of the Earth (as seen from Venus), but we need to convert it to degrees. We are given that the parallax is 62.8″, so we convert to degrees:

$$62.8'' \times \frac{1'}{60''} \times \frac{1°}{60'} = 1.74 \times 10^{-2} \text{ degrees}$$

So we calculate the distance from Venus to Earth:

$$\text{distance} = \frac{360° \times (12{,}756 \text{ km})}{2\pi \times (1.74 \times 10^{-2} \text{ degrees})}$$

$$= 4.20 \times 10^{7} \text{ km}$$

So the distance should be about 4.2×10^7 km. We can convert this to the more useful unit of astronomical units:

$$4.2 \times 10^{7} \text{ km} \times \frac{1 \text{ AU}}{1.496 \times 10^{8} \text{ km}} = 0.28 \text{ AU}$$

which corresponds to the difference between Earth's and Venus' semimajor axes.

Chapter 8. Formation of the Solar System

The chapter presents a single, unified theory for solar system formation. It essentially "builds" the solar system piece by piece so that the following chapters can look in detail at how those pieces work and interact with one another.

Note that, while there is some value in getting students to learn our modern theory of solar system formation itself, in the long run it is more important that they get a feel for what a theory is good for and how a theory should be judged. For example, while we emphasize the evidence that supports our modern theory, we do not emphasize the contorted path followed in its development; you may therefore wish to emphasize the unresolved issues that we point out in this chapter and others. This will also set the stage for the changes to the theory necessitated by the discovery of extrasolar planets.

As always, when you prepare to teach this chapter, be sure you are familiar with the relevant media resources (see the complete, section-by-section resource grid in Appendix 3 of this Instructor's Guide) and the online quizzes and other study resources available on the Mastering Astronomy Web site.

What's New in the Fourth Edition That Will Affect My Lecture Notes?

As everywhere in the book, we have added learning goals that we use as subheadings, rewritten to improve the text flow, improved art pieces, and added new illustrations. The art changes, in particular, will affect what you wish to show in lecture.

- The topic of extrasolar planets has blossomed to the point where it requires its own chapter. So all the material on this subject that formerly appeared at the end of this chapter has been moved and greatly expanded in Chapter 13. The coverage of solar system formation here in Chapter 8 is updated but otherwise not significantly changed.

Teaching Notes (By Section)

Section 8.1 The Search for Origins

This section sets the stage for the rest of the chapter by highlighting the key characteristics that must be answered by any theory that successfully explains the formation of the solar system. The four key characteristics are a review of those introduced in Chapter 7. We then outline a bit of the history of solar system formation theories, including an explanation of why the nebular theory gained credence over competing theories.

Section 8.2 The Birth of the Solar System

This section begins with a brief overview of how the solar nebula came to exist, emphasizing galactic recycling. It then goes through the processes that caused the solar nebula to become a flattened, spinning disk. Thus, in essence, this section explains the first of the four features mentioned above for our solar system.

- We discuss just enough about galactic recycling so that students can understand the origin of the solar nebula in a general sense. More detailed discussion of the galactic processes comes in Part V of the textbook.
- Useful classroom demos include: (1) conservation of angular momentum, using a swivel chair or large turntable; (2) a "pepper-and-water" demonstration of random motions averaging out into regular motions through collisions. The latter demonstration can be done with a clear plastic tub on a viewgraph machine. Be sure to refocus the projector onto the pepper flakes on the water's surface. Be sure to practice to find out the right amount of "random stirring" needed to create a perceptible rotational motion and to get a feeling for how long it takes the random motions to settle into the anticipated pattern.

Section 8.3 The Formation of Planets

We now discuss how the planets formed in the disk through the process of accretion, which gives us the opportunity to explain the next feature of the solar system that we introduced in Section 8.2.

- Students often have difficulty with the concept of a frost line and how it influences the eventual size of planets. You are encouraged to verify your students' understanding of this key point before moving on. Also, despite our repeated emphasis of the point, when discussing accretion students sometimes lose sight of the fact that the vast majority of the material in the solar nebula is hydrogen and helium throughout. Take the opportunity to remind them of this fact.
- The discussion of the four kinds of materials in the solar nebula—metal, rock, hydrogen compounds, and hydrogen and helium gas—will surface repeatedly in Part III, so it bears emphasis.
- Planetary scientists often refer to materials such as methane, ammonia, and water as "ices." However, it is difficult to get students to think of ice as something that is not actually in the solid state. Therefore, we prefer the more accurate term *hydrogen compounds* for these materials. On a related note, we use the term *rocky material* to refer to elements and compounds that make rock when solid, but that can be in the gas state at high temperatures.
- Students are always excited to see meteorites, and this section begins to give them an appreciation of their importance. If you have access to meteorites, either pass them around or invite students to see them after class. For notes on acquiring meteorites for your school, see Section 12.1 in this Instructor's Guide.

Section 8.4 The Aftermath of Planet Formation

This section discusses the final two features, both of which relate to the many small bodies in the solar system. First, we present a short discussion of the origin of asteroids and comets, with more detail to come in Chapter 12. Then we discuss impacts and the role they play in leading to "exceptions to the rules." Note that this is also where we discuss the giant impact hypothesis for the formation of our Moon.

- The fact that Earth probably did not form with much water helps students appreciate the importance of asteroids and comets. We've found that students love the idea that the water they drink likely came from comets.
- The giant impact hypothesis for the formation of the Moon is generally a favorite topic for many students. It's also a great place to discuss the nature of science, since science leads us to an idea that we may never be able to test directly.

Section 8.5 The Age of the Solar System

This section addresses the important question of the age of the solar system and why we think we know it so well.

- This question can be answered only with a digression into radioactive decay. This section offers three ways of learning the concepts: text, graphs, and equations. While some students may appreciate the Mathematical Insight 8.1 the text and graphs will stand on their own for other students.
- The techniques of radiometric dating are far more sophisticated than described here, but they depend on a deeper understanding of chemistry than we can present at this level. Nevertheless, the principles are easy to understand, and should lend students some confidence in the idea that we really can know how old things are.

Answers/Discussion Points for Think About It Questions

The Think About It Questions are not numbered in the book, so we list them in the order in which they appear, keyed by section number.

Section 8.2

- (p. 229) According to our theory, planets could not have formed along with the first generation of stars after the Big Bang. The Big Bang did not make the elements necessary for planets like ours.
- (p. 230) The random "strokes" in the water represent the gravitational influence of nearby stars stirring up the cloud from which we formed. The demonstration makes two main points: First, collisions average out the small random motions into large uniform motions. Second, it's very unlikely for random influences to exactly cancel out; the final rotation is always a bit one way or the other. Note that the experiment is not meant to reflect the shrinking of the cloud that formed the solar nebula, though it is likely that collisions were not able to average out random motions in the cloud until the cloud shrank. [Don't let students be confused by the clumping of pepper grains (surface tension) or the slowing of the rotation (friction with the tub).] We must always be on the lookout for aspects of the demonstration that might detract from the main point.

Section 8.3

- (p. 232) At 1,300 K, almost all (at least 99.8%) of the nebula is gaseous. Only some of the different kinds of metals could condense, and they make up a maximum of 0.2% of the nebula. At 100 K, 2% of the nebula will be in solid form: metals, rocks, and hydrogen compounds. The remaining 98% is hydrogen/helium gas that never condenses. The 100 K region is farther from the Sun.

Section 8.4

- (p. 237) Without Jupiter, comets and asteroids (which contain the ingredients for our oceans and atmosphere) would not have been deflected into the inner solar system. Earth would be made only of "dry" planetesimals that formed in our region of the solar system—no oceans or atmosphere.

Section 8.5

- (p. 241) In another 4.5 billion years, another half-life will have elapsed, for a total of about two half-lives. Only $\frac{1}{4}$ of the original uranium will exist, and $\frac{3}{4}$ will have been converted to lead.

Solutions to End-of-Chapter Problems (Chapter 8)

1. There are four properties to our solar system that provide us with clues to how it formed. The first is the fact that the motions in our solar system are orderly. The planets orbit in the same plane, on nearly circular orbits, and in the same direction around the Sun, and most of them spin in that same sense. Most of the major moons of our solar system also follow these rules when orbiting around their planets.

The second property of the solar system that provides clues about the formation is the existence of two kinds of planet, jovian and terrestrial. The two types of planets are found in different parts of the solar system (outer and inner, respectively) and are generally very different in nature. (Masses, densities, and compositions are especially significant differences.)

The third property we use to understand the formation of our solar system is the existence of the smaller bodies such as the comets and asteroids. Asteroids are rocky or metallic bodies that are usually found between the orbits of Jupiter and Mars. Comets are icy bodies found farther out into the solar system. Comets are found in two regions, the Oort cloud and the Kuiper belt.

Finally, there are exceptions to the rules. Venus spins backward, Uranus and Pluto spin on their sides, Earth has an incredibly large moon for a terrestrial planet, and a number of smaller moons (plus the large moon Triton) orbit their planets backward.

2. The nebular theory states that the solar system formed from a collapsing cloud of gas and dust billions of years ago. It is widely accepted by scientists today because it explains the properties of the solar system remarkably well. However, there is not observational evidence of other forming solar systems to support this theory.

3. The solar nebula was the cloud of gas and dust that formed our solar system. Much of the material in the clouds—everything heavier than helium—came from the interiors of stars that had exploded, spreading their material into the galaxy.

4. This cloud that formed the solar system began by collapsing under its own gravity. As it did so, to conserve angular momentum it spun faster. When clumps of gas and dust collided, both took velocities nearer the average. Since the average velocity tends to point along a flattened disk, the solar nebula eventually took on a disk shape. Evidence for these processes can be found in the fact that the planets orbit the Sun in nearly the same plane and in the same direction and that they spin in this same direction.

5. The four major components to the solar nebula were:
    - Gases (hydrogen and helium)—These were the least-dense part of the nebula but the most abundant.
    - Hydrogen compounds (water, ammonia, methane)—Compounds made up of the very abundant hydrogen and other, relatively abundant elements (oxygen, carbon, and nitrogen) are the second-most-abundant component of the solar nebula. These components are denser than the gases, but less dense than the other two components.
    - Silicates (rocky stuff)—Dust made of silicates is denser than the hydrogen compounds, but also less abundant.
    - Metals—Metals are the densest component of all and also the least abundant.

    The terrestrial planets are mainly made up of silicates and metals, which explains their high densities. The jovian planets are mainly made up of the gases and the hydrogen compounds, which explains both their lower densities and why they are so large: There was so much more material for them to be built from.

6. The frost line is the point moving away from the Sun where water is cool enough to freeze. Since the solar nebula was hotter near the center of the disk, water in the inner solar system stayed a gas. Outside of the frost line, it froze. As a solid, it was able to participate in building planets. Since there was so much more water (and, farther out, solid ammonia and methane) in the nebula, the planets past the frost line are much larger than the ones inside that limit.

7. The terrestrial planets are thought to have been formed by solid bits of silicates and metals colliding and sticking together in the solar nebula. The silicates and metals were able to condense in the hotter inner part of the disk where the temperature was too high for the hydrogen compounds to be solids. As the silicates form larger clumps, their gravity begins to allow them to gather up mass more efficiently. The biggest bodies, now called *planetismals*, are able to collect mass the fastest. Eventually, the biggest planetismals gather up the others and become planets.

8. The formation of the jovian planets is similar to the terrestrial planets in the early stages with the major exception that the jovian planets are able to use solid hydrogen compounds ("ices") to build up their masses. Since these ices are far more abundant than the silicates and metals, the jovian planets were able to build faster and larger than the terrestrial planets. Eventually, they became large enough to even hold on to nebular gases (hydrogen and helium). In this respect their formation is quite different from that of the terrestrial worlds. Once the jovian planets could collect gas, it was easy for them to grow very large in size. (Jupiter and Saturn in particular were able to grow extremely massive relative to the terrestrial planets.)

   Jovian planets have large retinues of moons because the material falling toward the jovian planets would have formed miniature disks around the planets just like the solar nebula formed a disk around the Sun. And just like planets formed out of the solar nebula disk, moons formed out of the disks around the planets.

9. The solar wind is a stream of particles that comes off of the Sun at all times. This is important to the planet formation process because it probably ended that process. When the Sun finally became large enough to start its own fusion reaction and create a solar wind, this wind would have blown away most of the gas and dust in the solar nebula. Deprived of material to use for building, planet formation would have slowed and stopped.

10. Asteroids and comets are essentially leftovers from planet formation. They are bits of material that were never swept up into planets, but probably could have been. We find asteroids in the inner solar system (between Mars and Jupiter, mainly) because near the center of the disk it was too hot for the hydrogen compounds to freeze and participate in planet formation. Farther out, where the ices were able to condense, the small bodies were icy. And so we have comets. The comets in the Kuiper belt probably formed in that general area of the solar system. However, the comets in the Oort cloud were almost certainly flung out into that area from much closer to the Sun. This tells us that the forming of the early solar system was a dynamic place full of collisions and close interactions between these small bodies and planets.

11. The heavy bombardment was a period when many comets or asteroids were striking the planets. This bombardment occurred near the end of planet formation or early in the solar system's life, within the first few hundred million years.

12. We think that the Moon formed when a large impactor struck the Earth with a glancing blow. Such a collision would have blown a lot of material into orbit around the Earth, temporarily forming a disk around our planet. From this disk, the Moon formed. Evidence in favor of this theory includes the chemical composition of the Moon: It is similar to Earth's mantle (which is where the material probably came from) and it is deficient in easily vaporized compounds like water.

It is not really surprising that such an event could have affected Earth. There are other objects in the solar system that seem to show the effects of similar collisions. Uranus is tipped over. Pluto is both tipped over and has a large moon like we do. Mercury lacks most of the outer layers we would expect it to have, which might have been blasted off in a collision.

13. Radiometric dating is a way to determine the age of something. It uses the fact that some isotopes of some elements are unstable and that they decay at a certain rate. Over some length of time, given a collection of atoms of this isotope, we know that about half of them will have decayed. This length of time is called the *half-life*. If we have a sample of rock and know what fraction of the original atoms of a given element have decayed, then we can figure out the age of the rock by knowing the element's half-life.
14. The solar system is about 4.55 billon years old. We determine this most accurately with meteorites, which have not been altered since the formation of the solar system. (Rocks on Earth have been recycled, so their ages do not tell us when the Earth formed.)
15. *A solar system has five terrestrial planets in its inner solar system and three jovian planets in its outer solar system.* This hypothetical solar system is consistent with our theory of solar system formation. The numbers of planets are different from our solar system, but that difference is not fundamental.
16. *A solar system has four large jovian planets in its inner solar system and seven small planets made of rock and metal in its outer solar system.* This discovery would be surprising, because solid objects that form beyond the jovian planets should include a great deal of ice, according to our current theories.
17. *A solar system has 10 planets that all orbit the star in approximately the same plane. However, five planets orbit in one direction (e.g., counterclockwise), while the other five orbit in the opposite direction (e.g., clockwise).* The nebular theory cannot explain a "backward" planet, as all planets should form from the "forward"-moving nebula. A planet could conceivably be captured into such an orbit, if a planet ejected from another star system somehow entered our own solar nebula and was slowed down. This process is analogous to the capture of a retrograde satellite in a jovian nebula. However, it is extremely unlikely that half the planets in a solar system would have been captured in this way.
18. *A solar system has 12 planets that all orbit the star in the same direction and in nearly the same plane. The 15 largest moons in this solar system orbit their planets in nearly the same direction and plane as well. However, several smaller moons have highly inclined orbits around their planets.* This statement makes sense and in fact is similar to the situation in our solar system.
19. *A solar system has six terrestrial planets and four jovian planets. Each of the six terrestrial planets has at least five moons, while the jovian planets have no moons at all.* This statement does not make sense. Although the number of planets is not an issue, the presence of so many moons around terrestrial planets is hard to explain, as is the lack of moons around jovian planets. The capture of nebular gas should result in more jovian-planet satellites.
20. *A solar system has four Earth-size terrestrial planets. Each of the four planets has a single moon that is nearly identical in size to Earth's Moon.* This would be surprising. The formation of our Moon is thought to be the result of a rare random event, so it would not be expected to occur for all terrestrial planets.

21. *A solar system has many rocky asteroids and many icy comets. However, most of the comets orbit in the inner solar system, while the asteroids orbit in far-flung regions much like the Kuiper belt and Oort cloud of our solar system.* This would be a surprise. Icy objects should appear outside the jovian planets, and rocky objects inside.
22. *A solar system has several planets similar in composition to the jovian planets of our solar system but similar in mass to the terrestrial planets of our solar system.* This would be a surprise. We know of no process by which to make a jovian planet—with abundant hydrogen and helium gas—without first making a massive core of hydrogen compounds. It may be possible to have planets similar in size to terrestrial planets but made of hydrogen compounds.
23. *A solar system has several terrestrial planets and several larger planets made mostly of ice. (Hint: What would happen if the solar wind started earlier or later than in our solar system?)* Either answer is acceptable if well argued. Although no such planets have been detected, it may be possible for the formation process of jovian planets to be interrupted by solar wind turn-on before the capture of nebular gases, leaving a real "ice giant."
24. *Radiometric dating of meteorites from another solar system shows that they are a billion years younger than rocks from the terrestrial planets of the same system.* This would not be surprising—there is no reason that other solar systems should be the same age as ours.
25. c; 26. b; 27. c; 28. c; 29. b;
30. b; 31. a; 32. c; 33. b; 34. b.
35. This question asks students to briefly restate and explain ideas taken directly from the reading. The key in grading is to make sure that students demonstrate that they understand the concepts about which they are writing, and that they have used their own words and not simply copied from the text.
36. Ices would have condensed in the inner solar system, significantly increasing the size and mass (or possibly number) of terrestrial planets. Water and other hydrogen compounds would be much more abundant.
37. Without capture of nebular gas, jovian planets would not accumulate substantial amounts of material into the planets themselves or into the disks surrounding them. The jovian planets would consist of icy cores without the envelopes of light gases (H, He). No satellites would form, as no disk would have formed.
38. More than one correct answer: On one hand, more angular momentum might force more material into the solar nebula (instead of into the protosun). In this scenario, more planets or more massive ones might form. On the other hand, too much angular momentum might result in two protostars, possibly destabilizing the planet-forming process. With no angular momentum, the entire nebula would collapse into a star, with no disk from which to make planets.
39. See comments for Problem 35.
40. See comments for Problem 35.
41. Hydrogen was created in the Big Bang, In the solar nebula, hydrogen appears in gaseous form and also in hydrogen compounds. Earth's H_2O came from ice-rich comets or asteroids that were flung our way by the jovian planets. Oxygen was not created in the Big Bang, but inside stars due to fusion. Stellar explosions spewed oxygen and other heavy elements out to mix in the giant clouds. Oxygen combined with hydrogen to make water, and the cloud collapsed to form the solar nebula.

42. If wandering planetesimals hadn't been flung out of their formation region, the following differences would be apparent: (1) little or no cratering on the terrestrial planets or moons; (2) little or no water or other hydrogen compounds for the terrestrial planets; (3) no comets in our skies; (4) asteroids might have formed a planet; (5) no Oort cloud, and jovian planets might have accreted more ice.
43. Pluto is basically an icy planetesimal, like so many that formed in the region of the jovian planets and beyond. It resembles the formation of terrestrial planets in that it's a solid object formed by accretion of flakes in the solar nebular—but mostly of ices instead of just rock and metal. Pluto's formation is identical to the initial stages of jovian planet formation—but without accreting with other planetesimals and capturing nebular gas.
44. The meteorite, being older, would have a larger proportion of argon, since more of the potassium would have undergone the radioactive decay that produces argon.
45. a. The equal amounts of the parent and daughter isotopes mean that the age of the rock is one half-life of potassium–40, or 1.25 billion years.
    b. The ratio of potassium–40 to argon–40 is 1 to 3, which means 1/4 of the original potassium–40 remains. Two half-lives have passed, so the rock is 2.5 billion years old.
46. a. Following the method of Mathematical Insight 8.1, we calculate the age of the rock from the lunar highlands to be:

$$t = T_{\text{half}} \times \frac{\log_{10}\left(\frac{\text{current amount}}{\text{original amount}}\right)}{\log_{10}(1/2)}$$

$$= 4.5 \text{ billion yr} \times \frac{\log_{10}(0.55)}{\log_{10}(1/2)} = 3.9 \text{ billion yr}$$

    b. The age of the rock from the lunar maria is:

$$t = T_{\text{half}} \times \frac{\log_{10}\left(\frac{\text{current amount}}{\text{original amount}}\right)}{\log_{10}(1/2)}$$

$$= 4.5 \text{ billion yr} \times \frac{\log_{10}(0.63)}{\log_{10}(1/2)} = 3.0 \text{ billion yr}$$

47. a. Based on the radioactive carbon content, the time since the cloth was painted is:

$$t = T_{\text{half}} \times \frac{\log_{10}\left(\frac{\text{current amount}}{\text{original amount}}\right)}{\log_{10}(1/2)}$$

$$= 5{,}700 \text{ yr} \times \frac{\log_{10}(0.77)}{\log_{10}(1/2)} = 2150 \text{ yr}$$

b. Based on the radioactive carbon content, the time since the wood was cut is:

$$t = T_{\text{half}} \times \frac{\log_{10}\left(\frac{\text{current amount}}{\text{original amount}}\right)}{\log_{10}(\frac{1}{2})}$$

$$= 5{,}700 \text{ yr} \times \frac{\log_{10}(0.062)}{\log_{10}(\frac{1}{2})} \approx 23{,}000 \text{ yr}$$

c. Carbon-14 is not useful for establishing the age of the Earth because its half-life is too short in comparison to the Earth's age; essentially all C-14 present when the Earth formed would have decayed long ago.

48. The age of this meteorite is:

$$t = T_{\text{half}} \times \frac{\log_{10}\left(\frac{\text{current amount}}{\text{original amount}}\right)}{\log_{10}(\frac{1}{2})}$$

$$= 14 \text{ billion yr} \times \frac{\log_{10}(0.94)}{\log_{10}(\frac{1}{2})} = 1.25 \text{ billion yr}$$

Because this time is much less than the age of the solar system, the meteorite cannot be a leftover from the solar system formation. The rock must have formed more recently, indicating that Mars must have had geological activity relatively recently.

49. If the star has the same mass as the Sun, we do not need Newton's version of Kepler's third law. In this case, we simply use Kepler's third law:

$$p^2 = a^3$$

where p is (as usual) the period in years and a is the semimajor axis in AU. We can solve this for the period by taking the square root of both sides:

$$p = a^{3/2}$$

and we are told that the semimajor axis is 50,000 AU. Putting in the number, we get:

$$p = (50{,}000 \text{ AU})^{3/2}$$
$$= 1.1 \times 10^7 \text{ yr}$$

Finally, we are told that the collapse time is about half of this:

$$\frac{1}{2} \times 1.1 \times 10^7 \text{ yr} = 5.6 \times 10^6 \text{ yr}$$

The collapse time for the cloud was about 5.6 million years.

50. We will assume that the ratio of the mass of the icy Earth to the mass of the actual Earth is the same as the ratio of the total abundance of ices, rock, and metals (out of which the icy Earth would form) to just rock and metals (out of which the real Earth formed). We will need to know what the relative abundances are. Table 8.1 provides this information: Metals made up 0.2% of the nebula, silicates 0.4%, and

hydrogen compounds 1.4%. The ratio of the mass of the icy Earth to the mass of the actual Earth is:

$$\frac{\text{mass of icy Earth}}{\text{mass of real Earth}} = \frac{\text{abundance of ices, rocks, and metals}}{\text{abundance of rocks and metals}}$$
$$= \frac{1.4\% + 0.4\% + 0.2\%}{0.4\% + 0.2\%}$$
$$= 3.3$$

Earth would be 3.3 times larger if it could have used ices.

We can make the same calculation if we ask what Earth would be like if it could have also used the gases, hydrogen and helium. In this case, we will make the same assumption as before, except that we will assume that the ratio of the mass of the gaseous Earth to the mass of the real Earth will be equal to the ratio of the total abundance of everything in the disk (100%, of course) to the abundance of rocks and metals. So:

$$\frac{\text{mass of gaseous Earth}}{\text{mass of real Earth}} = \frac{\text{abundance of gases, ices, rocks, and metals}}{\text{abundance of rocks and metals}}$$
$$= \frac{100\%}{0.4\% + 0.2\%}$$
$$= 170$$

Our planet would be 170 times more massive if it had been able to build itself out of everything in the disk and not just rocks and metals.

51. Since density is mass over volume, it serves as a conversion factor between the mass of an object and its volume. And because the cores of the planets are spherical, we can use the formula for the volume of a sphere to find the radius. First, we find the volume of the ice ball by converting from the mass (10 times the mass of Earth), using the density (2 g/cm^3). We'll have to figure out what the mass is in grams first, though. Looking up the mass of the Earth in Appendix E, we see that it is 5.98×10^{24} kg. Ten Earth masses converted to grams is:

$$10 \ \cancel{\text{Earth masses}} \times \frac{5.98 \times 10^{24} \ \cancel{\text{kg}}}{1 \ \cancel{\text{Earth mass}}} \times \frac{1000 \text{ g}}{1 \ \cancel{\text{kg}}} = 5.98 \times 10^{28} \text{ g}$$

Now, we convert to volume using the density:

$$5.98 \times 10^{28} \ \cancel{\text{g}} \times \frac{1 \text{ cm}^3}{2 \ \cancel{\text{g}}} = 2.99 \times 10^{28} \text{ cm}^3$$

Now we go from volume, V, to the radius, r. To do this, recall that the formula for the volume of a sphere is:

$$V = \frac{4}{3}\pi r^3$$

Solving this for the radius with some algebra:

$$r = \sqrt[3]{\frac{3V}{4\pi}}$$

So, we use the volume we already determined to get the radius:

$$r = \sqrt[3]{\frac{3(2.99 \times 10^{28}\ \text{cm}^3)}{4\pi}}$$
$$= 1.9 \times 10^9\ \text{cm}$$

This is nice, except that we probably should convert that to kilometers, which we know how to do:

$$1.9 \times 10^9\ \cancel{\text{cm}} \times \frac{1\ \cancel{\text{m}}}{100\ \cancel{\text{cm}}} \times \frac{1\ \text{km}}{1000\ \cancel{\text{m}}} = 19{,}000\ \text{km}$$

The icy core has a radius of about 19,000 kilometers. We can convert this to Earth radii by looking in Appendix E to find that Earth's radius is 6,378 kilometers:

$$19{,}000\ \text{km} \times \frac{1\ \text{Earth radius}}{6378\ \text{km}} = 3.0\ \text{Earth radii.}$$

This ice ball would have a radius about three times that of Earth.

52. Ranking the moons by radius is easy enough, once we look in Appendix E. From largest to smallest, they are: Ganymede (2,634 kilometers in radius), Titan (2,575 kilometers), the Moon (1,738 kilometers), and Charon (635 kilometers).

To rank them by ratio to the size of their planets, we'll need the planet radii. Appendix E once again comes to our aid with the radii, in the same order: 71,492 kilometers (Jupiter), 60,268 kilometers (Saturn), 6,378 kilometers (Earth), and 1,160 kilometers (Pluto).

Taking the ratios of the moons' radii to their planets:

$$\frac{\text{radius of Ganymede}}{\text{radius of Jupiter}} = \frac{2634\ \text{km}}{71{,}492\ \text{km}} = 0.037$$

$$\frac{\text{radius of Titan}}{\text{radius of Saturn}} = \frac{2725\ \text{km}}{60{,}268\ \text{km}} = 0.045$$

$$\frac{\text{radius of the Moon}}{\text{radius of Earth}} = \frac{1738\ \text{km}}{6378\ \text{km}} = 0.27$$

$$\frac{\text{radius of Charon}}{\text{radius of Pluto}} = \frac{635\ \text{km}}{1160\ \text{km}} = 0.55$$

Ranking the moons in the order of the ratio of their radii to the radii of their planets then goes: Charon, the Moon, Titan, and Ganymede. This is the reverse of what we found above for the order of their sizes. From this we can conclude that while the largest planets tend to have the largest moons, the smaller planets sometimes have moons that are out of proportion to their size. (Note that both Charon and the Moon are believed to have been formed in giant impacts. Thus, it is not entirely surprising that they are both bigger than we would expect given their planets' sizes.)

53. The probability of one planet traveling around in one direction and not the other is $\frac{1}{2}$. If the probabilities really did not depend on each other, then the probability that all nine would end up orbiting in this same direction is $(1/2)^9 = 1.9 \times 10^{-3}$. However, we need to allow for either orbital direction; (All of the planets merely have to go in the same direction, it does not matter which of the two.) We should take twice this value, or 3.9×10^{-3}. To get this in percent probability, we multiply by 100%:

$$3.9 \times 10^{-3} \times 100\% = 0.39\%$$

The probability of the planets all orbiting the Sun in the same direction is 0.39%.

54. The formula for angular momentum is (from Section 4.3) $m \times v \times r$, where m is the mass, v is the velocity, and r is the radius. We know that this quantity is conserved, so:

$$m_{\text{large}} \times v_{\text{large}} \times r_{\text{large}} = m_{\text{small}} \times v_{\text{small}} \times r_{\text{small}}$$

The mass of the objects is the same before and after the collapse, so $m_{\text{large}} = m_{\text{small}}$ and we can cancel the masses out and get:

$$v_{\text{large}} \times r_{\text{large}} = v_{\text{small}} \times r_{\text{small}}$$

We are asked to find the velocity in the uncollapsed (large) stage, so we can also solve for v_{large}:

$$v_{\text{large}} = v_{\text{small}} \times \frac{r_{\text{small}}}{r_{\text{large}}}$$

We are told that material near Pluto orbits at 5 km/sec. We know, from Appendix E, that Pluto orbits the Sun at an average of 39.5 AU, and we are told that the original size of the solar nebula was 40,000 AU. We put in those values:

$$v_{\text{large}} = 5 \text{ km/sec} \times \frac{39.5 \text{ AU}}{40{,}000 \text{ AU}}$$

$$= 4.9 \times 10^{-3} \text{ km/sec}$$

We might want to convert this to meters per second since it is so small. Converting:

$$4.9 \times 10^{-3} \cancel{\text{km}}\text{/sec} \times \frac{1000 \text{ m}}{1 \cancel{\text{km}}} = 4.9 \text{ m/sec}$$

Chapter 9. Planetary Geology: Earth and the Terrestrial Worlds

This chapter begins our true comparative planetology with a general introduction to planetary geology. Note that, while the initial focus is on processes, we still use features on Earth (and occasionally on other worlds) to give concrete examples of each process at work. Once students learn the basic ideas, we apply them to each of the terrestrial planets in the "geological tour" sections (9.3–9.6) that conclude the chapter.

- If you are more accustomed to teaching with a planet-by-planet approach than the comparative planetology approach, you may wish to emphasize Sections 9.3 through 9.6, which go through a geological tour of the terrestrial worlds. Indeed, you can feel free to use the earlier sections only as references, focusing on the worlds individually if you wish.
- Throughout this chapter, it is useful to ask "what if" questions in class. For example, ask how terrestrial planets would be different if they were larger, older, younger, and so on.
- Note that we do not introduce jargon that will not be useful elsewhere. For example, we do not use the terms *scarp, graben,* or *regolith*—instead, we describe such things using familiar words from everyday English.

As always, when you prepare to teach this chapter, be sure you are familiar with the relevant media resources (see the complete, section-by-section resource grid in Appendix 3 of this Instructor's Guide) and the online quizzes and other study resources available on the Mastering Astronomy Web site.

What's New in the Fourth Edition That Will Affect My Lecture Notes?

As everywhere in the book, we have added learning goals that we use as subheadings, rewritten to improve the text flow, improved art pieces, and added new illustrations. The art changes, in particular, will affect what you wish to show in lecture.

- This chapter consolidates all of the material that appeared in Chapters 10 (geology of the terrestrial worlds) and part of Chapter 14 (planet Earth) of the third edition. Thus, we no longer have a separate chapter about Earth. Earth is discussed here as the culmination of understanding planetary geology.
- We include expanded coverage of the latest results (as of fall 2005) from the *Spirit* and *Opportunity* Mars rovers, and their new view of a warmer, wetter past for Mars.

Teaching Notes (By Section)

Section 9.1 Connecting Planetary Interiors and Surfaces

This section presents the key features of planetary interiors that are important to understanding surface geology.

- Note that most students are familiar with the Earth's structure in terms of core, mantle, and crust, but are unfamiliar with the important term *lithosphere.* You should check to be sure your students understand this term clearly before continuing.
- Most students have heard of seismology but do not know how it can be used to explore a planet's interior. This is discussed in a Special Topic box.
- Note that this is the first section in which we introduce the idea of convection—a topic that will come up again and again throughout the remainder of the book.
- The discussion of planetary cooling might benefit from a simple demonstration of large and small objects—potatoes, rocks, ice cubes, or whatever is handy—cooling down or warming up.
- The discussion of magnetic fields also lends itself to a demonstration with iron filings andmagnets.

Section 9.2 Shaping Planetary Surfaces

This section introduces the four basic geological processes—impact cratering, volcanism, tectonics, and erosion—with concrete examples of each. A key intention in this section is to enable students to connect the presence or absence of the geological processes to basic planetary properties such as size, distance from the Sun, and rate of rotation.

Sections 9.3–9.5 Geological Tours: The Moon and Mercury, Mars, and Venus

These sections provide an up-to-date and fairly comprehensive geological tour of each of the terrestrial worlds, except Earth. Please remember that our key objective is for students to understand the features we see on each world, not to memorize the names or characteristics of these features.

Section 9.6 The Unique Geology of Earth

This final section covers Earth's geology, with particular emphasis on the specific ways it differs from the other planets. The topic of plate tectonics is well covered, so that students are well prepared to appreciate its importance in maintaining Earth's atmosphere as covered in the next chapter.

Answers/Discussion Points for Think About It Questions

The Think About It questions are not numbered in the book, so we list them in the order in which they appear, keyed by section number.

Section 9.1

- (p. 254) The smallest: It has lost the most interior heat, and thus its lithosphere has thickened the most. Note that a smaller planet will have a thicker lithosphere both in relative terms (percentage of radius) and in kilometers.
- (p. 256) Faster rotation.

Section 9.2

- (p. 256) Typical globes have relief of one-to-several millimeters, which is out of proportion to their size—a factor of several too large. If the bumps were to scale, they would hardly be discernable.

Section 9.3

- (p. 263) Mercury, with more ancient craters, has an older surface. The planets are the same age—i.e., the planets as a whole formed roughly simultaneously. Geological processes have been active on Venus, remaking the surface over billions of years.

Section 9.4

- (p. 268) Size. Earth's large size has sustained widespread, global, and ongoing volcanism and tectonics, preventing any part of the surface from still showing the scars of the heavy bombardment. Erosion, indirectly a result of size, has been just as important in keeping the surface "young looking."

Section 9.5

- (p. 273) Volcanism and tectonics ought to be essentially unchanged, as they depend primarily on planetary size. But fast rotation would create winds in Venus's thick atmosphere, leading to more erosion.
- (p. 274) This asks the students about the status of ESA's *Venus Express* mission.

Section 9.6

- (p. 278) Earthquakes are common in these areas because they are near plate boundaries, where plates interact with one another and can therefore slip in ways that cause earthquakes.
- (p. 280) This question will require students to investigate some local geology. You may wish to do this for your school locality, and discuss notable geological formations in class.

Solutions to End-of-Chapter Problems (Chapter 9)

1. Planetary geology is the extension of geology on the Earth: It's the study of the surface and interior of any solid world. A brief summary of the geology of the solid worlds:

   Mercury—Heavily cratered with some evidence for volcanism and tectonics.

   Venus—Young surface, with much volcanism tectonics but little cratering or erosion present.

   Earth—Young surface, with evidence for tectonics, volcanism, and erosion. Little impact cratering apparent.

   The Moon—Much like Mercury, with many craters and some evidence for volcanism in the past.

   Mars—Lots of evidence of volcanism and easily visible evidence of tectonics. Lots of craters in the southern hemisphere particularly. Some erosion apparent.
2. Differentiation is the process by which dense material sinks to the bottom of a fluid and the less dense material floats up to the top. Since the solid worlds were all probably fluid early on in their lives, differentiation likely happened on all of them. This leads to dense, metallic cores and less dense, rocky mantles. The crusts on some of the worlds (like Earth) are made up of mantle material that was even less dense than the average so that it floated to the top.
3. A lithosphere is a planet's outer layer of relatively rigid rock. We did not list it in the core/mantle/crust distinction because the lithosphere is composed of the crust and part of the mantle so that it is a separate distinction.

   Among the five terrestrial worlds, the largest, Venus and Earth, have the thinnest lithospheres. The smallest, Mercury and the Moon, have the thickest lithospheres. Mars lies in between.
4. Planets can heat up in three ways: differentiation, accretion, or radioactivity. In differentiation, the gravitational potential energy released when the denser material sinks is converted into heat. Accretion also uses gravitational potential energy, but in this case the energy comes from bodies falling down onto the planet from outside. Radioactivity uses the nuclear potential energy released when unstable isotopes of some elements decay.

Planets can get the heat out in three ways as well: conduction, convection, and radiation. In conduction, the heat is transferred by physical contact. In convection, the hotter material in the Earth rises, carrying the heat with it, and the cooler material sinks. And in radiation, the heat is taken away by photons due to black body emission.

Because all of the heat must be radiated away from the surface of the planets eventually, the rate at which a planet can get rid of its heat is proportional to its surface area. But the amount of heat that a planet has is proportional to its volume. Since volume is proportional to radius cubed and surface area goes like radius squared, large planets have to get rid of more heat per square meter of surface, so they take longer to cool.

5. Earth's magnetic field requires that its interior be made of an electrically conducting fluid that convects and rotates. Apart from Venus, most of the other terrestrial worlds should have cooled to the point where their interiors are no longer convection fluids, so we would not expect to see magnetic fields there. In the case of Venus, the planet spins very slowly, which probably keeps any magnetic field from occurring on that planet. (Why Mercury has a magnetic field at all is something of a mystery. Apparently, the core is still molten.)
6. The four geological processes are volcanism, tectonics, impact cratering, and erosion. Volcanism is the process in which hot material from inside a planet is leaked out onto the surface. The Hawaiian Islands are an example of this process. Tectonics is a disruption of the planet's surface from stresses. An example of tectonics in action is the Rocky Mountains. Impact cratering is the process by which bowl-shaped craters are created due to outside impacts. We can see craters all over the Moon's surface. Finally, erosion is the process by which features are worn down by wind or water. The Grand Canyon on Earth is an example of such a feature.
7. Outgassing is the process in which gases from a planet's interior are expelled through volcanism. This process is very important to our existence since it probably supplied most of the water on our surface and in our atmosphere.
8. Since impact cratering occurs on planets at known rates, we can tell the age of a planetary surface by how many craters it has. If there are more craters, we know that the surface is older. Thus, the fact that the Moon is much more heavily cratered than the Earth tells us that the Moon's surface is much older. This is because the Moon is long dead geologically, except for impact cratering. Earth, however, continues to have volcanism, tectonics, and erosion. All of these processes renew the surface and erase impact craters, making the surface young and relatively uncratered.
9. Size: Larger planets are able to stay warm inside longer than smaller planets. Thus, on these worlds we see volcanism and tectonics for much longer into their lives than the smaller worlds. Larger planets are also better able to retain atmospheres, making erosion more important there.

   Distance from the Sun: This affects only erosion. The closer a planet is to the Sun, the warmer it will be and therefore the more weather it should have. Also, planets that are warm enough to have liquid water will have much more erosion than ones that are not that warm, since water can do a lot of erosion (assuming everything else is the same, of course).

   Rotation rate: Planets that spin faster have faster winds. This results in more erosion.
10. The Moon's history begins with its birth 4.55 billion years ago. At this time, it was still hot enough from accretion that it must have had a liquid interior. Thus, it was capable of volcanism and tectonics. Early in its life, the Moon experienced many

large impacts. These impacts left enormous craters and also triggered volcanism. The runny lava that seeped into the giant craters filled the craters in, forming the lunar maria. The maria also show evidence of tectonic stress features, the only tectonic features known on the Moon.

Even though the heavy bombardment has ended, impact cratering continues on the Moon. Since the Moon is a small world, most of its heat was quickly lost and the interior became solid. At this time, volcanism and tectonics stopped. And with no atmosphere to drive erosion, there has been nothing to erase the craters. The Moon's surface today shows many craters.

11. When Mercury was young, it still retained heat from its accretion. Thus, volcanism and tectonics were still possible. Indeed, we do see evidence of lava flows, although they are not as large as the lunar maria. During this time, the planet cooled and contracted. As it did so, the surface also had to shrink, meaning some parts of it had to shift to compensate. In these areas we see long cliffs. Since it cooled, Mercury has become a dead world, with no volcanism or tectonics left. Because it has never had an appreciable atmosphere, erosion has never been a major factor in Mercury's geology. Like the Moon, this left impact cratering as the dominant geological process for most of its life, leaving us with the heavily cratered world we see today.

12. Olympus Mons—A large stratovolcano formed from extensive, long-term volcanic activity.

    Tharsis Bulge—A volcanic feature formed by the extensive and long-term volcanic activity. Evidence of its volcanic nature can be found in the presence of three large volcanoes on the bulge.

    Valles Marineris—A tectonic feature formed from the stress due to the heavy Tharsis Bulge sitting on the planet at the west end.

    Southern hemisphere—Heavily cratered due to impacts. This part of the planet was not resurfaced like the north was, so the craters are much more obvious here.

    Pathfinder landing site—Appears to be an eroded area where water may have once flowed out of the canyon system to the south.

13. Liquid water is not stable on the surface of Mars today because of the cold temperatures and the low atmospheric pressure. The low temperature means that water is almost always frozen on Mars today. However, if it did warm enough to thaw, the low pressure would cause the water to almost immediately evaporate off of the surface. Water might exist below the surface, however, where the planet is still warm enough for it to exist in liquid form.

    We see lots of evidence for liquid water in the past. Evidence includes networks of what look like river channels, lots of erosion of features like craters, layers in rocks seen by landers, and chemical evidence indicating that some of the rocks were formed in water.

14. Coronae—These circular tectonic features were probably formed by rising mantle plumes pushing on the crust from below.

    Shallow-sided volcanoes—Formed by runny lava.

    Steep-sided volcanoes—Formed by thicker lava.

    Craters—Due to impact cratering, of course. However, the small number of craters points to a young surface. There is also a lack of small craters due to the thick atmosphere destroying them before they hit the surface.

15. We think most of Venus's surface may have been repaved around 750 million years ago because most of the surface appears to be this age.

    Venus's lack of plate tectonics is perhaps due to its lithosphere being stronger than Earth's. A stronger lithosphere would be harder to break up into plates, preventing plate tectonics. We think that Venus's lithosphere may be stronger because of the lack of water in the rocks due to the extreme surface temperature, since water tends to soften rocks.

16. Earth's lithosphere is broken up into plates that float on the underlying mantle. The tops of the mantle's convection cells drag the bottom of the plates, making them move. This movement causes spreading in some places (where the plates are moving apart) and subduction in others (where one plate is diving under another). The plate tectonics force the surface material to be reprocessed and expelled back out onto the surface in volcanic eruptions. Every time a bit of rock is subducted and expelled, its chemistry changes, making it lighter. The lighter rock becomes the continental plates, while the relatively unprocessed rocks are the seafloor plates.

17. Seafloors—Formed in the spreading zones where upwelling material pushes out and forces two seafloor plates apart.

    Continents—Formed by crustal material that has been recycled, forming less dense rocks.

    Islands—Formed by undersea volcanoes erupting over many years. This builds up undersea mountains that eventual protrude above sea level.

    Mountain ranges—Formed where two plates are colliding. The stress on the surface causes the crust to buckle upward, forming a mountain range.

    Rift valleys—Formed by two continental plates pulling apart.

    Faults—Places where two plates are sliding sideways relative to each other. These are areas that are often prone to earthquakes.

18. The geological controlling factors (size, distance from the Sun, and rotation rate) probably set the planet's geology almost entirely. There is some question about why Earth has plate tectonics and Venus does not, but that may well be explained by Venus's distance from the Sun.

19. *The next mission to Mercury photographs part of the surface never seen before and detects vast fields of sand dunes.* This discovery would be a big surprise, because Mercury never had an atmosphere sufficient for significant erosional activity.

20. *Seismographs placed on the surface of Mercury record frequent and violent earthquakes.* Thiswould be surprising, as the kind of tectonic activity responsible for earthquakes occurred very long ago.

21. *A future orbiter observes a volcanic eruption on Venus.* This would not be surprising. Venus is thought to be sufficiently active for new eruptions to occur and create lava flows.

22. *A Venus radar mapper discovers extensive regions of layered sedimentary rocks, similar to those found on Earth.* This discovery would be very surprising. Erosion is virtually negligible on Venus, due to the lack of liquid water and significant winds. Even if Venus had more water early in its existence, more recent geological activity has wiped out every trace of such ancient surfaces.

23. *Radiometric dating of rocks brought back from one lunar crater shows that the crater was formed only a few tens of millions of years ago.* This would not be surprising. Craters are continuing to form throughout the solar system, as there are

still plenty of impactors around. Meteor Crater is an example of a recent crater—only a few tens of thousands of years old.

24. *New orbital photographs of craters on Mars that have gullies also show pools of liquid water to be common on the crater bottoms.* This would be surprising. Under the current conditions of Mars's atmosphere, pools of liquid water should rapidly freeze and/or evaporate.

25. *Drilling into the Martian surface, a robotic spacecraft discovers liquid water deep beneath the slopes of a Martian volcano.* This would be exciting, but not surprising. "Geothermal" heat from Martian volcanoes may well be enough to melt water under the Mars surface.

26. *Clear-cutting in the Amazon rain forest on Earth exposes vast regions of ancient terrain that is as heavily cratered as the lunar highlands.* This would be surprising. Erosion has been so strong in that region that no ancient terrain would be recognizable. Furthermore, Earth had little continental crust so long ago that South America wouldn't even have existed as a large land mass at that time.

27. *Seismic studies on Earth reveal a "lost continent" that held great human cities just a few thousand years ago but that is now buried deep underground off the western coast of Europe.* This is not plausible. Plates move only a few centimeters per year, so a continent could not be subducted in a few thousand years. Neither could erosional processes bury a continent on the time scale of human civilization.

28. *We find a planet in another solar system that orbits at the same distance from a Sun-like star as Earth orbits the Sun and has Earth-like plate tectonics, but it is only the size of the Moon.* This would be surprising because we expect only a larger world to have plate tectonics. However, it might be possible if the planet is young and still hot inside.

29. *We find a planet in another solar system that is as large as Earth but as heavily cratered as the Moon.* This would be surprising. Such a large planet would be expected to have extensive geological activity from volcanism, tectonics, and probably erosion.

30. *We find a planet in another solar system with an Earth-like seafloor crust and continental crust but that apparently lacks plate tectonics or any other kind of crustal motion.* This would be surprising—we do not know of any process for creating different kinds of crust that does not involve plate tectonics.

31. b; 32. b; 33. a; 34. c; 35. b; 36. c; 37. a; 38. b; 39. c; 40. b.

41. Radiometric dating is usually considered more reliable than measuring crater abundances, partly because it is much more precise. In addition, cratering is somewhat random and more likely to be misleading. Moreover, the precise time at which the early bombardment ended is not well known. Crater abundances are easier to measure on other planets, because it is much cheaper to take photographs than to land on the surface and analyze rocks for radioactivity—either with an intelligent robot or by returning the sample to Earth.

42. Mars has had the greatest erosional activity, because it once had liquid water on its surface and it now has wind and dust storms. Its size—large compared to the three other words considered—is the main reason. Mars outgassed more, and was able to retain its atmosphere due to stronger gravity.

    Mercury has a negligible atmosphere from the point of view of erosion, primarily due to its high temperature related to its distance from the Sun. Its relatively small size also led to only a small amount of outgassing to form an atmosphere in the first place.

The Moon also has a negligible atmosphere, primarily related to the inability of such a small world to create or retain an atmosphere.

Venus has a great deal of atmosphere but very little erosion. Water erosion doesn't occur because the planet is too hot, a condition related to its distance from the Sun. More straightforwardly, it lacks significant wind erosion because its slow rotation rate leads to very slow winds.

43. If Mars were smaller, it would have undergone less volcanic and tectonic activity because its interior would have cooled more. With less atmosphere from less outgassing, it is likely that erosion would be less important as well. As a result, craters would be more widespread on the Martian surface. With less atmosphere, Mars would have been a less hospitable place for life.
44. Essay question. Grading should emphasize logical discussion of observed features on the Martian surface, the known geological processes, the history of water, and time scales for change on Mars.
45. Sample solution: If Earth had been closer to the Sun, the warmer temperatures might have forced more water vapor into the atmosphere. This would have increased the greenhouse effect, leading to further warming, evaporation of the oceans, and possible escape of all the water to space (as with Venus). Without liquid water on the surface, life as we know it would not have arisen on Earth.
46. Extrapolating from the Moon to Earth and beyond, we would predict very high rates of volcanic and tectonic activity that would have completely erased evidence of past cratering. We'd expect it to have a substantial atmosphere, and with Earth's rotation rate there should at least be substantial wind erosion. Water erosion might be possible, given the distance, but students could argue that the extra atmosphere might make the planet too hot for liquid water.
47. a. The spacecraft should include a magnetic field detector, because the size and rapid rotation of the planet would be expected to cause a magnetic field if the core is metallic. The spacecraft should also measure the gravitational pull of the planet on the spacecraft (giving the mass) and the size of the planet; together these quantities provide the planet's density.
    b. Given the planet's size and rotation rate, erosional features should be present if there is an atmosphere.
48. The pressure from the textbook will squash a 2-centimeter ball of Silly Putty at room temperature to about 1 centimeter in 5 seconds. A warmer ball of Silly Putty will squash to about 0.5 centimeter in the same amount of time, and a chilled ball to about 1.5 centimeters. Actual measurements may vary due to the temperatures used and the size of the ball. Opinions may vary as to whether the differences are large or small effects, though we feel they are large given the small range in temperatures used. The geological connection is that warmer rocks deform more easily than cooler rocks.
49. Answers will vary depending on the size and shape of the containers used. Complete freezing will normally take hours. Measurement of the "lithospheric thickness" will be difficult in some cases. The main result—that larger containers take longer to freeze—should be quite obvious, and its relationship to small bodies cooling off faster than larger bodies should be clear.
50. Observations of the Moon should be feasible with virtually any small telescope. Answers will vary depending on the phase of the Moon and the interest of the student, but virtually any attempt that has engaged the student's mind in active observation should be considered a success.

51. First, we will need an expression for the surface area-to-volume ratio. For spherical objects, this is just:

$$\frac{4\pi r^2}{\frac{4}{3}\pi r^3}$$

We can simplify this by canceling terms to get:

$$\frac{3}{r}$$

For Mars, where $r = 3{,}397$ km (from Appendix E), the surface area-to-volume ratio is 8.83×10^{-4} km^{-1} while for the Moon, where $r = 1{,}738$ km (Appendix E), the ratio is 1.73×10^{-3} km^{-1}. Sincc thc ratc of cooling of a planct is proportional to this ratio, we would expect Mars's interior to be much warmer than the Moon's.

52. First, we will need an expression for the surface area-to-volume ratio. For spherical objects, this is just:

$$\frac{4\pi r^2}{\frac{4}{3}\pi r^3}$$

We can simplify this by canceling terms to get:

$$\frac{3}{r}$$

For Venus, where $r = 6{,}051$ km (from Appendix E), the surface area-to-volume ratio is 4.96×10^{-4} km^{-1} while for Earth, where $r = 6{,}378$ km (Appendix E), the ratio is 4.70×10^{-3} km^{-1}. Since the rate of cooling of a planet is proportional to this ratio and since these numbers are pretty close, we would expect that the two planets should have comparable interior temperatures.

53. a. Waist size is a length, so it should be proportional to my height. I would expect that doubling my height should double my waist size.

b. Since clothes cover my body area, I would expect the amount I need to increase like my surface area. My surface area should increase like my height squared (just like a sphere's surface area is proportional to the radius squared). Doubling my height should increase my area by a factor of $2^2 = 4$.

c. Weight is proportional to volume, which depends on my height cubed (just like a sphere's volume depends on the radius cubed). Doubling my height should increase my volume (and thus my mass) by $2^3 = 8$.

d. We are told that the pressure on my joints goes like the total mass over the area of the joints. Now, the area of the joints should increase by the same factor as the surface area. We already worked out that the mass increases by a factor of 8 while the area increases by a factor of 4. The pressure increases by a factor of $8/4 = 2$. By doubling my size, I've doubled the pressure on my joints.

(This is why larger animals like elephants require such massive legs relative to their bodies while small animals like insects have tiny little legs.)

54. We will have to begin by finding how many micrometeorites hit every square centimeter of the Moon's surface. We are told that 25 million micrometeorites hit the Moon's surface each day, so we just need the Moon's surface area to get the rate per area. We know that the surface area of a sphere is:

$$\text{surface area} = 4\,\pi r^2$$

and that the Moon's radius is 1,738 kilometers (Appendix E). We can calculate the surface area of the Moon. First, however, we should convert the radius to centimeters since we want the rate per square centimeter:

$$1738\ \cancel{\text{km}} \times \frac{1000\ \cancel{\text{m}}}{1\ \cancel{\text{km}}} \times \frac{100\ \text{cm}}{1\ \cancel{\text{m}}} = 1.738 \times 10^8\ \text{cm}$$

Applying the formula for surface area:

$$\begin{aligned}\text{surface area} &= 4\,\pi(1.738 \times 10^8\ \text{cm})^2 \\ &= 3.80 \times 10^{17}\ \text{cm}^2\end{aligned}$$

The impact rate is given by:

$$\begin{aligned}\text{impact rate per area} &= \frac{\text{25 million impacts per day}}{3.80 \times 10^{17}\ \text{cm}^2} \\ &= 6.59 \times 10^{-11}\ \text{impact/day/cm}^2\end{aligned}$$

Now, we are told that it will take 20 impacts to destroy a footprint. We just need to know the area of an astronaut's footprint. We will approximate their boots as rectangles that are 10 centimeters by 30 centimeters. The area, therefore, is 10 cm × 30 cm = 300 cm^2. How long does it take to get 20 impacts in that area? The rate of impacts in 300 cm^2 is given by the rate per area times the area, so:

$$\begin{aligned}\text{rate of impacts} &= (\text{rate per area}) \times (\text{area}) \\ &= (6.59 \times 10^{-11}\ \text{impact/day/cm}^2) \times (300\ \text{cm}^2) \\ &= 1.98 \times 10^{-8}\ \text{impact/day}\end{aligned}$$

There are 1.98×10^{-8} impact/day in the footprint.

Finally, we can find how long it takes for 20 impacts to hit this footprint. Using the usual rule for rate:

$$\text{impacts} = \text{rate} \times \text{time}$$

we solve for time and use our values for rate and number of impacts:

$$\begin{aligned}\text{time} &= \frac{\text{impacts}}{\text{rate}} \\ &= \frac{20\ \text{impacts}}{1.98 \times 10^{-8}\ \text{impact/day}} \\ &= 1.01 \times 10^9\ \text{days}\end{aligned}$$

This is clearly a long time, a billion days. But we had better convert to years since this number is difficult to gauge:

$$1.01 \times 10^{9} \ \cancel{\text{days}} \times \frac{1 \text{ yr}}{365 \ \cancel{\text{days}}} = 2.77 \times 10^{6} \text{ yr}$$

It would take about 2.77 million years to obliterate those footprints on the Moon.

55. To solve this problem we will use the formula for kinetic energy:

$$\text{kinetic energy} = \frac{1}{2} mv^{2}$$

where m is the mass and v is the velocity. We need to get the mass of the asteroid and we will probably need to convert the velocity to meters per second. Starting with the easier of these two, we convert the velocity:

$$20 \ \cancel{\text{km}} \text{/s} \times \frac{1000 \text{ m}}{1 \ \cancel{\text{km}}} = 20{,}000 \text{ m/s}$$

The impact speed is 20,000 m/s.

Now for the mass of the impactor. Since we are given the radius, we can estimate the volume by assuming that the asteroid is spherical. To get the mass, we will assume that the density is about 3 g/cm^3, pretty average for something rocky. First computing the volume, we use the volume of a sphere:

$$\text{volume} = \frac{4}{3} \pi r^{3}$$

The asteroid is 1 kilometer in diameter, or half a kilometer in radius. Converting this to centimeters (to match the density):

$$0.5 \ \cancel{\text{km}} \times \frac{1000 \ \cancel{\text{m}}}{1 \ \cancel{\text{km}}} \times \frac{100 \text{ cm}}{1 \ \cancel{\text{m}}} = 5 \times 10^{4} \text{ cm}$$

The volume is:

$$\begin{aligned} \text{volume} &= \frac{4}{3}\pi \ (5 \times 10^{4} \text{ cm})^{3} \\ &= 5.2 \times 10^{14} \text{ cm}^{3} \end{aligned}$$

Multiplying by the density gives the mass of the body:

$$\begin{aligned} \text{mass} &= (5.2 \times 10^{14} \text{ cm}^{3}) \times (3 \text{ g/cm}^{3}) \\ &= 1.6 \times 10^{15} \text{ g} \end{aligned}$$

Converting this back into kilograms:

$$1.6 \times 10^{15} \ \cancel{\text{g}} \times \frac{1 \text{ kg}}{1000 \ \cancel{\text{g}}} = 1.6 \times 10^{12} \text{ kg}$$

Now we are ready to find the kinetic energy. Plugging into the formula above:

$$\text{kinetic energy} = \frac{1}{2}(1.6 \times 10^{12}\ \text{kg})(20{,}000\ \text{m/s})^2$$
$$= 3.2 \times 10^{20}\ \text{J}$$

It is difficult for most of us to understand this number, since we have little intuitive sense of how large a joule is, to say nothing of such a large number of joules. Let us convert to megatons of TNT. We are told that a megaton is 4×10^{15} joules, so:

$$3.2 \times 10^{20}\ \cancel{\text{J}} \times \frac{1\ \text{megaton}}{4 \times 10^{15}\ \cancel{\text{J}}} = 7.9 \times 10^{4}\ \text{megatons}$$

This impact would be like nearly 80 thousand megatons of TNT, much larger than the largest weapon in the human arsenal.

56. To get the average power emitted per unit area due to internal heat, we will need to compute the surface area of the Earth. To do this, we will use the relation:

$$\text{surface area} = 4\ \pi r^2$$

The radius of the Earth is 6,378 kilometers (from Appendix E). Converting this to meters:

$$6{,}378\ \cancel{\text{km}} \times \frac{1000\ \text{m}}{1\ \cancel{\text{km}}} = 6.378 \times 10^{6}\ \text{m}$$

The surface area is:

$$\text{surface area} = 4\ \pi(6.378 \times 10^{6}\ \text{m})^2$$
$$= 5.11 \times 10^{14}\ \text{m}^2$$

We are told that 3 trillion watts of energy leak out of Earth due to internal heat. Dividing the energy leak rate by the area, we get the power area:

$$\frac{3 \times 10^{12}\ \text{W}}{5.11 \times 10^{14}\ \text{m}^2} = 5.87 \times 10^{-3}\ \text{W/m}^2$$

5.87×10^{-3} W/m^2 of internal heat energy escape the Earth. The power per area received by the Earth due to the Sun is much larger than this, telling us that the Earth's surface temperature is set by solar radiation and is not internal. However, the internal heat has to work its way out of the Earth. It is in this working its way out of the Earth's interior that heat drives geological activity. (Solar heating occurs at the surface, so it does not have to work its way there to escape. Thus, it does not drive tectonic and volcanic activity.)

57. This is a simple rate problem. We are told that the continents are 3,000 kilometers apart and that they move at 1 centimeter per year. Using:

$$\text{distance} = \text{rate} \times \text{time}$$

and solving for time, we get:

$$\text{time} = \frac{\text{distance}}{\text{rate}}$$

All that we need to do is convert either the rate or the distance so that they have the same units. We will convert the distance to centimeters, although it does not really matter which we do:

$$3{,}000\ \cancel{\text{km}} \times \frac{1000\ \cancel{\text{m}}}{1\ \cancel{\text{km}}} \times \frac{100\ \text{cm}}{1\ \cancel{\text{m}}} = 3 \times 10^8\ \text{cm}$$

Applying our formula for the time:

$$\text{time} = \frac{3 \times 10^8\ \text{cm}}{1\ \text{cm/yr}}$$
$$= 3 \times 10^8\ \text{yr}$$

The two continents would take about 300 million years to collide.

58. This is another rate problem. The area produced in some time is:

$$\text{area} = \text{rate} \times \text{time}$$

We will need the rate of crust production. We are told that the spreading center spreads about 1 centimeter per year over its 2,000-kilometer length. First converting centimeters to kilometers so that we can find the area per year:

$$1\ \cancel{\text{cm}} \times \frac{1\ \cancel{\text{m}}}{100\ \cancel{\text{cm}}} \times \frac{1\ \text{km}}{1000\ \cancel{\text{m}}} = 1 \times 10^{-5}\ \text{km}$$

The area is the width of new crust times the length:

$$\text{area of new crust} = (2{,}000\ \text{km}) \times (1 \times 10^{-5}\ \text{km})$$
$$= 1 \times 10^{-2}\ \text{km}^2$$

The rate is $1 \times 10^{-2}\ \text{km}^2/\text{year}$. Thus, in 100 million years, we get:

$$\text{area} = (1 \times 10^{-2}\ \text{km}^2/\text{yr}) \times (1 \times 10^8\ \text{yr})$$
$$= 1 \times 10^6\ \text{km}^2$$

The spreading center produces 1 million square kilometers of new crust every 100 million years.

59. a. This is clearly a ratio problem. Density is mass over volume. The volume of a planet is proportional to $\frac{4}{3}\pi r^3$, so the density is:

$$\text{density} = \frac{m}{\frac{4}{3}\pi r^3}$$

where m is the mass of the planet.

The ratio of the densities is:

$$\frac{\dfrac{m_{\text{Bearth}}}{\frac{4}{3}\pi r_{\text{Bearth}}^3}}{\dfrac{m_{\text{Earth}}}{\frac{4}{3}\pi r_{\text{Earth}}^3}}$$

Canceling like terms in the numerator and denominator:

$$\frac{\dfrac{m_{\text{Bearth}}}{r_{\text{Bearth}}^3}}{\dfrac{m_{\text{Earth}}}{r_{\text{Earth}}^3}}$$

Finally, we can use a little algebra to rearrange things to:

$$\frac{m_{\text{Bearth}}}{m_{\text{Earth}}}\left(\frac{r_{\text{Earth}}}{r_{\text{Bearth}}}\right)^3$$

At this point, we need numbers. By definition, radius is half the diameter, so if the Bearth has twice the diameter of Earth, then it has twice the radius as well. Mathematically: $r_{\text{Bearth}} = 2\ r_{\text{Earth}}$. Meanwhile, Bearth's mass is eight times Earth's, so $m_{\text{Bearth}} = 8\ m_{\text{Bearth}}$. The ratio becomes:

$$\frac{8\ m_{\text{Earth}}}{m_{\text{Earth}}}\left(\frac{r_{\text{Earth}}}{2\ r_{\text{Earth}}}\right)^3 = 8\left(\frac{1}{2}\right)^3 = \frac{8}{8} = 1$$

Bearth has the same density as Earth.

b. Surface area goes like $4\pi r^2$, so the ratio of surface areas is:

$$\frac{4\ \pi r_{\text{Bearth}}{}^2}{4\ \pi r_{\text{Earth}}{}^2}$$

Canceling leaves the simpler expression:

$$\left(\frac{r_{\text{Bearth}}}{r_{\text{Earth}}}\right)^2$$

From part (a), we know that $r_{\text{Bearth}} = 2\ r_{\text{Earth}}$, so the ratio of the surface areas is:

$$\left(\frac{2\ r_{\text{Earth}}}{r_{\text{Earth}}}\right)^2 = (2)^2 = 4$$

Bearth has four times the surface area of Earth.

c. Bearth is probably made of the same material as Earth, so it probably produces about as much internal heat per mass as Earth. However, with eight times the mass but only four times the surface area, Bearth has a harder time getting rid of its heat. The interior is probably hotter and it probably has more tectonic and volcanic activity as a result.

Chapter 10. Planetary Atmospheres: Earth and the Terrestrial Worlds

This chapter continues our focus on the terrestrial worlds, this time with emphasis on their atmospheres. As in the previous chapter, we introduce important ideas and processes with concrete examples from Earth. The final sections contrast how the climate histories of Venus, Earth, and Mars have differed, and present current understanding of why these differences arose. Again, if you are more comfortable with a planet-by-planet approach than a comparative planetology approach, you could consider focusing on Sections 10.3–10.6 first and bringing in supporting material from other sections as needed.

As always, when you prepare to teach this chapter, be sure you are familiar with the relevant media resources (see the complete, section-by-section resource grid in Appendix 3 of this Instructor's Guide) and the online quizzes and other study resources available on the Mastering Astronomy Web site.

What's New in the Fourth Edition That Will Affect My Lecture Notes?

As everywhere in the book, we have added learning goals that we use as subheadings, rewritten to improve the text flow, improved art pieces, and added new illustrations. The art changes, in particular, will affect what you wish to show in lecture. We have not made any substantial content or organizational changes to this chapter.

- Sections have been consolidated and renumbered, but the same topics are covered in the same order.
- We build on coverage of the latest results from the *Spirit* and *Opportunity* Mars rovers relating to the climate history of Mars.
- Earth's atmosphere is covered in depth in this chapter, as there is no longer a separate chapter on Earth. This allows a more immediate comparison between Earth and the other planets, and a better sense of Earth's unique atmosphere. The prior coverage of the greenhouse effect prepares students for an enhanced discussion of global warming with new graphics.

Teaching Notes (By Section)

Section 10.1 Atmospheric Basics

We begin with an overview of the terrestrial atmospheres, focusing on the effects that atmospheres have on planets. This should provide motivation for why the material in this chapter is important. We also cover the important role of the greenhouse effect.

- The concept of the "no greenhouse" temperature—the temperature a planet would have in the absence of greenhouse gases—is introduced qualitatively in the text and in a more mathematical way in Mathematical Insight 10.1. This concept helps students understand the importance of the greenhouse effect.

This section focuses on how interactions between light and gases determine a planet's basic atmospheric structure. The ideas involve some physics, but once students understand them, they should be able to follow the idea of atmospheric structure easily. Students may find the discussion of why the sky is blue to be particularly interesting.

Section 10.2 Weather and Climate

After distinguishing between weather and climate, the section proceeds to delve into both. In terms of material that will be important to understanding later sections, one key topic is the short discussion of long-term climate change. This sets the stage for understanding the mechanisms of long-term atmospheric gain and loss, and for understanding the climate histories of the planets that follow in the final sections. This section also covers the processes that add or remove gas from planetary atmospheres, which are very important to understand the climate histories of the terrestrial worlds. Note that this section includes our brief discussion of the exospheres of the Moon and Mercury.

Section 10.3–10.5 The Atmospheric Histories of Mercury and the Moon, Mars, and Venus

These sections explore how these worlds have ended up so completely different from each other, setting the stage for understanding how Earth has turned out differently still.

Section 10.6 Earth's Unique Atmosphere

Parelleling the last section of the previous chapter, this section culminates with an appreciation of how unique Earth's atmosphere is—arguably even "more unique" than our geology. It is these features that make Earth such a pleasant place to live.

- For many students, the discussion of human impacts, especially global warming, will be of the greatest interest. We urge you to cover this topic even if you do not have time to cover other topics in this section or chapter. Students have gained a sense of how planets work, giving them a much better appreciation for possible human influence on the planet.
- Figure 10.36, in particular, should open students' eyes. It clearly shows that current CO_2 levels are well above anything that Earth has experienced in the past 400,000 years. Although this does not prove that anything will happen as a result, it certainly suggests that we should be aware of the human impact and its potential effects.

Answers/Discussion Points for Think About It Questions

The Think About It questions are not numbered in the book, so we list them in the order in which they appear, keyed by section number.

Section 10.1

- (p. 291) Lower. Gravity is weaker on Mars, so the same amount of air would weigh less and press down less hard, leading to lower pressure.
- (p. 293) On clear nights, the thermal radiation from the surface can escape to space, cooling the planet rapidly. If it's cloudy, the thermal radiation is reflected back down and keeps the troposphere warmer.
- (p. 293) The high-reflectivity (white) shirt absorbs less sunlight, so you'll be cooler than the person in black.

- (p. 294) Since nitrogen and oxygen make up most of the Earth's atmosphere, the greenhouse effect would be far stronger if they were greenhouse gases, and so the Earth would be much hotter.
- (p. 298) Moon: Yes, since there's not enough atmosphere to absorb solar X rays. Mars: No, because X rays are absorbed in the thermosphere.

Section 10.2

- (p. 300) It will move generally to the west.
- (p. 302) Aim to the left. Put yourself in the place of the person near the outer perimeter of the merry-go-round in Figure 10.14. To end up "straight north,"—i.e., along the line to the pole at the center—you'll have to aim a bit left.

Section 10.4

- (p. 312) Releasing CO_2 in to the Martian atmosphere would strengthen the greenhouse effect and warm the planet. However, it could not restore ancient Martian oceans (if they existed) because most of the water was probably permanently lost to space.

Section 10.5

- (p. 315) Not really. If we moved Venus to Earth's distance, it would still have its thick CO_2 atmosphere and strong greenhouse effect. Because there is no water on Venus to form oceans, there would be no way to dissolve the atmospheric CO_2 and substantially reduce the greenhouse effect.

Section 10.6

- (p. 316) Oxygen would eventually be used up in chemical reactions with the surface, and animals could not survive. This process might take millions of years.
- (p. 317) Lots of possible answers for everyday examples of feedback. For most students, the easiest examples to cite will be "people" ones: e.g., positive or negative reinforcement of behaviors.
- (p. 319) Without plate tectonics, carbonates would still form on the seafloor, but they would not be recycled into the mantle, so less CO_2 would be outgassed. This kind of outgassing is the only way out of "snowball Earth." Thus, without plate tectonics, Earth could not recover from a snowball phase.
- (p. 321) This is a subjective question that should engender lively discussion and debate.

Solutions to End-of-Chapter Problems (Chapter 10)

1. Mercury's atmosphere has an extremely low pressure and is made up of helium, oxygen, and silicon. The temperatures vary from very hot to very cold and there is no weather to speak of.

   Venus has a very dense atmosphere made mainly of carbon dioxide. It is a very, very hot planet (470°C) due to an extreme greenhouse effect and there are thick clouds made of sulfuric acid that sometimes drop an acidic rain. There is, however, little wind and no violent storms.

Earth's atmosphere, while not as dense as Venus's, is still dense by planetary standards. Earth's atmosphere is composed mainly of nitrogen and oxygen and has a mild temperature of about 15°C on the average. Earth's atmosphere also possesses copious weather, with violent storms, rain, and variable cloud coverage.

The Moon's atmosphere is akin to Mercury's, except that it's much cooler.

Finally, Mars's atmosphere is low density compared to Earth or to Venus, but still much higher than Mercury or the Moon. Like Venus, the atmosphere is mostly carbon dioxide, but it is much cooler at –50°C. And unlike Venus, Mars has winds and dust storms on the surface. Like Earth, Mars has patchy clouds, although the clouds on Mars are composed of water vapor and carbon dioxide.

2. Gas pressure comes from collisions of atoms or molecules of gas. For example, air molecules in a balloon hit the walls of the balloon, pushing it outward. Since more molecules in a given volume means more collisions, increasing the density increases the pressure of the gas. Another way to increase the pressure is to raise the temperature, since temperature is a way of measuring the kinetic energy of individual molecules. The harder the molecules hit the walls of the balloon, the more pressure they exert.

   Atmospheric pressure decreases as we climb in altitude because there is less air above us to push down on us. One bar of pressure is roughly the air pressure at sea level on Earth.

3. There is some atmosphere at the orbital altitude of the Space Station. Atmospheres do not abruptly end: They fade away with altitude. So while the air at the Space Station's altitude is extremely thin, it is present and causes drag on the Station, which means that the Station must periodically be boosted back up to keep it from falling to Earth.

4. The greenhouse effect causes the surfaces of planets to be warmer than they would be without an atmosphere. The effect occurs because the light coming down to a planet's surface is visible and generally passes through atmospheres with little absorption. However, the planet's surface, being cooler than the Sun's, emits infrared radiation that can be absorbed by certain molecules called *greenhouse gases*. The atmosphere then radiates the energy in all directions, some of it back down to the planet again, reheating the surface a bit.

5. If there were no greenhouse effect, a planet's temperature would be set by its distance from the Sun (since more distant planets get less solar energy per square meter) and the planet's reflectivity (since more reflective planets absorb less energy from the Sun). In reality, the "no greenhouse" temperatures of the planets and their actual temperatures differ. For Mercury and the Moon, the "no greenhouse" temperature falls between the day and night extreme temperatures, which we would expect for planets with too little atmosphere to cause a greenhouse effect. Mars's surface temperature is quite close to the predicted "no greenhouse" temperature, which makes sense as it has little atmosphere to cause a greenhouse effect. However, Earth's temperature is significantly higher than its predicted "no greenhouse" temperature due to its atmosphere. Venus's atmosphere is even further off the prediction, thanks to its extreme greenhouse. Both planets have substantial atmospheres so that it is not surprising that the "no greenhouse" predictions fail.

6. The lowest level of Earth's atmosphere is the troposphere. The troposphere is heated from below by the Earth's surface, which gets its energy by absorbing incoming solar radiation. The heat is then radiated back upward in the infrared,

which is absorbed by the greenhouse gases in the troposphere. Since the troposphere is heated from below, the temperature decreases with altitude. Above the troposphere is the stratosphere, which is heated by absorption of ultraviolet light. The temperature in the stratosphere first rises and then falls with altitude. The next layer up is the thermosphere, which is heated by X rays. And finally, the top layer of the atmosphere is the exosphere, which is hot from absorbing ultraviolet and X rays. The exosphere is also so thin that fast-moving molecules can escape to space completely before striking other molecules.

7. The sky is blue because air molecules scatter blue light much more than they scatter red light. So much of the blue light takes meandering paths to our eyes, and reaches us from all directions. So our sky appears blue. On the other hand, at sunrise or sunset, when we look at the Sun, we see red. This is because the Sun's light has to pass through much more atmosphere than at noon and the blue light has been largely scattered in other directions by the time that the light reaches us. The blue light that was scattered out to make our red sunset has gone to make someone else's daytime sky blue.
8. We get convection in the troposphere because it is heated from below. Convection allows this heat to escape upward when radiation is too inefficient. Radiation is inefficient in the troposphere because of the greenhouse gases, which try to stop the ultraviolet radiation from escaping. However, in the stratosphere the heating occurs everywhere, although it is concentrated near the middle of the layer. With a thinner atmosphere in that layer, there is much less to keep the infrared from escaping outward, so convection is not needed to help the heat escape.
9. Ozone is a molecular form of oxygen with three oxygen atoms rather than the usual two. Ozone happens to be very good at absorbing ultraviolet light. This property allows it to protect us on the Earth's surface and heat the stratosphere, where most of the ozone is concentrated. Since Venus and Mars lack ozone, they do not possess stratospheres.
10. A magnetosphere is a protective bubble created by a planet's magnetic field. It contains trapped ions and deflects most of the solar wind around the planet. However, some of the particles can make it to the planet's surface near the poles. Where these high-energy particles enter our atmosphere, they usually collide with air molecules. This causes the molecules to gain energy, which they radiate away, giving us beautiful auroras.
11. *Weather* describes the atmospheric conditions at a given time in a given place. This includes air pressure, cloud cover, temperature, winds, and precipitation. Weather changes rapidly, over hours or days. *Climate* is a long-term average of the weather and changes much more slowly, usually over decades or longer.
12. Earth has alternating bands of winds that predominantly blow toward the west or toward the east. These bands are due to the circulation cells in which the hot air from the equator rises and then spreads toward the colder regions of the Poles. However, because of Earth's rapid rotation, the airflow is deflected by the Coriolis effect, keeping the air from making it to the poles. So rather than get two large cells (one in each hemisphere), we get six smaller ones. In each cell, the Coriolis effect deflects the surface winds, which try to get heat to either the Poles or the equator, eastward or westward. The deflection causes the surface winds to tend to flow eastward or westward in a given band.

13. Clouds on Earth are made of tiny droplets of water or ice flakes, although on other planets they can be made of droplets or ice flakes of other compounds (such as carbon dioxide on Mars). If the ice flakes or water droplets grow too large to be held up by the convection currents, they fall to the ground as precipitation.
14. There are four factors that can lead to climate change. The first is due to the Sun's getting brighter as it ages, causing more solar energy to be delivered to the planets. This tends to increase temperatures on the planets.

    The second cause of climate change is changes in axis tilt. If the tilt changes, the seasons are affected: A smaller tilt leads to weaker seasons while a larger tilt causes stronger seasons. This means that different parts of the planet would tend to get warmer or cooler if we average the tilt over a year. (For example, with no tilt and no seasons, the polar regions will never really get much sunlight and will therefore be even colder than they are now.)

    A third cause of climate change is changes in the reflectivity of the planet. Darker planets absorb more of the Sun's light and therefore tend to be hotter than similar, brighter planets. Changes in reflectivity can be caused by changes in the cloud cover, dust particles, or changes in the surface like deforestation or paving a large area with asphalt.

    The final cause of climate change is changes in the abundance of greenhouse gases in the atmosphere. Adding more greenhouse gases leads to a stronger greenhouse effect and more warming. Conversely, decreasing the greenhouse gas content decreases the temperature.
15. Atmospheres can gain gases in three ways. The first is outgassing, the emission of gases from the interior of a planet into the atmosphere through volcanic activity. The second process that adds gases to the atmosphere is evaporation or sublimation, the process where the solid or liquid states change to gases. Finally, on planets with thin atmospheres, micrometeorites, high-energy particles, and high-energy photons can strike the planet's surface and vaporize rock.

    Planets can lose atmospheric gases five ways. The first is thermal escape: Particles gain enough energy near the top of the atmosphere to escape the planet's gravity. Second, particles can condense back into a liquid or solid state onto the surface, leaving the atmosphere. Third, atmospheric gases can have chemical reactions with surface materials and leave the atmosphere. Fourth, particles can be stripped away from the planet's upper atmosphere by collisions with solar wind particles. And, finally, planets can lose atmospheric gases through large impacts blasting gases into space.

    The factors that determine what gases will be lost through thermal escape are the planet's escape speed (a higher escape speed means that molecules need more energy to escape), the temperature (hotter gases have more energy per molecule, so the molecules are more likely to be able to escape), and the mass of the molecule (heavier molecules move more slowly at the same temperature and so don't escape as easily).
16. The Moon and Mercury have little atmospheric gas because they are small worlds with low escape speeds. It is therefore easy for most molecules to gain enough energy to escape these worlds. But in their polar regions where some craters may have areas that are permanently in shadow, the surfaces may be cold enough that water could stay frozen and not escape.

17. Mars has seasons for two reasons. The first is the tilt of its axis, which causes seasons there for the same reason that Earth has seasons. The second is that its orbit is much more eccentric than Earth's, meaning that the distance from the Sun *is* important on Mars. As things work out, the southern hemisphere has more extreme seasons: Summers are shorter and hotter than in the northern hemisphere and winters are longer and colder.
18. Mars may have lost atmospheric gases through solar wind stripping. Early in Mars's life this effect was probably not important, thanks to Mars's magnetic field. However, as the planet cooled and convection stopped, the magnetic field would have disappeared, leaving the planet vulnerable to the solar wind particles. With little or no outgassing to replenish the atmosphere (because there would be no internal convection to drive volcanism), the atmosphere would have disappeared. If this scenario is correct, it is the small size of Mars that is to blame for the loss of atmosphere. Had Mars been a larger planet, it would have stayed hotter longer and convection would have continued.
19. A runaway greenhouse occurs when a planet gets so hot that it cannot keep liquid water stable on the surface. As the planet heats up, more water is vaporized into the air. A potent greenhouse gas, the water vapor raises the temperature further. This evaporates more water, causing the cycle to continue. This did not occur on Earth, but did happen to Venus, because Earth is farther from the Sun and therefore was not hot enough to start this cycle.
20. Earth's atmosphere is different from the atmospheres of the other planets in several ways. One is the composition: Earth's atmosphere is composed mainly of nitrogen and oxygen, and there is relatively little carbon dioxide. This composition keeps our greenhouse effect much milder than Venus's and prevents our planet from being too hot for life. Earth is also unique in that it has an ozone layer, which protects us from the Sun's damaging ultraviolet light. Earth has kept its water, which is vital for the chemistry of life.
21. The carbon dioxide cycle is the cycle in which carbon dioxide is dissolved from the atmosphere into rain. This rain is mildly acidic and dissolves some of the rocks onto which it falls. The water (with dissolved rocks) flows to the oceans, where it forms carbonate rocks. These rocks are subducted into the Earth due to plate tectonics. Eventually, volcanism releases the carbon dioxide back into the atmosphere, completing the cycle. Because this cycle keeps the level of carbon dioxide in Earth's atmosphere stable, it tends to regulate our temperature and keep Earth hospitable for life rather than becoming too hot or too cold.
22. The debate over global warming centers around the question: Are humans causing the Earth to warm up? Over the past few decades, it has become clear that the Earth's temperature is rising. We also know that the amount of carbon dioxide in our atmosphere is increasing. Since humans burn fossil fuels for energy, a process that releases carbon dioxide, it is considered likely by many people that we are responsible for the rising levels of carbon dioxide. Since carbon dioxide is a greenhouse gas, we might therefore be causing the Earth to warm. However, since the climate is a complex system with many feedbacks and drivers, it is difficult to prove that we are the cause of the warming.

    If the Earth is really warming, the effect would be devastating to humans. Sea levels would rise significantly, flooding coastal areas. Global warming would also alter weather patterns, which could make some currently food-producing areas deserts.

We might also see intensified storms due to global warming. In addition, there are secondary effects such as changes in ecologies or ocean currents. These are difficult to predict, unfortunately.

23. *If Earth's atmosphere did not contain molecular nitrogen, X rays from the Sun would reach the surface.* False. Almost any atmospheric ingredient will stop X rays.
24. *If the molecular oxygen content of Earth's atmosphere increases, it will cause our planet to warm up.* False. Oxygen is not a greenhouse gas.
25. *Earth's oceans must have formed at a time when no greenhouse effect operated on Earth.* False. Oceans always put water vapor, which is a very strong greenhouse gas, into the atmosphere.
26. *In the distant past, when Mars had a thicker atmosphere, it also had a stratosphere.* False. Not all atmospheric ingredients can absorb UV as ozone does on Earth.
27. *If Earth rotated faster, hurricanes would be more common and more severe.* True. Weather gets much of its energy from Earth's rotation.
28. *Mars would still have seasons even if its orbit around the Sun were perfectly circular rather than elliptical.* This statement is true. Mars's tilt is comparable to Earth's, so seasons would still be important. A circular orbit would have the effect of making seasons similar in the two hemispheres.
29. *Mars once may have been warmer than it is today, but it could never have been warmer than Earth because it is farther from the Sun than is Earth.* This statement is incorrect. With enough greenhouse gasses, or even reflectivity differences, it's possible that Mars could have been warmer than Earth. Temperature depends on more than distance.
30. *If the solar wind were much stronger, Mercury might develop a carbon dioxide atmosphere.* False. CO_2 comes from volcanic outgassing, not bombardment of a rocky surface.
31. *If Earth had as much carbon dioxide in its atmosphere as Venus, our planet would be too hot for liquid water to exist on the surface.* True.
32. *A planet in another solar system has an Earth-like atmosphere with plentiful oxygen but no life of any kind.* This would be very hard to explain, since oxygen is not stable in our atmosphere without continuous resupply by living things.
33. a; 34. a; 35. c; 36. b; 37. a; 38. c; 39. b; 40. c; 41. c; 42. c.
43. a. Venus's cloudiness causes it to reflect so much sunlight that it actually absorbs less than the Earth.
    b. In the absence of clouds, Venus's dark surface would lead to the absorption of much more sunlight and higher temperatures. Without performing a calculation, it's debatable whether Venus would be warmer with the low reflectivity or the greenhouse effect. (Acalculation would show that the low reflectivity would not warm the planet as much as the current greenhouse effect.)
    c. The clouds contain sulfuric acid probably derived from volcanic outgassing. If outgassing ceased, the amount of sulfur compounds—and therefore clouds—would probably decrease.
44. a. With no greenhouse gases, the troposphere would not be warmer at the bottom.
    b. Without UV light, no stratosphere would form.
    c. With greater X-ray output, the thermosphere and exosphere would be warmer.
45. No. Mercury would lose its new atmosphere for exactly the same reasons it lost whatever it originally outgassed and whatever is generated nowadays by

bombardment: Its low gravity makes it hard to hold on to an atmosphere, and its closeness to the Sun makes thermal escape rapid.

46. If Venus rotated faster, it would have (1) more erosion from faster winds, (2) circulation patterns like Earth's, and (3) a magnetic field (probably) that could keep the solar wind at a greater distance, so (4) atmospheric escape through nonthermal processes would probably be reduced.
47. In the daytime, the warmer air over land rises and draws air from the sea to replace it. This leads to winds from sea to shore. At night, the air over the sea is warmer, so the circulation is reversed. The circulation pattern resembles Hadley circulation, with the warmer region (sea or shore) corresponding to Earth's equatorial region.
48. Answers will depend on the students' choice of processes. For the example given (bombardment as a source process): Bombardment is important only on planets for which other processes are unimportant, so it mainly occurs on small planets for which outgassing is negligible. It can be more important for planets closer to the Sun, where the solar wind is stronger. Bombardment might depend on the planet's composition or rotation rate, but this is beyond the scope of this text.
49. a. The bottle contracts quite noticeably. Microscopically, the molecules inside are slowed as the air cools. They collide with the walls less frequently and with less force. Molecules striking the bottle wall from the inside are no longer balancing those striking from the outside, so the bottle contracts.
    b. The accumulation of frost is condensation, and the disappearance of ice cubes is sublimation. Whether either of these occurs in your home depends on your local humidity and the operation of your freezer. These processes are important on Mars at the polar caps.
50. Earth and Venus were both large enough for substantial outgassing of water, CO_2, and N_2. But Earth formed far enough from the Sun that the water remained liquid and formed oceans, which then drew CO_2 from the atmosphere to make carbonate rocks. Venus, too close to the Sun, left H_2O in the atmosphere, so CO_2 remained there as well, causing the runaway greenhouse.
51. Sample solution: If Earth had been closer to the Sun, the warmer temperatures might have forced more water vapor into the atmosphere. This would have increased the greenhouse effect, leading to further warming, evaporation of the oceans, and possible escape of all the water to space (as with Venus). Without liquid water on the surface, life as we know it would not have arisen on Earth.
52. Higher temperatures increase the formation rate of carbonate rocks, reducing atmospheric CO_2 and the greenhouse effect. This is an example of negative feedback—a stabilizing effect.
53. Earth would not become Mars. Its large size, warm interior, and ensuing strong magnetic field would retain the atmospheric ingredients. It's possible that the feedback mechanisms of the CO_2 cycle would increase the atmospheric CO_2 enough to maintain temperate conditions. Even if the oceans froze, ongoing plate tectonics would continue to release CO_2, perhaps enough to remelt the oceans. Earth might well go through freezing cycles, reach a cooler equilibrium, or possibly become too frozen for the CO_2 cycle to melt.
54. Essay question—answers will vary.

55. Since we are told that there are 10,000 kilograms of air pushing down on every square meter of the Earth, we can convert this to the total mass of the atmosphere by finding Earth's surface area. Surface area is proportional to $4\pi r^2$ and the radius of Earth (given in Appendix E) is 6,378 kilometers. Converting this to meters:

$$6{,}378 \cancel{\text{km}} \times \frac{1000 \text{ m}}{1 \cancel{\text{km}}} = 6.378 \times 10^6 \text{ m}$$

So we can find the surface area:

$$\begin{aligned} \text{surface area} &= 4\pi r^2 \\ &= 4\pi (6.378 \times 10^6 \text{ m})^2 \\ &= 5.11 \times 10^{14} \text{ m}^2 \end{aligned}$$

So to get the total mass of the air, we multiply the mass per square meter by the surface area:

$$\begin{aligned} \text{mass} &= (10{,}000 \text{ kg/m}^2) \times (5.11 \times 10^{14} \text{ m}^2) \\ &= 5.11 \times 10^{18} \text{ kg} \end{aligned}$$

The mass of Earth's atmosphere is about 5.11×10^{18} kg.

56. To solve this problem, we reverse the logic used in Problem 55. First, we will find the total mass of Earth's oceans. We are told that they cover 75% of the planet's surface. We can find the surface area of the Earth with the formula:

$$\text{surface area} = 4\pi r^2$$

Earth's radius (given in Appendix E) is 6,378 kilometers. Converting this to meters:

$$6{,}378 \cancel{\text{km}} \times \frac{1000 \text{ m}}{1 \cancel{\text{km}}} = 6.378 \times 10^6 \text{ m}$$

So we can find the surface area:

$$\begin{aligned} \text{surface area} &= 4\pi r^2 \\ &= 4\pi (6.378 \times 10^6 \text{ m})^2 \\ &= 5.11 \times 10^{14} \text{ m}^2 \end{aligned}$$

Since the oceans cover only 75% of the Earth's surface, we multiply this by 0.75 to get the surface area that they cover:

$$5.11 \times 10^{14} \text{ m}^2 \times 0.75 = 3.83 \times 10^{14} \text{ m}^2$$

To get the volume of the oceans, we need the average depth. Conveniently, this is provided for us as 3.5 kilometers. Inconveniently, we have to convert this to meters:

$$3.5 \cancel{\text{km}} \times \frac{1000 \text{ m}}{1 \cancel{\text{km}}} = 3{,}500 \text{ m}$$

The volume is the depth times the surface area (we are ignoring Earth's curvature, which is OK since the oceans' depth is much less than Earth's radius). So we have:

$$\begin{aligned} \text{volume} &= (3{,}500 \text{ m}) \times (3.83 \times 10^{14} \text{ m}^2) \\ &= 1.35 \times 10^{18} \text{ m}^3 \end{aligned}$$

Now we almost have the mass of the Earth's oceans. To get this, we just need the density, given as 1,000 kg/m^3. So:

$$\begin{aligned} \text{mass of oceans} &= (1{,}000 \text{ kg/m}^3) \times (1.35 \times 10^{18} \text{ m}^3) \\ &= 1.35 \times 10^{21} \text{ m}^3 \end{aligned}$$

The pressure is spread over the Earth's total surface area, so we divide into this (calculated above) to give the mass over each square meter of Earth:

$$\begin{aligned} \text{mass per area} &= \frac{1.35 \times 10^{21} \text{ kg}}{5.11 \times 10^{14} \text{ m}^2} \\ &= 2.63 \times 10^{6} \text{ kg/m}^2 \end{aligned}$$

In Problem 55 we were told that Earth's atmospheric pressure (about 1 bar) is 10,000 kg/m^2. So we can convert to Earth atmosphere pressures:

$$2.63 \times 10^{6} \text{ kg/m}^2 \times \frac{1 \text{ Earth atmosphere pressure}}{10{,}000 \text{ kg/m}^2} = 263 \text{ Earth atmosphere pressures}$$

This early steam atmosphere would have exerted 263 times the pressure of Earth's present atmosphere.

57. From Mathematical Insight 10.1, we know that the "no greenhouse" temperature formula is:

$$T = 280 \text{ K} \sqrt[4]{\frac{1 - \text{reflectivity}}{d^2}}$$

where the distance is in AU. For a totally black planet, the reflectivity is 0. We are told to assume a distance of 1 AU, so we can calculate the maximum temperature:

$$\begin{aligned} T &= 280 \text{ K} \sqrt[4]{\frac{1-0}{(1 \text{ AU})^2}} \\ &= 280 \text{ K} \end{aligned}$$

So a totally black planet at 1 AU would have a surface temperature of 280 K. We can convert this to degrees Celsius by subtracting 273°C as described in the chapter to get a temperature of 7°C.

On the other hand, if the planet were totally white, the reflectivity would be 1. In this case, we would have:

$$\begin{aligned} T &= 280 \text{ K} \sqrt[4]{\frac{1-1}{(1 \text{ AU})^2}} \\ &= 0 \text{ K} \end{aligned}$$

Converting to degrees Celsius by subtracting 273°C, we get –273°C. So a white planet would be at absolute zero temperature, or –273°C.

Now, to keep a planet at exactly freezing we require the temperature to be exactly 273 K (the freezing point of water). In other words:

$$273\text{ K} = 280\text{ K}\sqrt[4]{\frac{1-\text{reflectivity}}{(1\text{ AU})^2}}$$

We can divide both sides by 280 K and then take the fourth power to remove the root:

$$\left(\frac{273\text{ K}}{280\text{ K}}\right)^4 = 1-\text{reflectivity}$$

A little more algebra lets us solve for the reflectivity and then calculate it:

$$\text{reflectivity} = 1-\left(\frac{273\text{ K}}{280\text{ K}}\right)^4$$
$$= 0.096$$

So a reflectivity of 0.096 would be required to keep the surface temperature at exactly freezing. However, this is well below Earth's actual reflectivity, implying that Earth should be below freezing were it not for our atmosphere and the greenhouse effect. (Indeed, our "no greenhouse" temperature is below freezing.)

58. From Mathematical Insight 10.1, we know that the "no greenhouse" temperature formula is:

$$T = 280\text{ K}\sqrt[4]{\frac{1-\text{reflectivity}}{d^2}}$$

where the distance is in AU. From Appendix E, we know that Venus is at 0.723 AU from the Sun and we are told to assume the same reflectivity as the Earth. The reflectivity of Earth is given in Table 10.2 as 0.29, so we can calculate the "no greenhouse" temperature:

$$T = 280\text{ K}\sqrt[4]{\frac{1-0.29}{(0.723)^2}}$$
$$= 302\text{ K}$$

We can convert this to degrees Celsius by subtracting 273°C to get 29°C. So in the absence of clouds, if Venus had the same reflectivity as Earth, Venus's "no greenhouse" temperature would be 29°C.

If Venus had a greenhouse effect and if it changed the temperature by the same amount as it does today (510°C, given in Table 10.2), we could find the temperature by adding the greenhouse effect's contribution to the "no greenhouse" temperature: 29°C + 510°C = 539°C. So Venus would be a very hot 539°C if its clouds did not reflect most of the sunlight that hits the planet.

59. From Mathematical Insight 10.1, we know that the "no greenhouse" temperature formula is:

$$T = 280\text{ K}\sqrt[4]{\frac{1-\text{reflectivity}}{d^2}}$$

where the distance is in AU. Table 10.2 gives Mars's reflectivity as 0.16, and we are told that at its closest to the Sun, Mars is at 1.38 AU and at its farthest is at 1.66 AU. So we can find the temperatures of Mars at these two points in its orbit:

$$T_{\text{closest}} = 280\text{ K}\sqrt[4]{\frac{1-0.16}{(1.38\text{ AU})^2}} = 228\text{ K}$$

$$T_{\text{farthest}} = 280\text{ K}\sqrt[4]{\frac{1-0.16}{(1.66\text{ AU})^2}} = 208\text{ K}$$

We can convert these to degrees Celsius by subtracting 273°C to find that the temperature at Mars's closest approach to the Sun is –45°C and that when Mars is farthest from the Sun, its temperature is –65°C. So Mars's "no greenhouse" temperature varies by 20°C over the course of a Martian year.

60. a. In Mathematical Insight 5.3 we learn that we can find the escape speed via:

$$v_{\text{escape}} = \sqrt{\frac{2\,GM}{r}}$$

where G is Newton's gravitational constant, M is the planet's mass, and r is the distance from the planet's center. For Venus, $M = 4.87 \times 10^{24}$ kg according to Appendix E. We are told to calculate the escape speed for the exosphere, 200 kilometers above the planet's surface. Appendix E tells us that the planet's radius is 6,051 kilometers, so $r = 6{,}251$ km for our purposes. Now we need to convert this to meters:

$$6{,}251\ \cancel{\text{km}} \times \frac{1000\text{ m}}{1\ \cancel{\text{km}}} = 6.251 \times 10^6\text{ m}$$

Finally, knowing that $G = 6.67 \times 10^{-11}\,\dfrac{\text{m}^3}{\text{kg} \times \text{sec}^2}$, we can find the escape speed:

$$v_{\text{escape}} = \sqrt{\frac{2\left(6.67 \times 10^{-11}\,\dfrac{\text{m}^3}{\text{kg} \times \text{s}^2}\right)(4.87 \times 10^{24}\text{ kg})}{(6.251 \times 10^6\text{ m})}}$$
$$= 10{,}200\text{ m/sec}$$

So the escape speed from Venus's exosphere is 10,200 m/s.

b. From Mathematical Insight 10.2 we know how to calculate the thermal speed of molecules in the atmosphere:

$$v_{\text{thermal}} = \sqrt{\frac{2\,kT}{m}}$$

where k is Boltzmann's constant (1.38×10^{-23} joules/Kelvin), T is the temperature, and m is the mass of the molecule. For this situation, we are told that hydrogen has a mass of 1.67×10^{-27} kg and that the temperature is 350 K. So we find the thermal speed:

$$v_{\text{thermal}} = \sqrt{\frac{2(1.38 \times 10^{-23} \text{ joules/Kelvin})(350 \text{ K})}{1.67 \times 10^{-27} \text{ kg}}}$$

$$= 2{,}400 \text{ m/sec}$$

Deuterium is twice the mass of hydrogen, so the mass is $2 \times (1.67 \times 10^{-27} \text{ kg}) = 3.34 \times 10^{-27}$ kg. We can find the thermal speed for deuterium under the same conditions:

$$v_{\text{thermal}} = \sqrt{\frac{2(1.38 \times 10^{-23} \text{ joules/Kelvin})(350 \text{ K})}{3.34 \times 10^{-27} \text{ kg}}}$$

$$= 1{,}700 \text{ m/sec}$$

So ordinary hydrogen atoms move in Venus's exosphere at a speed of about 2,400 m/s while the heavier deuterium moves at about 1,700 m/s.

c. The thermal speed of hydrogen in Venus's exosphere is about one-fourth the escape speed from that location. We are told that if the thermal speed is more than about 20% the escape speed, most of the atoms will have escaped over the age of the solar system. Since the hydrogen moves faster than this, Venus should have lost most of its hydrogen (and therefore water) by now. Note that the deuterium does not exceed one-fifth of the escape speed, so it might still be found in Venus's atmosphere.

61. a. In Mathematical Insight 5.3 we learn that we can find the escape speed via:

$$v_{\text{escape}} = \sqrt{\frac{2\,GM}{r}}$$

where G is Newton's gravitational constant, M is the planet's mass, and r is the distance from the planet's center. For Jupiter, $M = 1.90 \times 10^{27}$ kg according to Appendix E. We are told to calculate the escape speed for the exosphere, 1,000 kilometers above the planet's surface. Appendix E tells us that the planet's radius is 71,492 kilometers, so $r = 72{,}492$ km for our purposes. Now we need to convert this to meters:

$$72{,}492 \cancel{\text{km}} \times \frac{1000 \text{ m}}{1 \cancel{\text{km}}} = 7.2492 \times 10^{7} \text{ m}$$

Finally, knowing that $G = 6.67 \times 10^{-11} \dfrac{\text{m}^3}{\text{kg} \times \text{sec}^2}$, we can find the escape speed:

$$v_{\text{escape}} = \sqrt{\frac{2\left(6.67 \times 10^{-11} \dfrac{\text{m}^3}{\text{kg} \times \text{s}^2}\right)(1.90 \times 10^{27} \text{ kg})}{(7.2492 \times 10^{7} \text{ m})}}$$

$$= 55{,}900 \text{ m/sec}$$

So the escape speed from Jupiter's exosphere is 55,900 m/s.

b. From Mathematical Insight 10.2 we know how to calculate the thermal speed of molecules in the atmosphere:

$$v_{\text{thermal}} = \sqrt{\frac{2\,kT}{m}}$$

where k is Boltzmann's constant (1.38×10^{-23} joules/Kelvin), T is the temperature, and m is the mass of the molecule. For this situation, we are told that hydrogen has a mass of 1.67×10^{-27} kg and that the temperature is 800 K. So we find the thermal speed:

$$v_{\text{thermal}} = \sqrt{\frac{2(1.38 \times 10^{-23} \text{ joules/Kelvin})\ (350 \text{ K})}{1.67 \times 10^{-27} \text{ kg}}}$$
$$= 3{,}600 \text{ m/sec}$$

So ordinary hydrogen atoms move in Jupiter's exosphere at a speed of about 3,600 m/s.

c. The thermal speed of hydrogen in Jupiter's exosphere is high, but the escape speed is much, much larger (more than 10 times larger, in fact). Since we are told that most of the atoms will have been lost over the age of the solar system if the thermal speed is more than about 20% of the escape speed and since the thermal speed of hydrogen in this case is well below the escape speed, Jupiter should not have lost most of its hydrogen, which is comforting to conclude, since observations show that the planet is made mostly of hydrogen.

62. a. First, we need to compute what fraction of the Earth's atmosphere is carbon dioxide. From Figure 10.35, carbon dioxide appears to currently make up about 370 ppm of Earth's atmosphere. This is parts per million and we want parts per hundred, often called *percent*. There are 10,000 ppm in 1%, so we can convert:

$$370 \cancel{\text{ppm}} \times \frac{1\%}{10{,}000 \cancel{\text{ppm}}} = 0.037\%$$

Gas	Nitrogen	Oxygen	Argon	Water	Carbon Dioxide
Abundance	77%	21%	1%	1%	0.037%
Greenhouse Strength	X	X	X	1	2
Importance	X	Needed for animal life. Necessary for ozone creation.	X	Needed for life. Warms planet by greenhouse effect. Controls temperature with the carbon dioxide cycle and with clouds (which change the reflectivity)	Helps control the planet's temperature with the carbon dioxide level. Needed for plant life.

b. We are told in the chapter that single atoms (like argon) and two-atom molecules (especially ones with two of the same atom) are poor greenhouse gases. So that leaves water and carbon dioxide. Water is the most potent greenhouse gas in the atmosphere, so it is first. Carbon dioxide is therefore number 2 on our table.
   Note in our table that the least-abundant gases do the most for the greenhouse effect. This suggests that the amount of the gas is not as important as its infrared-absorbing properties are.
c. Listed on table.

Chapter 11. Jovian Planet Systems

This chapter covers the four jovian planets, their satellites, and their rings. It is possible to cover this broad subject range in a single chapter because we emphasize the general properties of these objects, highlighting their differences in cases where these differences have meaningful interpretations.

As always, when you prepare to teach this chapter, be sure you are familiar with the relevant media resources (see the complete, section-by-section resource grid in Appendix 3 of this Instructor's Guide) and the online quizzes and other study resources available on the Mastering Astronomy Web site.

What's New in the Fourth Edition That Will Affect My Lecture Notes?

As everywhere in the book, we have added learning goals that we use as subheadings, rewritten to improve the text flow, improved art pieces, and added new illustrations. The art changes, in particular, will affect what you wish to show in lecture.

- We have streamlined the organization of material into just three sections: 8.1 on the jovian planets themselves, 8.2 on the jovian moons, and 8.3 on the jovian rings.
- Jargon reduction: We no longer emphasize the terms *zones* and *belts* for Jupiter's atmosphere, instead just focusing on the fact that we see alternating bands of color and what causes them.
- All the latest *Cassini/Huygens* results have been included, including text and image updates to almost all the Saturn satellites, and dramatically better images of the rings. The discovery that Enceladus is a geologically active world should surprise and interest students.

Teaching Notes (By Section)

Section 11.1 A Different Kind of Planet

The section opens with a discussion of how we learned that the jovian planets are so different from Earth and reminds students of those differences. A few slides will help drive this home. Many students fail to appreciate how fundamentally different the jovian planets are, and still try to imagine craters and a rocky surface below the clouds.

As we did in explaining geological processes, we turn first to the interiors of the planets. The subsection Inside Jupiter is intended to help build intuition with respect to this nonintuitive phenomenon before moving on to interior comparisons.
This section builds directly on the concepts covered in Chapter 10 on terrestrial planet atmospheres. As in the previous chapter, we begin with the case of Jupiter before moving on to a comparative study of the jovian atmospheres. If you are short on time, you may wish to gloss over the details in this section so that you can move on to the jovian moons and rings.

- We suggest emphasizing that the same atmospheric processes govern structure and circulation on both terrestrial and jovian planets.
- Many students are interested in the colors of the jovian planets, which are discussed in this section.
- Students might wonder whether jovian planets undergo atmospheric evolution through the same processes discussed in Chapter 10. The short answer is "no" (because nothing can escape their strong gravity), but the question can make for a useful discussion.

This section also discusses jovian planet magnetospheres; it may be skipped if time limitations are a factor. However, the relationship between interiors, magnetic fields, and charged particles is a recurring theme throughout the book.

Section 11.2 A Wealth of Worlds: Satellites of Ice and Rock

This section covers the diverse subject of icy satellites. It opens with some new physical ideas necessary to an understanding of the behavior and geology of the satellites: tidal heating and ice geology. Then the major satellites are discussed from the jovian system outward. We do not cover all satellites, sticking instead to the most important or most curious. In many cases the geology of the satellites is not understood, and we do not consider it worthwhile to have students memorize facts for which there is no explanation. In addition to exploring the most interesting cases of Io, Europa, Titan, Miranda, and Triton, students should mainly carry away the idea that even frigid, icy bodies undergo a surprising amount of geological activity.

Section 11.3 Jovian Planet Rings

This section covers planetary rings. The two key points to emphasize are (1) how rings work, including tidal forces and orbital resonances, and (2) origin of the rings. Note our emphasis on the idea of resonances in helping to explain ring systems, and on the surprising result that Saturn's rings are not as old as the solar system.

- The rings are very compelling visually, and a slide show may provide additional motivation for students to learn these concepts.
- Resonances are a recurrent theme in the outer solar system and will come up again in Chapters 12 and 13.

Answers/Discussion Points for Think About It Questions

The Think About It questions are not numbered in the book, so we list them in the order in which they appear, keyed by section number.

Section 11.1

- (p. 331) This question asks students to think about the nature of planets and gravity. If Saturn were a solid object with a density less than that of water, then it would float. But it is not solid, and the presence of a downward gravity would cause Saturn to spread out over the ocean like a popped water balloon. At that point, dense materials would sink and gases would rise upward into the atmosphere of the gigantic planet.

- (p. 335) Jupiter has plenty of explosives but no free oxygen with which to burn them.

Section 11.2

- (p. 347) Like all bodies in the solar system, Titan should have been heavily cratered when the solar system was young. So the question is whether the craters have been erased. In Titan's case, erosion has certainly been an important process. The search for volcanic and tectonic structures has been suggestive—but not definitive. Proof may come after the book has been published, so it's a good idea to stay up-to-date through the *Cassini* mission Web site.

Section 11.3

- (p. 351) Ring particles at the inner edge travel faster, just as planets closer to the Sun travel faster.

Solutions to End-of-Chapter Problems (Chapter 11)

1. The jovian planets were all formed from similar planetesimals made of hydrogen compounds mixed with rock and metals. But Jupiter and Saturn captured more hydrogen and helium than Uranus and Neptune, which is why Jupiter and Saturn are larger and richer in those gases, while Uranus and Neptune are composed of a much higher fraction of hydrogen compounds.
2. Jupiter is denser than Saturn due to the "pillow effect." As extra mass is added to the planets, the lowest layers of gases compress more under the added weight. So as we add gas to the jovian planets, they can grow denser as Figure 11.2 indicates. Figure 11.2 also shows that if a planet were to grow more than about three times more massive than Jupiter, adding more gas would actually result in making the planet shrink in radius in order to accommodate the extra weight.
3. Jupiter has a gaseous envelope on the outside where the pressure is low. However, as we travel downward into Jupiter, the pressure increases and the hydrogen atmospheres are squeezed closer together. Eventually, we get far enough down that the hydrogen becomes a liquid. Below the liquid hydrogen, the pressure increases so much that the hydrogen behaves like a metal, creating the metallic hydrogen layer. Finally, deep in the interior lies a relatively small core of hydrogen compounds, silicates, and metals.

   Other jovian planets have similar structures, except for the metallic hydrogen layer. Uranus and Neptune are too small to generate the pressures needed for this layer, while Saturn's layer is smaller than Jupiter's because Saturn does not generate the pressure needed for metallic hydrogen in as much of its volume as Jupiter does.
4. Jupiter seems to generate its extra internal heat by contracting slowly, suggesting that it is not totally finished forming yet. Saturn probably generates its heat with a helium rain in its atmosphere. This is just the heat of differentiation, really. Neptune, like Jupiter, might still be contracting, although we are not sure how this can happen. Uranus does not generate extra internal heat.

5. Jupiter has a hot thermosphere that is low-density and is heated by X rays. Below this, Jupiter has a stratosphere that is heated by molecules absorbing ultraviolet light, although unlike Earth, the absorbers are *not* ozone. Finally, the lowest layer is the troposphere, where the temperature rises as we go farther down due to greenhouse gases trapping heat. Atmospheric structure on the other planets is similar, although the others are cooler.

   Jupiter has several cloud layers, each one due to different gases condensing. The lowest is water, followed by ammonium hydrosulfide, and ammonia at the top. Each cloud layer condenses around the lowest altitude where the temperature is low enough for it to do so, which is why there are distinct layers. Saturn has the same cloud layers. Uranus and Neptune might also possess these layers, but we cannot see them because they would be under the methane clouds that we need to see below in order to do so.

6. The different cloud layers are made of different chemicals and therefore apparently reflect light differently. This explains why Jupiter has different-color clouds in its atmosphere. Saturn's clouds also vary in color, but because they are deeper in the atmosphere and under more haze, the differences are more subdued.

   In contrast to colorful Jupiter and subdued Saturn, Uranus and Neptune are both blue. This is because the methane in these planets' atmospheres absorbs red light. So any light that enters the atmospheres, hits the clouds, and reflects back to us will lose most of its red light along the way, leaving us with blue-looking planets.

7. Jupiter has bands of alternating winds like the Earth, but Jupiter has many more bands because of the planet's rapid rotation and great size. Jupiter also features powerful storms. These include the famous Great Red Spot, a spinning storm system that is rather like Earth's hurricanes, except that it spins in the wrong direction. Unlike terrestrial hurricanes, the Great Red Spot has lasted for several centuries.

   Saturn and Neptune also feature similar patterns. Uranus, though, appeared bland when it was visited by *Voyager 2* in 1986, with virtually no banding or visible clouds. However, recent observations show Uranus's atmosphere coming alive again, perhaps as a result of seasonal changes.

8. Jupiter has a strong magnetic field because it has a fast rotation and a very large volume of metallic hydrogen by which to generate its field. Jupiter's magnetosphere is similar to Earth's in many ways, but there are some very important differences. One of these is that the charged particles in Jupiter's field come not from the solar wind, but from the volcanic moon Io. Because of this source embedded inside the magnetosphere, there is a donut-shaped region of charged particles in the magnetosphere called the *Io plasma torus*, something Earth does not have. The magnetosphere also interacts with other moons, bombarding them with high-energy particles and creating thin atmospheres.

   The other jovian planets have magnetospheres as well, although their magnetic fields are much weaker. None of the planets has any moon quite like Io, so they do not have as many charged particles in their magnetospheres as Jupiter does. Another difference between their magnetospheres is that the more distant planets feel less pressure from the solar wind. These planets can therefore have larger magnetospheric bubbles than they would if they were at Jupiter's distance from the Sun. A final interesting difference between the magnetospheres is that neither the magnetic field of Uranus nor the magnetic field of Neptune is closely aligned with the planets' spin axes, differing by 60° and 46°, respectively.

9. We categorize moons by their sizes into three groups: small, medium, and large. The large moons are over 1,500 kilometers in diameter, the medium moons are between 300 and 1,500 kilometers in diameter, and the small moons are below 300 kilometers in diameter. Most of the large and medium moons were probably formed in disks around the planets while the small moons were probably mostly captured asteroids or comets.
10. The key feature of Io is that it is extremely volcanically active due to tidal heating. Europa, Ganymede, and Callisto all seem to have subsurface oceans. Europa's surface is young, with many features indicating geological activity. Ganymede's surface is young in places and much older in others, which also indicates that tectonic and volcanic processes may have recently been active. Callisto, by comparison, looks heavily cratered and appears to be geologically dead.

    The inner three of these moons (Io, Europa, and Ganymede) get their heat from tidal heating. This comes about because their orbits are not quite circular, which means the moons are constantly being flexed by changing tidal forces as they orbit. While the tidal heating tries to circularize their orbits, they are trapped in an orbital resonance with each other that keeps their eccentricities larger than they would otherwise be. This explains why these moons show geological activity when we expected them to be quiet and nearly dead.
11. The atmosphere of Titan is the only other one in the solar system that is made of nitrogen gas like Earth's. The surface pressure is also comparable to Earth, at 1.5 bars. However, Titan's atmosphere is also very cold (well below the freezing point of water) and contains no oxygen. It also has dense methane clouds, which may create methane, or ethane rain, which could fall onto the surface.

    The *Huygens* probe has shown us that Titan appears to have valleys carved by rain, a sort of dirt on the surface (which is probably precipitated smog particles), and a slush of water ice and ammonia that functions like lava on Earth.
12. Many of the medium-size moons of Saturn and Uranus seem to have had geological activity in the recent past. Evidence for this includes bright surfaces (suggesting recent resurfacing); grooves like those on Ganymede, which appear on Enceladus; a large ridge on Iapetus; and tectonic features on Miranda. Mysteries of these moons include the source of the dark material on Iapetus, the origin of Iapetus's ridge, and why Uranus's moons vary so much in their amount of geological activity.
13. Triton is the large moon of Neptune. It is heavily cratered, but also shows evidence for geological activity in the recent past. Part of the surface is covered with wrinkles called "cantaloupe terrain," there is evidence of volcanic activity, and it has enough of an atmosphere to have left wind streaks. What makes Triton especially intriguing is that we are pretty certain that it is a captured body. It orbits Neptune backward, something that our model for planet and moon formation does not allow unless the moon is captured.
14. Ice moons can have geological activity at smaller sizes than rocky worlds because the ice of which they are made melts at much lower temperatures. Because of this, it takes much less heat to create geological activity. Combined with the heat source of tidal heating, this has allowed for a surprising amount of geological activity in the outer solar system.
15. Planetary rings are made up of countless small icy particles on similar orbits. Saturn's rings are the most impressive, being visible from Earth. Jupiter's rings are dusty and almost invisible. Uranus and Neptune both have narrow, bright, and

dense rings with very sparse, dusty rings in between. Neptune's dense, brighter rings also appear in arcs around the planet rather than making complete circles. Gap moons are small moons that orbit inside the rings. Their gravity nudges the ring particles away from themselves, clearing gaps in the rings. Other, more distant moons also affect the rings via resonances. Resonances can cause the ring particles to clear large gaps from their repeated nudges on the ring particles. Resonances can also launch waves in the rings.

16. Because collisions are constantly occurring within the ring systems, ring particles are continually being ground down to dust. Dust cannot survive long in ring systems because sunlight pressure makes them slowly fall into the planet. So we think that ring systems cannot survive very long and thus that ring particles must be replenished. Over time, the jovian planets probably will have varying ring characteristics. While Saturn has the most impressive ring system now, a billion years ago it might have been Neptune or Jupiter that had the biggest rings. A billion years from now, Saturn may not have any rings at all.
17. *Saturn's core is pockmarked with impact craters and dotted with volcanoes erupting basaltic lava.* This is not a possible discovery. The core lies below tens of thousands of kilometers of dense gas and even metallic hydrogen. No impactors reach that depth, and the conditions at the core are so extreme that nothing resembling familiar basaltic volcanism could occur.
18. *Neptune's deep blue color is not due to methane, as previously thought, but instead is due to its surface being covered with an ocean of liquid water.* This is not plausible. Neptune does have a great deal of water, but temperatures at the visible surface would be far too cold for it to be liquid.
19. *A jovian planet in another star system has a moon as big as Mars.* This is plausible. We know of no reason why larger satellites could not exist around extrasolar planets. After all, some of them are much larger than Jupiter.
20. *An extrasolar planet is discovered that is made primarily of hydrogen and helium. It has approximately the same mass as Jupiter but the same size as Neptune.* This is unlikely. A Jupiter-mass planet might be larger if it were hotter, but we know of no means to make such a planet much smaller than Jupiter's size.
21. *A new small moon is discovered orbiting Jupiter. It is smaller than Jupiter's other moons but has several large, active volcanoes.* This is not plausible. Tidal forces are too weak on objects so far away, and they are not forced into elliptical orbits by resonances. It would also be too small to have volcanoes from radioactive heating.
22. *A new moon is discovered orbiting Neptune. The moon orbits in Neptune's equatorial plane and in the same direction that Neptune rotates, but it is made almost entirely of metals such as iron and nickel.* This is not plausible. Solid objects at those distances are largely icy and rocky. While some asteroids have metallic composition, they would not be captured by Neptune.
23. *An icy, medium-size moon is discovered orbiting a jovian planet in a star system that is only a few hundred million years old. The moon shows evidence of active tectonics.* A part from the difficulty of observing moons around extrasolar planets, there is nothing unusual about tectonic activity on medium-size icy satellites.
24. *A jovian planet is discovered in a star system that is much older than our solar system. The planet has no moons at all, but it has a system of rings as spectacular as the rings of Saturn.* This is not plausible. With no moons of any size, there would be no source of ring particles.

25. *Radar measurements of Titan indicate that most of the moon is heavily cratered.* Surprising (but debatable). The initial observations show the presence of erosional features, and relatively few craters. But *Cassini* hasn't seen the whole surface at sufficient resolution!
26. *The Cassini mission finds 20 more moons of Saturn.* Not surprising. In fact, in the first year of the mission, *Cassini* has already found a few.
27. c; 28. b; 29. c; 30. c; 31. b; 32. c; 33. a; 34. b; 35. b; 36. c.
37. If the Jupiter-forming nebula had come together with no rotation:
    - All the material would have fallen into Jupiter, leaving nothing to make satellites.
    - With no satellites, no rings will form.
    - If Jupiter didn't rotate, its weather patterns wouldn't be smeared into belts and zones.
    - If Jupiter didn't rotate, it wouldn't have a magnetic field.
    - If Jupiter didn't rotate, it would be spherical instead of "squashed."
38. Its dark appearance in the infrared indicates that the Great Red Spot is cooler and therefore higher than neighboring clouds.
39. a. The gravitational attraction of Saturn on Mimas is less than that of Jupiter on Amalthea; therefore, Saturn's mass must be less than Jupiter's.
    b. If Saturn is less massive but almost as large as Jupiter, its density must be lower.
40. Without ingredients besides hydrogen and helium, the jovian planets would all be gray in color, and there would be no clouds or precipitation.
41. This question basically asks what a typical jovian planet system would be like if moved closer to the Sun. A jovian planet orbiting close to its star would have a much higher atmospheric temperature, though probably little difference in its interior. The hotter atmosphere would likely support more active weather. The heat would have the effect of vaporizing all the cloud-making materials. Other materials could condense at higher temperatures, making other kinds of clouds. Note that it would not be likely for the increased heat to evaporate the planet itself. The moons too would be much warmer. Since our jovian planet moons are ice-rich, they might turn into water worlds, with atmospheres—or conceivably be boiled away. Students may propose alternate scenarios given the tremendous change suggested, which should receive credit if well argued.
42. Essay question, answers may vary. Given the prevalence of erosional processes on Titan, "Earth" or "Mars" will be common answers.
43. The Galilean moons decrease in density with distance from Jupiter, the same trend seen in planets. This implies that the nebula surrounding Jupiter was warmer near its center. The fact that all of Jupiter's moons are less dense than their "counterparts" in the planets, it's clear that the proportion of ices must be higher, suggesting that the jovian nebula must have been cooler.
44. This question asks for a great deal of speculation. A wide variety of answers should be accepted if well supported.
    a. SuperJupiter would actually be smaller than Jupiter, and therefore much denser in every layer. The core would be even more compressed than Jupiter's (a smaller fraction of the size), and the metallic and liquid hydrogen layers would reach even closer to the cloud tops.

b. The cloud layers would not necessarily be that different, though students could argue that the weather might be more extreme given their logic in part (e).
c. Wind speeds could well be larger, especially if the smaller size has led to a faster rotation, or if internal heat is greater.
d. The magnetic field should be much stronger due to the greater metallic hydrogen layer. If students mentioned faster rotation elsewhere in their answer, it should be mentioned here as well.
e. All other things being equal, the greater mass, and hence stronger gravity, should produce more thermal energy through gravitational contraction.

45. Observing project; answers will vary. Consult the latest *Sky and Telescope* magazine (or many Web sites) for tabulations of the location of the jovian satellites. Good observers may be able to record the orbital resonances between the three inner satellites.

46. Observing project; answers will vary. Saturn is an exciting object to observe, and students should be encouraged to do so.

47. a. First, we convert a loss rate of a ton of sulfur dioxide per second into units of kilograms per year:

$$1000 \frac{\text{kg}}{\cancel{\text{s}}} \times \frac{3600\ \cancel{\text{s}}}{1\ \cancel{\text{hr}}} \times \frac{24\ \cancel{\text{hr}}}{1\ \cancel{\text{day}}} \times \frac{365\ \cancel{\text{days}}}{1\ \text{yr}} = 3.15 \times 10^{10} \frac{\text{kg}}{\text{yr}}$$

Over 4.5 billion years, the total mass loss is then:

$$\text{mass lost} = 3.15 \times 10^{10} \frac{\text{kg}}{\cancel{\text{yr}}} \times 4.5 \times 10^{9}\ \cancel{\text{yr}} = 1.4 \times 10^{20}\ \text{kg}$$

Thus, the fraction of Io's mass lost in 4.5 billion years is:

$$\text{fraction of mass lost} = \frac{\text{mass lost}}{\text{mass}} = \frac{1.4 \times 10^{20}\ \cancel{\text{kg}}}{893 \times 10^{20}\ \cancel{\text{kg}}} = 0.0016 = 0.16\%$$

Note that this is a fairly insignificant overall mass loss.

b. If sulfur dioxide is 1% of Io's mass, then the current total mass of sulfur dioxide is:

$$1\% \times 893 \times 10^{20}\ \text{kg} = 8.93 \times 10^{20}\ \text{kg}$$

Given the mass loss rate from part (a), Io would run out of sulfur dioxide in a time of:

$$\text{time to run out} = \frac{\text{mass available}}{\text{mass loss rate}} = \frac{8.93 \times 10^{20}\ \cancel{\text{kg}}}{3.15 \times 10^{10}\ \cancel{\text{kg}}/\text{yr}} = 2.8 \times 10^{10}\ \text{yr} = 28\ \text{billion yr}$$

Thus, Io will not run out of sulfur dioxide for quite some time.

48. With one collision every 5 hours, the total number of collisions suffered by a ring particle is given by the following equation:

$$\#\ \text{collisions} = \text{collision rate} \times \text{time} = \frac{1\ \text{collision}}{5\ \text{hr}} \times \text{time}$$

A time of 4.5 billion years is equivalent to:

$$\text{time} = 4.5\times10^{9}\ \cancel{\text{yr}}\times\frac{365\ \cancel{\text{days}}}{1\ \cancel{\text{yr}}}\times\frac{24\ \text{hr}}{1\ \cancel{\text{day}}} = 3.94\times10^{13}\ \text{hr}$$

Thus, the total number of collisions suffered by a ring particle in 4.5 billion years would be:

$$\#\ \text{collisions} = \frac{1\ \text{collision}}{5\ \text{hr}}\times3.94\times10^{13}\ \text{hr} = 7.8\times10^{12}\ \text{collisions}\approx 8\ \text{trillion collisions}$$

49. a. Newton's version of Kepler's third law states that period (p), semimajor axis (a), and mass of the central body (M) are related via:

$$p^2 = \frac{4\pi^2}{GM}a^3$$

We can solve this for p by taking a square root:

$$p = \left(\frac{2\pi}{\sqrt{GM}}\right)a^{3/2}$$

Since both Prometheus and Pandora are orbiting the same planet, we can save ourselves some trouble by calculating the part in parentheses right now and using the result for both moons. The mass of Saturn is given in Appendix E as 5.69×10^{26} kg, so we get:

$$p = \left(3.225\times10^{-8}\frac{\text{s}}{\text{m}^{3/2}}\right)a^{3/2}$$

We have to convert the semimajor axes for Prometheus and Pandora to meters to use this. We are told that the semimajor axes are 139,350 kilometers, and 141,700 kilometers, respectively. So converting:

$$139{,}350\ \cancel{\text{km}}\times\frac{1000\ \text{m}}{1\ \cancel{\text{km}}} = 1.3935\times10^{8}\ \text{m}$$

and:

$$141{,}700\ \cancel{\text{km}}\times\frac{1000\ \text{m}}{1\ \cancel{\text{km}}} = 1.417\times10^{8}\ \text{m}$$

So now we can find their orbital periods. First, Prometheus:

$$p = \left(3.225\times10^{-8}\frac{\text{s}}{\text{m}^{3/2}}\right)(1.3935\times10^{8}\ \text{m})^{3/2}$$

$$= 5.305\times10^{4}\ \text{s}$$

Converting this to hours, which would probably be a more suitable unit:

$$5.305\times10^{4}\ \cancel{\text{s}}\times\frac{1\ \cancel{\text{min}}}{60\ \cancel{\text{s}}}\times\frac{1\ \text{hr}}{60\ \cancel{\text{min}}} = 14.7\ \text{hr}$$

Similarly, Pandora:

$$p = \left(3.225 \times 10^{-8} \frac{\text{s}}{\text{m}^{3/2}}\right)(1.417 \times 10^{8}\,\text{m})^{3/2}$$

$$= 5.440 \times 10^{4}\,\text{s}$$

Converting this to hours, which would probably be a more suitable unit:

$$5.440 \times 10^{4}\,\cancel{\text{s}} \times \frac{1\ \cancel{\text{min}}}{60\ \cancel{\text{s}}} \times \frac{1\ \text{hr}}{60\ \cancel{\text{min}}} = 15.1\ \text{hr}$$

So Prometheus has an orbital period of 14.7 hours and Pandora has a period of 15.1 hours. The percent difference in the periods is given by:

$$\frac{15.1\ \text{hr} - 14.7\ \text{hr}}{15.1\ \text{hr}} \times 100\% = 2.00\%$$

b. Prometheus finishes each of its orbit 15.1 hr – 14.7 hr = 0.4 hr before Pandora does. So we can find out how long it takes Pandora to fall behind by a full orbit. This is just a rate problem: Pandora loses 0.4 hr/orbit and we want to know how long it takes to fall behind by a full 1.51 hr. The old rule:

$$\text{distance} = \text{rate} \times \text{time}$$

can be solved for time to give us:

$$\text{time} = \frac{\text{distance}}{\text{rate}}$$

Putting in our numbers, we get:

$$\text{time} = \frac{15.1\ \text{hr}}{0.4\ \text{hr/orbit}}$$

$$= 37.8\ \text{orbits}$$

So Prometheus will lap Pandora every 37.8 orbits. Since a Prometheus orbit is 14.7 hours, we can convert orbits to hours:

$$37.8\ \text{orbits} \times \frac{14.7\ \text{hr}}{1\ \text{orbit}} = 555\ \text{hr}$$

We should probably convert hours to days, since 555 hours is hard to understand:

$$555\ \cancel{\text{hr}} \times \frac{1\ \text{day}}{24\ \cancel{\text{hr}}} = 23.1\ \text{days}$$

So Prometheus laps Pandora every 23.1 days.

50. We can get the orbital periods from Appendix E. Titan's period is 15.945 days while Hyperion's is 21.277 days. Taking a ratio of the two periods will give us an idea of the resonance they are in. The ratio is:

$$\frac{21.277\ \text{days}}{15.945\ \text{days}} = 1.33$$

We recognize that 1.33 is about 4/3. So Titan and Hyperion are in a 4-to-3 resonance. We are also asked to find which medium-size moon is in a 2-to-1 resonance with Enceladus. From Appendix E, Enceladus has a period of 1.370 days. A moon with a 2-to-1 resonance with Enceladus would have to have either twice the period or half the period. Twice this period is 2.740 days, while half the period is 0.685 day. Looking down the list of orbit periods of the moons in Appendix E, we see that Dione has a period of about 2.737 days. No medium moons have periods around 0.685 day, so Dione is the moon that is in the 2-to-1 resonance with Enceladus.

51. From Appendix E, the mass of Titan is 1.3455×10^{23} kg. We can add the masses of the rest of the moons of Saturn that appear in Appendix E to get that they total 5.7920×10^{21} kg. So the ratio of the mass of Titan to the mass of all of the rest of Saturn's moons is:

$$\frac{1.3455 \times 10^{23} \text{ kg}}{5.7920 \times 10^{21} \text{ kg}} = 23.23$$

Titan is more than 23 times more massive than all of the rest of Saturn's moons combined. Another way to compare Titan to the other moons is to compare the surface gravities. We recall that we can find the acceleration due to gravity with the formula:

$$g = \frac{GM}{R^2}$$

where g is the acceleration due to gravity, G is Newton's gravitational constant, M is the mass of the body, and R is the radius of the body. For Titan, we have already looked up the mass (1.3455×10^{23} kg) and we can also use Appendix E to find that the radius is 2,575 kilometers. Converting this to meters:

$$2575 \not{\text{km}} \times \frac{1000 \text{ m}}{1 \not{\text{km}}} = 2.575 \times 10^6 \text{ m}$$

We can also use Appendix E to find the mass and radius of Mimas: 3.70×10^{19} kg and 199 kilometers. Converting the radius into meters:

$$199 \not{\text{km}} \times \frac{1000 \text{ m}}{1 \not{\text{km}}} = 1.99 \times 10^5 \text{ m}$$

So we can find the accelerations due to gravity on both worlds. First, Titan:

$$g_{\text{Titan}} = \frac{\left(6.67 \times 10^{-11} \frac{\text{m}^3}{\text{kg} \times \text{s}^2}\right)(1.3455 \times 10^{23} \text{ kg})}{(2.575 \times 10^6 \text{ m})^2}$$
$$= 1.35 \text{ m/s}^2$$

So Titan's acceleration due to gravity is 1.35 m/sec^2. We can do the same for Mimas:

$$g_{\text{Mimas}} = \frac{\left(6.67 \times 10^{-11} \frac{\text{m}^3}{\text{kg} \times \text{s}^2}\right)(3.70 \times 10^{19} \text{ kg})}{(21.99 \times 10^5 \text{ m})^2}$$
$$= 0.062 \text{ m/s}^2$$

So the acceleration due to gravity on Mimas is $0.062\ \text{m/s}^2$. The ratio of these two accelerations is:

$$\frac{1.35\ \text{m/sec}^2}{0.062\ \text{m/sec}^2} = 21.8$$

So the acceleration due to gravity on Titan is 21.8 times the acceleration due to gravity on Mimas.

52. To solve this problem, we need to find both the escape speed from the top of Titan's atmosphere and the thermal speed of the hydrogen atoms. We recall that escape speed is calculated via:

$$\text{escape velocity} = \sqrt{\frac{2GM}{r}}$$

where G is Newton's gravitational constant, M is the mass of the body, and r is the distance from the center of the body. Appendix E tells us that Titan has a mass of 1.3455×10^{23} kg and a radius of 2,575 kilometers. Since we seek the escape speed at 1,400 kilometers up in the atmosphere, r is the sum of the radius and the altitude, or 3,975 kilometers. Converting this altitude to meters so we can compute the escape speed we get:

$$3{,}975\ \cancel{\text{km}} \times \frac{1000\ \text{m}}{1\ \cancel{\text{km}}} = 3.975 \times 10^6\ \text{m}$$

So now we can compute the escape speed:

$$\text{escape speed} = \sqrt{\frac{2\left(6.67 \times 10^{-11} \dfrac{\text{m}^3}{\text{kg} \times \text{s}^2}\right)(1.3455 \times 10^{23}\,\text{kg})}{3.975 \times 10^6\,\text{m}}}$$
$$= 2{,}125\ \text{m/s}$$

The escape speed from the top of Titan's atmosphere is 2,125 m/s.
Moving to the thermal speed, we also recall that the thermal speed is given by:

$$\text{thermal speed} = \sqrt{\frac{2kT}{m}}$$

where k is Boltzmann's constant, T is the temperature, and m is the mass of the molecule or atom we are interested in. We are told that hydrogen has a mass of 1.67×10^{-27} kg and that the temperature is 200 K. So we can calculate the thermal speed:

$$\text{thermal speed} = \sqrt{\frac{2\left(1.38 \times 10^{-23} \dfrac{\text{joules}}{\text{Kelvin}}\right)200\ \text{K}}{1.67 \times 10^{-27}\ \text{kg}}}$$
$$= 1{,}960\ \text{m/s}$$

The thermal speed is more than half of the escape speed. We were told in Chapter 10 that if the thermal speed is more than $\frac{1}{5}$ of the escape speed most of that molecule or atom will have escaped the atmosphere through thermal escape over the age of the solar system. Since hydrogen on Titan *is* moving faster than this limit, we expect to see little hydrogen on Titan.

53. We recognize this as a conversion problem. We are told that the rings, 270,000 kilometers in diameter, are equivalent to a 6.6-centimeter dollar bill in our model. So we can convert the 50-meter thickness (in real life) to the model thickness. But first, we have to convert meters to kilometers so our units work out:

$$50 \cancel{m} \times \frac{1 \text{ km}}{1000 \cancel{m}} = 0.05 \text{ km}$$

Now we can convert from kilometers in the real rings to centimeters in our model:

$$0.05 \cancel{km} \times \frac{6.6 \text{ cm}}{270{,}000 \cancel{km}} = 1.22 \times 10^{-6} \text{cm}$$

So in the model, the rings would only be 1.22×10^{-6} cm thick, much thinner than a dollar bill.

54. a. We recall that density is defined as:

$$\text{density} = \frac{\text{mass}}{\text{volume}}$$

and we know that the volume of a planet is:

$$\frac{4}{3}\pi r^3$$

where *r* is the radius. So we can solve for the radius of the planet given the mass and density. Starting with:

$$\text{density} = \frac{\text{mass}}{\frac{4}{3}\pi r^3}$$

we can rearrange to get the radius by itself:

$$r^3 = \frac{3 \text{ (mass)}}{4 \pi \text{ (density)}}$$

Now we just take the cube root of both sides and we have solved for the radius:

$$r = \sqrt[3]{\frac{3 \text{ (mass)}}{4 \pi \text{ (density)}}}$$

We are told in the problem that the planet has a mass five times that of Jupiter but with the same density. From Appendix E, Jupiter's density is 1.33 g/cm^3 and its mass is 1.90×10^{27} kg. The mass of this new planet is therefore:

$$5 \times (1.90 \times 10^{27} \text{kg}) = 9.50 \times 10^{27} \text{kg}$$

We need to convert this to grams so that the units match with the density:

$$9.50 \times 10^{27} \cancel{kg} \times \frac{1000 \text{ g}}{1 \cancel{kg}} = 9.50 \times 10^{30} \text{ g}$$

So we can calculate the radius of this planet:

$$r = \sqrt[3]{\frac{3(9.50 \times 10^{30}\ \text{g})}{4\ \pi(1.33\ \text{g/cm}^3)}}$$
$$= 1.19 \times 10^{10}\ \text{cm}$$

We had probably better convert this to kilometers:

$$1.19 \times 10^{10}\ \cancel{\text{cm}} \times \frac{1\ \cancel{\text{m}}}{100\ \cancel{\text{cm}}} \times \frac{1\ \text{km}}{1000\ \cancel{\text{m}}} = 119{,}000\ \text{km}$$

So the planet has a radius of 119,000 kilometers. The diameter is twice this, 239,000 kilometers.

b. Mathematical Insight 6.1 tells us that we can calculate the angular size of an object via the formula:

$$\text{angular size} = 206{,}265'' \times \frac{\text{physical size}}{\text{distance}}$$

We are told that the planet is 230-light-years distant, but we will need to convert this to kilometers so that the units work out in the above formula. Luckily, Appendix E tells us that there are 9.46×10^{12} km in a light-year. So we can convert:

$$230\ \cancel{\text{light-years}} \times \frac{9.46 \times 10^{12}\ \text{km}}{1\ \cancel{\text{light-year}}} = 2.18 \times 10^{15}\,\text{km}$$

Now it is easy to calculate the angular size:

$$\text{angular size} = 206{,}265'' \times \frac{119{,}000\ \text{km}}{2.18 \times 10^{15}\ \text{km}}$$
$$= 1.13 \times 10^{-5}\,\text{arcseconds}$$

So the planet would appear to be 1.13×10^{-5} arcseconds from here on Earth.

Chapter 12. Remnants of Rock and Ice: Asteroids, Comets, and Kuiper Belt Objects

Small bodies in the solar system are important for two reasons: what they tell us about the formation of the solar system, and what small bodies can do to planets if they collide. This chapter connects these two subjects with an understanding of how comets and asteroids have been nudged by the jovian planets ever since formation. While our understanding of asteroids has improved steadily in recent years, the study of comets has undergone a major breakthrough in the discovery of Kuiper belt objects. This discovery has helped put Pluto and "Planet X" in context as the largest free-roaming Kuiper belt objects.

The topic of "cosmic collisions" offers a real opportunity to demonstrate the relevance of astronomy to our existence. Hollywood movies have tried to make this point through the possible threat of an impact, but the deeper connection goes back to the role that impacts have played in biological evolution and our very presence on the planet.

As always, when you prepare to teach this chapter, be sure you are familiar with the relevant media resources (see the complete, section-by-section resource grid in Appendix 3 of this Instructor's Guide) and the online quizzes and other study resources available on the Mastering Astronomy Web site.

What's New in the Fourth Edition That Will Affect My Lecture Notes?

As everywhere in the book, we have added learning goals that we use as subheadings, rewritten to improve the text flow, improved art pieces, and added new illustrations. The art changes, in particular, will affect what you wish to show in lecture. We have not made any substantial content or organizational changes to this chapter.

- The previous edition's sections 13.1–13.3 have been merged into a single Section 12.1 to better match the learning goals. The same content is still covered.
- The discovery of "Planet X" is thoroughly integrated into this chapter. It bears noting that since the first edition of *The Cosmic Perspective,* the discovery of an object rivaling Pluto has been predicted, so its discovery was straightforward to incorporate into the text. In addition to the complete coverage in this chapter, the topic is also mentioned in Chapters 1 and 7. This chapter includes all of the basic properties of the planet known at the time of publication. The name, however, had not been approved by the IAU. If you are teaching from an early printing of the fourth edition, be aware that the text refers to it by its "minor planet" designation "2003 UB313," or "Planet X."
- The topic of meteor showers is now covered with comets in Section 12.2.

Teaching Notes (By Section)

Section 12.1 Asteroids and Meteorites

This section reminds students of the motivation for studying the small objects and introduces the minimum number of new terms needed for the chapter. After an overview of asteroid properties, this section explains how orbital resonances are responsible for both the existence of asteroids and their occasional collisions with planets. Students are usually amazed to learn that meteorites are pieces of asteroids and that asteroids may themselves be pieces of shattered planets. This section emphasizes the relationship between rocks that you can hold, the processes that brought them to Earth, and what we can learn about the asteroids themselves.

- Note that we do not use the term *meteoroid,* but do expect students to distinguish between meteors and meteorites.
- Note also that we do not use the jargon of *chondrites versus achondrites*, instead calling them *primitive versus processed meteorites*. This puts the emphasis on the origin of these two categories of meteorites rather than on their visible morphology.
- Learning about meteorites can be all the more effective if you have samples for students to view or to hold. Meteorite dealers can be found easily on the Web or in popular astronomical magazines. One well-established dealer is Robert Haag, PO Box 27527, Tucson, AZ 85726, 520-882-8804 or http://www.meteoriteman.com. Etched iron meteorites are particularly impressive: Their visual appeal and high

density immediately attract students. The inescapable story they tell (from the core of a shattered "mini-planet") challenges a lot of preconceived notions about the permanence of celestial bodies. Meteorites are normally sold by the gram, with common varieties currently priced around $1 a gram for small specimens.

Section 12.2 Comets

This section covers the appearance and origin of comets. Because most students have little idea what they're actually seeing when they look at a comet, we cover this topic first. Emphasize how small the nucleus is and how large and insubstantial the tails are. The different effects of the Sun on atoms, dust, and large grains in the tail make a useful sequence and help students understand meteor showers.

- We do not classify comets using an arbitrary period cutoff (long period versus short period), instead classifying them by their place of origin (Oort cloud versus Kuiper belt). The two kinds of classification are basically the same, but the latter is more physically motivated.
- Note that we use the term *comet* to include all ice-rich objects of the outer solar system. Thus, we refer to the objects of both the Kuiper belt and the Oort cloud as "comets." We use the term *comet* even when objects are far from the Sun with no tail. Note that Gerard Kuiper, though originally Dutch, Americanized the pronunciation of his name to rhyme with "piper."
- When discussing the origin of comets, emphasize the parallels with asteroids: Jovian planets keep both kinds of objects from forming a single large object; jovian planets nudge their orbits through resonances; and both kinds of objects can collide with planets. The explanation for their different composition (ice versus rock) should be very familiar to students by now.
- Some students may have a hard time accepting that the Oort cloud is populated by comets ejected from much closer in, believing instead that they must have formed in place, at huge distances from the Sun. You should point out that densities in a solar nebula the size of the Oort cloud would have been much too low for material to accrete into comets. You should also point to the discussion of gravitational encounters in Part I.

Section 12.3 Pluto: Lone Dog No More

This section focuses on Pluto and "Planet X" and the question of whether they are true planets or merely large Kuiper belt comets. This is a good opportunity for class discussion, because there are many defensible viewpoints.

- FYI, at this time there is still no IAU-sanctioned definition for the "minimum planet." Some scientists are calling for definitions of various types. It's worth noting that one potential definition, promoted by some planetary scientists, calls for conferring planetary status on any object large enough for gravity to begin to shape it into a sphere. By this definition, there could be dozens of planets in our solar system, including Pluto, Quaoar, Sedna, and Ceres. Others argue that any definition has limitations, and suggest that the term *planet* will always have inherent fuzziness.

Section 12.4 Cosmic Collisions: Small Bodies versus the Planets

The discussion of cosmic collisions proceeds from the irrefutable (Shoemaker-Levy 9 impact on Jupiter, meteor showers) to the controversial (death of the dinosaurs, threats to civilization). This is a great opportunity to show connections between astronomy, geology, biology, and sociology.

- The question of how to deal with asteroid threats makes an excellent discussion topic—again with many defensible viewpoints and no right answer.
- Showing clips from popular movies such as *Deep Impact* and *Armageddon* would make an excellent starting point for a discussion on science fact versus science fiction concerning the asteroid impact threat.

Answers/Discussion Points for Think About It Questions

The Think About It questions are not numbered in the book, so we list them in the order in which they appear, keyed by section number.

Section 12.1

- (p. 361) The less reflective one must be closer.
- (p. 363) The gaps are visible in the distribution of semimajor axes—the *average* distance from the Sun. But asteroids travel on elliptical orbits, meaning that they rarely lie at exactly their average orbital distance. Although there may be very few asteroids with semimajor axes of 2.5 AU, for example, many asteroids with other semimajor axes will cross the distance of 2.5 AU on their elliptical orbits.

Section 12.2

- (p. 368) The question asks students to learn the status of the *Stardust* mission. Reminding students of the crash landing of the *Genesis* sample return capsule may underscore the difficulty of missions like these.
- (p. 370) Phaeton appears to be an extinct comet—one with all its ices used up or buried deep below acrust.

Section 12.3

- (p. 374) The definition of a planet makes an excellent classroom discussion, not just in developing the definition but also in evaluating the value of definitions. Encourage students to offer specific criteria (orbiting the Sun but not another planet, roughly spherical, larger than a certain size, etc.), and then see how many objects fit the definition. The test cases of Pluto, "Planet X," Charon, Ceres, and Triton may lead to answers other than "nine planets." Close with a discussion of whether the definition is useful or just a matter of semantics.

Section 12.4

- (p. 378) This is a subjective question that should generate interesting debate.

Solutions to End-of-Chapter Problems (Chapter 12)

1. Asteroids are small leftovers from solar system formation made of rock and metal. They are much smaller than the planets: The largest asteroid is less than 1,000 kilometers in diameter and most are much smaller than this. We have measured the masses of a few of the asteroids and found them to have densities of between 1 and 3 g/cm^3, consistent with rocky compositions or even rocky rubble.
2. Even the largest of asteroids, Ceres, is small compared to the planets. (Ceres's diameter is less than half that of Pluto.) So despite the large number of asteroids, their masses add up to a small fraction of the mass of any of the planets.

3. The asteroid belt exists because of orbital resonances with Jupiter. In that region of the solar system, Jupiter's resonances pump up the eccentricities of objects, making their collisions high speed. This keeps them from clumping together and forming a planet. In fact, it tends to make the asteroids break apart when they collide.
4. While the terms *meteor* and *meteorite* are often interchanged, they mean different things. A meteor is a piece of debris that is hitting our atmosphere and glowing from the heat of entry. A meteorite is what is left of the debris after it hits the ground. We can identify a meteorite several ways. One is that the crust tends to be pitted in an unusual way due to its entry into our atmosphere. Another feature common to many meteorites is a high metal content that causes them to attract magnets. The ultimate test, however, is a chemical analysis since meteorites tend to be rich in elements that are rare in the Earth's crust.
5. Primitive meteorites are intact leftovers of the solar nebula. They contain rocks and metals mixed together, as well as carbon compounds in some cases. Processed meteorites, on the other hand, have undergone differentiations in larger asteroids. As a result, they are made up of either metals or rocky material rather than a mix of the two.
6. When a comet is far from the Sun, it appears to be a dirty snowball. It is only when it approaches the Sun and warms up that it begins to sublimate, producing a coma of gas around it and its tails of gas and dust. But because the coma and tails are made of gas and dust from the comet's nucleus, each pass near the Sun reduces the mass of the comet. Eventually, the comet will become too small to hold itself together and will come apart. However, before this happens the comet might form a layer of rocky material like a crust, protecting the icy interior from the Sun. Such a comet would look much like an asteroid.
7. The coma and tails of a comet are produced by ices warming and sublimating off of the surface. As these ices turn to gas, they escape the nucleus (the solid icy part of the comet in the middle), carrying dust with them. The gas that escapes becomes ionized and is carried away from the Sun by the solar wind. The dust that escapes is pushed, more gently, away from the Sun by solar radiation pressure.
8. Meteor showers occur when Earth passes through a stream of pea-size debris left from a comet passage. Meteor showers occur at times of the year when the Earth is in parts of its orbit that intersect with the orbits of comets.
9. There are two reservoirs of comets in our solar system. The first is the Kuiper belt, which is similar to the asteroid belt except that it is beyond the orbit of Neptune and is filled with icy bodies rather than rocky and metallic ones. The other reservoir of comets is the Oort cloud, a spherical halo of comets well outside of the orbits of the planets. While comets in the Kuiper belt have orbits that are nearly in the plane of the solar system and go around the Sun in the "forward" direction, Oort cloud comets often orbit on highly inclined orbits in any direction. Comets that are now in the Oort cloud were probably flung there early in the solar system's life by Jupiter. Comets in the Kuiper belt probably formed in that region of the solar system, or close to it.
10. Pluto is significantly smaller than any of the other planets and is made of ice and rock. This means that it does not fit into either the jovian class or the terrestrial class of planets. Pluto possesses a thin atmosphere that may freeze to the surface as the planet moves farther from the Sun along its orbit. Its orbit is also very elliptical and inclined relative to the plane of the solar system. In spite of the fact that Pluto's orbit takes it inside the orbit of Neptune for part of the time, the two will never

collide because Pluto and Neptune are in a 3-to-2 resonance, which keeps Neptune far away when Pluto crosses Neptune's orbit. Pluto's moon, Charon, is slightly more than 1/10 the mass of Pluto, making it the largest moon-to-planet ratio in the solar system. Evidence suggests that Charon was formed in a giant impact, the same way we think Earth's Moon was formed.

11. Planet X, like Pluto, lies beyond the orbit of Neptune in the realm of Kuiper belt comets. They are icy and reflective, with composition similar to the other Kuiper belt comets. In addition, their orbits, while highly elliptical and inclined compared to the planets, is normal for a Kuiper belt object. In fact, many Kuiper belt comets are in the same resonance with Neptune that Pluto is.
12. Some time before we discovered Shoemaker-Levy 9, it had broken up into a string of smaller comet nuclei, probably due to tidal forces from Jupiter. As a result, we saw not one, but many impacts on Jupiter. Each impact released energies equivalent to a million hydrogen bombs and left a scar that was larger than the Earth and lasted for more than a month.
13. One piece of evidence we have that a large impact caused the extinction of the dinosaurs is a worldwide layer of iridium that was laid down at that time. Iridium is rare on Earth's surface but more common in comets and asteroids. Since the layer exists all around the globe, something had to have brought the iridium to the Earth's surface and spread it over the entire Earth. A large impact could have done exactly that.

    The layer that contains the iridium also contains other elements rare on Earth's surface, further suggesting an impact with a comet or asteroids. In addition, we see soot, droplets that appear to have been formed by solidifying molten rock, and "shocked quartz," crystals that show evidence of having been placed under high temperatures and high pressures. We find shocked quartz at known impact sites like Meteor Crater, so we associate it with impacts. The soot points to global fires that would likely have been caused by the hot impact debris.

    We have also now found a large crater, called *Chicxulub*, on the Yucatán Peninsula, which has the right size and age to be the result of the same impact event.
14. Impacts that would cause mass extinctions occur every 50 to 100 million years and are therefore considered very rare. Smaller impacts are more common, however. For example, 100-meter-size impacts (like the one that caused Meteor Crater) should occur every 10,000 years or so. These could cause local devastation and kill millions of people, but they would not cause mass extinction. So impacts pose a real threat, but there is little chance that we will have a large impact in our lifetimes.
15. *A small asteroid that orbits within the asteroid belt has an active volcano.* Surprising! Even the largest asteroids are too small to be geologically active now.
16. *Scientists discover a meteorite that, based on radiometric dating, is 7.9 billion years old.* Surprising! This would be older than our solar system, so it would be possible only if the meteorite originated in a different (and older) star system.
17. *An object that resembles a comet in size and composition is discovered to be orbiting in the inner solar system.* This is reasonable. Many comets from the outer solar system have their orbits altered so they become locked in the inner solar system. But exposed ices would be used up in a few years, so the object might not last long or might become covered in a protective layer of dust.

18. *Studies of a large object in the Kuiper belt reveal that it is made almost entirely of rocky (as opposed to icy) material.* This would be surprising. Objects forming that far out should contain mostly ices. We know of no way for a large asteroid to be flung into a Kuiper belt orbit.
19. *Astronomers discover a previously unknown comet that will produce a spectacular display in Earth's skies about 2 years from now.* This would not be surprising at all. We wish it happened more.
20. *A mission to Pluto finds that it has lakes of liquid water on its surface.* Surprising. Water would be frozen at Pluto's temperature, and we know of no extra heat sources.
21. *Geologists discover a crater from a 5-kilometers object that impacted the Earth more than 100 million years ago.* This would not be surprising. A slightly larger impact 65 million years ago is thought to have caused a mass extinction.
22. *Archaeologists learn that the fall of ancient Rome was caused in large part by an asteroid impact in southern Africa.* This would be surprising. An impact large enough to affect Roman civilization all the way from South Africa would have caused more widespread devastation and should have been noted in many historical and geological records.
23. *Astronomers discover three objects with the same average distance from the Sun (and the same orbital period) as Pluto.* This would not be surprising, as astronomers have already found many objects in the Kuiper belt with orbital properties similar to those of Pluto.
24. *Astronomers discover an asteroid with an orbit suggesting that it will impact the Earth in the year 2064.* This can be argued both ways: On the one hand, there are likely to be many undiscovered asteroids that could have orbits that cross Earth's path. On the other hand, it's unlikely that an impact will occur in any particular year.
25. b; 26. b; 27. c; 28. b; 29. c; 30. a; 31. c; 32. b 33. b; 34. b.
35. Answers will vary, but here are some possibilities:
    - The asteroids might have accreted into a single planet between Mars and Jupiter.
    - The asteroid belt would not have gaps.
    - Fewer comets would have been ejected into the Oort cloud, leaving many more that could potentially impact Earth.
36. The factual story goes as follows, though students are encouraged to add more creative touches: At first the iron atom floated in space in the molecular cloud. The cloud collapsed into the solar nebula. As the nebula cooled, iron atoms collected together and condensed into metallic flakes. As the nebula cooled further, rocky flakes formed, and then planetesimals formed by accreting the rocky and metallic flakes. A planetesimal grew into an asteroid large enough to melt, and the iron flowed to the core (differentiation). The asteroid collided with another asteroid, shattering it into pieces. A metallic piece was blasted away (and probably nudged by Jupiter) and collided with the Earth.
37. Because Albert emits a higher proportion of thermal emission (infrared) than of reflected light (visible), it must be warmer than Isaac. This means it must be darker (in order to be warmed by the Sun). Because Albert has a lower reflectivity than Jordan, it must be larger in order to appear as bright. Isaac would make a better candidate for mining metal, and Albert a better candidate for carbon-rich material.

38. Asteroids are rocky and close; comets are icy and distant. This relates to their formation in different regions of the solar nebula: asteroids inside the frost line, and comets outside.
39. Comets have tails due to the sublimation of ice caused by sunlight as they approach the Sun. The ejected gas is ionized and carried away in the ion tail by the solar wind—straight back from the Sun. Small solid particles drift away in the dust tail, which curves slowly away from the comet because the pressure of sunlight is so small compared to that from the solar wind. The third tail of larger particles is of interest because it's the source of meteor showers.
40. Kuiper belt objects are leftovers from the outer edges of the original solar nebula—they haven't moved much since their formation, and their (relatively) organized motions around the Sun support this conclusion. Oort cloud comets formed closer in, but were ejected by gravitational encounters with the jovian planets onto random orbits (highly eccentric and highly tilted) well beyond the Kuiper belt.
41. Observing project. Answers will vary.
42. Students should find that only a small amount of dark material is necessary in order to turn a snowball dark.
43. To estimate the size of a body composed of all of the asteroids, we will find the volume of 1 million 1-kilometer asteroids and find the size of a single body with the same volume. We know that the volume of a sphere is:

$$\text{volume} = \frac{4}{3}\pi r^3$$

So we can equate the total volume of 1 million 1-kilometer asteroids with the volume of a single object:

$$10^6 \times \left(\frac{4}{3}\pi r^3_{\text{1-km asteroids}}\right) = \frac{4}{3}\pi r^3_{\text{single body}}$$

(We could calculate the left-hand side first and then solve for radius in the equation for volume, but doing the problem this way lets us cancel out some constants and saves us extra number crunching.)

Canceling the $\frac{4}{3}$ and the π on both sides and solving for $r_{\text{single body}}$, we get:

$$r_{\text{single body}} = \sqrt[3]{10^6 \times r^3_{\text{1-km asteroids}}}$$

We can recall that a cube root is the $\frac{1}{3}$ power and if we remember the rules for exponents, we can simplify this further:

$$r_{\text{single body}} = 10^{6/3} \times r^{3/3}_{\text{1-km asteroids}}$$
$$= 10^2 \times r_{\text{1-km asteroids}}$$

(If we did not recall these rules, it would not have been terribly difficult to just crunch the numbers without this simplification. However, doing this little bit of algebra reduces the risk of making a mistake when punching the numbers into our calculators and it makes the result easy to check.)

Now we calculate $r_{\text{single body}}$. Recall that the diameter of the asteroids is 1 kilometer, so the radius must be half a kilometer. And so we calculate:

$$r_{\text{single body}} = 10^2 \times 0.5 \text{ km}$$
$$= 50 \text{ km}$$

So the radius of the single body would be 50 kilometers. This means that the diameter would be twice this, or 100 kilometers. This is still much, much smaller than any planet.

44. a. The comet's volume is:

$$V = \frac{4}{3} \times \pi \times (\text{radius})^3 = \frac{4}{3} \times \pi \times (1{,}000 \text{ m})^3 = \frac{4}{3} \times \pi \times 10^9 \text{ m}^3 = 4.2 \times 10^9 \text{ m}^3$$

so its mass (given in the problem) is:

$$m = V \times \text{density} = (4.2 \times 10^9 \text{ m}^3) \times 1{,}000 \frac{\text{kg}}{\text{m}^3} = 4.2 \times 10^{12} \text{ kg}$$

Therefore, its kinetic energy is:

$$KE = \frac{1}{2} mv^2 = \frac{1}{2} \times (4.2 \times 10^{12} \text{ kg}) \times \left(3 \times 10^4 \frac{\text{m}}{\text{s}}\right)^2 = 1.9 \times 10^{21} \text{ joule}$$

b. Converting this kinetic energy to units of megatons, we find that it is equivalent to:

$$\frac{1.9 \times 10^{21} \text{ joule}}{4.2 \times 10^{15} \dfrac{\text{joule}}{\text{megaton}}} = 450{,}000 \text{ megatons}$$

Clearly, such an impact could be devastating, though not enough to cause a massextinction.

45. Following the hint in the problem, we calculate the probability of impact by assuming a giant dartboard in which the bull's-eye is the area of the Earth's disk and the total dartboard area has a radius of 3 million kilometers:

$$\text{Probability} \approx \frac{\text{area of bull's-eye}}{\text{area of dartboard}}$$
$$= \frac{\text{area of Earth's disk}}{\text{area of target}}$$
$$= \frac{\pi \times (6371 \text{ km})^2}{\pi \times (3{,}000{,}000 \text{ km})^2}$$
$$= 4.5 \times 10^{-6} \text{(or about 4.5 chances in a million)}$$

Given that Toutatis was bound to come within 3 million kilometers of the Earth, the probability of a collision is still quite small. Nevertheless, the probability of nearly 5 in a million is greater than the probability of winning big in a lottery. The question of whether this probability should count as a "near miss" is a matter of opinion.

46. To find the average spacing between comets, we will take the cube root of the average volume that each comet occupies. To get the average volume, we will find the total volume of the Oort cloud and divide by the number of comets. The volume of the Oort cloud will be:

$$\text{volume} = \frac{4}{3}\pi r_{\text{Oort cloud}}^3$$

We are told that the radius of the Oort cloud is 50,000 AU, so we can find the volume pretty easily:

$$\begin{aligned}\text{volume} &= \frac{4}{3}\pi(50{,}000\ \text{AU})^3 \\ &= 5.24 \times 10^{14}\,\text{AU}^3\end{aligned}$$

So the total volume of the Oort cloud is 5.24×10^{14} AU^3.

The average volume occupied is the total volume divided by the number of bodies. We are told that there are a trillion (10^{12}) objects in the Oort cloud, so this average can be calculated:

$$\begin{aligned}\text{volume per object} &= \frac{\text{total volume}}{\text{number of objects}} \\ &= \frac{5.24 \times 10^{14}\ \text{AU}^3}{10^{12}} \\ &= 5.24 \times 10^{2}\ \text{AU}^3\end{aligned}$$

So each Oort cloud object has 5.24×10^2 AU^3 of space to itself.

So now we just take the cube root to get the average spacing:

$$\begin{aligned}\text{average spacing} &= \sqrt[3]{\text{average volume per object}} \\ &= \sqrt[3]{5.24 \times 10^{2}\ \text{AU}^3} \\ &= 8.06\ \text{AU}\end{aligned}$$

47. a. The ratio of the reflected light will be the ratio of the areas. We know that the area of a disk is πr^2, so the ratio of the light reflected by the two bodies is:

$$\begin{aligned}\text{ratio of reflected light} &= \frac{\pi r_{\text{Charon}}^2}{\pi r_{\text{Pluto}}^2} \\ &= \frac{r_{\text{Charon}}^2}{r_{\text{Pluto}}^2} \\ &= \left(\frac{r_{\text{Charon}}}{r_{\text{Pluto}}}\right)^2\end{aligned}$$

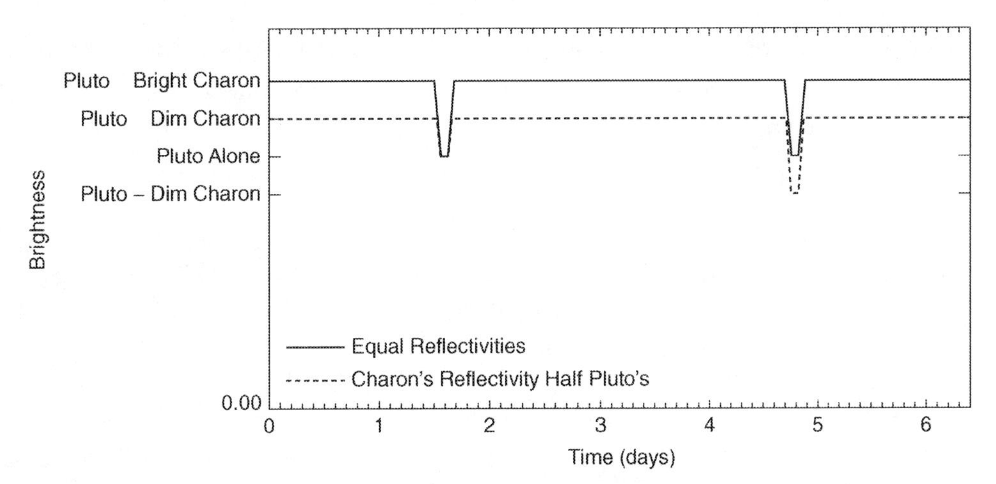

b. In this case, the two dips in brightness are different depths. When Charon passes behind Pluto, we lose some of the area reflecting light. But when Charon passes in front of Pluto, not only is there less area total, but Charon's darker surface is replacing part of the brighter surface of Pluto that we would otherwise see. So the dip corresponding to when Pluto is behind Charon is deeper than when Charon goes behind Pluto. In the plot, The dashed red line indicates this brightness, and the dip on the right corresponds to Charon passing in front of Pluto.

48. Following the method of Mathematical Insight 11.1, we find the "no greenhouse" temperature of a comet with reflectivity 0.03 at 50,000 AU to be:

$$T = 280\text{ K} \times \sqrt[4]{\frac{(1-\text{reflectivity})}{d^2}} = 280\text{ K} \times \sqrt[4]{\frac{(1-0.03)}{50{,}000^2}} = 1.24\text{ K}$$

At 3 AU, the "no greenhouse" temperature of this comet is:

$$T = 280\text{ K} \times \sqrt[4]{\frac{(1-\text{reflectivity})}{d^2}} = 280\text{ K} \times \sqrt[4]{\frac{(1-0.03)}{3^2}} = 160\text{ K}$$

At 1 AU, the "no greenhouse" temperature of this comet is:

$$T = 280\text{ K} \times \sqrt[4]{\frac{(1-\text{reflectivity})}{d^2}} = 280\text{ K} \times \sqrt[4]{\frac{(1-0.03)}{1^2}} = 278\text{ K}$$

Given that water sublimes at about 150 K, the temperature is high enough whenever the comet is within about 3 AU of the Sun, meaning that the comet should form a tail when it is in this region of the solar system (assuming it has enough volatiles remaining to form a tail).

49. This is another rate problem. We know that:

$$\text{rate} = \frac{\text{amount}}{\text{time}}$$

so we can solve for time to get:

$$\text{time} = \frac{\text{amount}}{\text{rate}}$$

The amount of material we want to fall to Earth is 0.1% of the mass of the Earth. From Appendix E we know that the mass of the Earth is 5.97×10^{24} kg. So the amount we want is:

$$0.001\times5.97\times10^{24}\text{ kg}=5.97\times10^{21}\text{ kg}$$

Now we find the rate. We are told that a few hundred tons of material fall to Earth per day, so we will say 200 tons for our estimate. A ton is 2,000 pounds and there are 2.204 pounds per kilogram. So the rate of material falling to Earth is:

$$200\frac{\text{tons}}{\text{day}}\times\frac{2000\text{ pounds}}{1\text{ ton}}\times\frac{1\text{ kg}}{2.204\text{ pounds}}=1.81\times10^{5}\frac{\text{kg}}{\text{day}}$$

With these two bits of data, we can find out how many days it takes for Earth to gain 0.1% of its mass:

$$\begin{aligned}\text{time}&=\frac{5.97\times10^{21}\text{ kg}}{1.81\times10^{5}\text{ kg/day}}\\&=5.30\times10^{16}\text{ days}\end{aligned}$$

So it takes 5.30×10^{16} days. But that number is hard to understand, so let us change it to years:

$$5.30\times10^{16}\cancel{\text{days}}\times\frac{1\text{ yr}}{365\cancel{\text{days}}}=9.04\times10^{13}\text{ yr}$$

So it would take 90.4 trillion years for Earth to gain even 0.1% of its mass from this falling dust. Clearly this extra mass is unimportant compared to Earth's existing mass.

Chapter 13. Other Planetary Systems: The New Science of Distant Worlds

This is an entirely new chapter, greatly expanding on the material on extrasolar planets in previous editions. This change was motivated not only by the sheer number of extrasolar planets now known, but also by our increasing knowledge about their properties. The discovery of new planets around other stars is one of the hottest topics in astronomy, and students will probably be eager to hear the latest news.

As always, when you prepare to teach this chapter, be sure you are familiar with the relevant media resources (see the complete, section-by-section resource grid in Appendix 3 of this Instructor's Guide) and the online quizzes and other study resources available on the Mastering Astronomy Web site.

What's New in the Fourth Edition That Will Affect My Lecture Notes?

As everywhere in the book, we have added learning goals that we use as subheadings, rewritten to improve the text flow, improved art pieces, and added new illustrations. The art changes, in particular, will affect what you wish to show in lecture.

- This is an entirely new chapter, greatly expanding on the material on extrasolar planets that previously appeared at the end of the "Formation of the Solar System" chapter.
- Techniques that were hypothetical—or not even imagined—have been proved successful since the previous edition. Specifics are given section by section below.
- We have added a large number of new explanatory figures and diagrams.
- The topic is now covered more completely and rigorously, including the addition of three new optional Mathematical Insights.
- A new Appendix E.4 was added to list the properties of the most interesting extrasolar planets.

Teaching Notes (By Section)

Section 13.1 Detecting Extrasolar Planets

This section goes through the many methods that have been (or will be) used to detect extrasolar planets. The difficulty of detecting planets directly bears repeating. It heightens the appreciation for the challenge and ingenuity of detecting them indirectly.

- New discoveries are reported virtually every month, with proved methods improving their sensitivity and other methods yielding their first detections. Since the first edition of this book, three methods have gone from potential to proved (transit, astrometry, gravitational lensing). It bears watching to see if any listed here as potential in fact have yielded recent detections.

Section 13.2 The Nature of Extrasolar Planets

The previous section is strictly limited to how planets can be detected, and not what the properties of the planets are. In some senses, this section is analogous to Chapter 7, which introduces the properties of the planets in our solar system.

- Again, new discoveries may stretch the boundaries of what is covered in this section. Keep an eye out for new discoveries of smaller, more distant planets. Similarly, ingenious techniques may be announced that reveal the properties of planets in new ways. For example, atmospheric information from transits and from eclipses are new capabilities since the previous edition of this textbook.

Section 13.3 The Formation of Other Solar Systems

Analogous to Chapter 8, this section covers what we deduce about solar system formation based on the observed properties of the planets.

- This is an important opportunity to draw attention to how science works. Specifically, the theory of solar system formation that was developed to explain our own planets completely failed to explain many of the new discoveries. Does it need to be modified, or completely thrown out? Are scientists "acting scientifically"? What would it mean for our solar system to form by exceptional rules, instead of the general rules once thought to apply? As with the previous sections, new discoveries continue to challenge even the revised theories. All in all, this section should be taught as "science in action" and not as "we have a complete understanding, and here it is."

Section 13.4 Finding New Worlds

Part III of the textbook closes with this forward-looking section on the future of planetary science as it extends to Earth-like planets around other stars. This major transformation should take place in the lifetimes of your students, and bring us closer to the central question of whether life in the universe is likely to be rare or common.

- You may wish to remind students how this latest phase of discovery is a continuation of the Copernican revolution. Not only is the Earth not the center of the Universe—it may not even be unusual or unique.
- You may wish to teach the “Life in the Universe” chapter (Chapter 24) next, which might change your emphasis in teaching this section.

Answers/Discussion Points for Think About It Questions

The Think About It questions are not numbered in the book, so we list them in the order in which they appear, keyed by section number.

Section 13.1

- (p. 390) a) Closer planets would make the wiggles larger and more closely spaced. b) More massive planets would only make the wiggles larger. These statements apply regardless of eccentricity or orientation.
- (p. 393) A large planet close to its star. A closer planet has a better chance of passing in front of its star, and will do so more frequently. A larger planet will make a larger drop in intensity.

Section 13.2

- (p. 397) No. A complete orbit of such planets takes decades, and we haven’t been able to detect planets for that long.

Section 13.3

- (p. 402) In science there is nothing wrong with being “just a theory.” Similarly, all theories are subject to modification as new discoveries are made. Just because Newton’s theory of gravitation had to be modified to account for relativity didn’t make it wrong—just incomplete. The same is true for the theory of solar system formation.

Section 13.4

- (p. 404) This question asks student to find out the status of the Kepler and COROT missions, which should be in orbit and operational before the next edition of the textbook.
- (p. 405) This opinion question should draw out good classroom discussion. For some, it may be the reason they took your class.

Solutions to End-of-Chapter Problems (Chapter 13)

1. Extrasolar planets are difficult to detect directly because they are small when viewed from Earth, many light-years away. In addition to their tiny sizes, the light from the stars near the planets is much, much brighter than the planets themselves. As a result, the tiny, dim planets get lost in the glare of the stars so that we have great difficulties seeing them.

2. There are three main ways used to look for extrasolar planets. The first way is the Doppler technique. This method uses the Doppler effect to detect the small wobble in a star's position due to the planet orbiting it. The astrometric technique is similar, except that rather than look for motions toward and away from us, it looks for tiny shifts in position. Finally, when a planet appears to cross in front of the star as seen from Earth, the transit method can be used to detect the planet's presence since the star's light is dimmed slightly.
3. The planet and the star both make orbits about a common center of mass rather than the planet alone orbiting the center of the star. The star, being much larger than the planet, has a much smaller orbit. But it does move slightly.

   Another way to look at this problem is conservation of momentum. As the planet changes direction in its orbit, something else has to move in the opposite direction to conserve momentum. The only other thing in the problem is the star, so we know that the star has to move in response to the planet.
4. The astrometric technique watches for tiny movements of the stars against the sky. However, this is very difficult to observe for two reasons. The first is that the amount of movement is very small and is difficult to detect even under good conditions. The second problem is that the time needed to see the movement is long: around 10 years for a planet with an orbit like Jupiter's. And, unfortunately, more distant (and therefore slower) planets cause the biggest changes of position in their stars.
5. The Doppler technique watches for movement in stars by looking for periodic Doppler shifts. These shifts in the frequency of the light we observe tell us that the star is moving toward or away from us and must therefore have something in orbit. Because the star's movement has the same period as the planet's orbit, watching for the period in the Doppler shift tells us the orbital period of the planet. The size of the shift tells us how fast the star is moving and, coupled with knowing the mass of the star and the distance to the planet (which we get from the mass of the star and the orbital period with Newton's version of Kepler's third law), we can find the mass of the planet inducing the wobble. Unfortunately, we really know the exact mass only if we also know the inclination of the orbit. The last thing we can learn from the Doppler data is the eccentricity of the orbit. Since the planet moves faster when it's closer to the star, the star must also move faster when the planet is closest. As a result, planets with eccentric orbits cause Doppler shifts that are not exactly symmetric.
6. Because the Doppler technique detects only movement toward and away from us, we do not know the planet's orbital inclination. A very massive planet with a large inclination will give the same tugs along our line of sight to its star as a smaller planet with no inclination. We can work out what the minimum mass of the planet would have to be if there were no inclination to the orbit, but we generally do not know if the mass is much larger than this since the planet could be on a very inclined orbit.

   That said, in general the planet masses should be quite close to these minimum masses. Statistically, we expect that $\frac{2}{3}$ of the planets will have masses less than twice the minimum mass. And 94% will have masses that are less than 10 times the minimum mass. While we cannot say for sure what a particular planet's mass is, we can say that most of the planets we have found so far are close to the minimum masses we calculate.
7. To use the transit method, we monitor the brightness of a star as the planet passes in front of the star's disk, much like when Venus or Mercury transit the Sun. Because the planet is dimmer than the star, the star's light dims slightly. From this, we can learn the relative sizes of the star and planet. (And since we usually have a good

estimate on the size of the star from stellar evolution theories, we know the size of the planet.) Unfortunately, this technique works only in the rare cases when the planet's orbit is nearly edge-on as seen from Earth, since otherwise the planet will pass to one side of the star as it orbits (much like how Mercury and Venus rarely transit the Sun as seen from Earth).

8. There is one candidate planet that was directly detected as of this writing. The (possible) planet is orbiting a brown dwarf and has a spectrum that indicates the presence of water in the atmosphere. This suggests that the object is a planet, although it is somewhat early to say for certain.
9. This table shows the properties we can measure and which of the three techniques we can use to measure each.

Property	**Doppler**	**Astrometric**	**Transit**
Radius			X
Mass*	X	X	
Eccentricity	X		
Orbital Distance	X	X	X
Density			X
Composition			X

* The mass derived depends on the inclination of the orbit.

10. The orbits of many known extrasolar planets are much more eccentric and much nearer their stars than the jovian planets of our solar system. This is surprising since our planet formation model suggests that planets should have nearly circular orbits and that jovian planets, which require ices to form, should form only farther out in the solar system.
11. There is reason to think that the extrasolar planets we have discovered so far are similar to the jovian planets of our solar system. For example, the masses of the planets are similar to the masses of the jovian planets. For cases where we have radii (when we have had transits), the radii are also consistent with jovian planets. Also, the spectroscopic analysis indicates that their atmospheres are similar in composition to those of the jovian planets.
12. "Hot Jupiters" are jovian planets that orbit much closer to their stars than the jovian planets in our solar system. These planets would be similar to the jovian planets, with a few interesting exceptions. Of course, they will be hotter than our jovian planets since they are closer to their stars. This added heat would keep them from contracting as far as our jovian planets have so that the hot Jupiters will be bigger in radius. (Theory and data both indicate that they should be about 50% larger.) Also, the planets will have different types of clouds in their atmospheres than our jovian planets do. Since they should be too hot for the hydrogen compounds to condense as in the jovian planets, these clouds will not be present. However, we expect to see clouds made up of silicates. These clouds might form bands on their planets like the clouds on the jovian planets, although the Coriolis effect will be weaker on the hot Jupiters since the planets should rotate more slowly with periods equal to their orbital periods (for the same reason that the Moon's spin period is equal to its orbital period: tidal forces).
13. Many extrasolar planets are much closer to their stars than we had expected. In fact, our formation model will not allow jovian planets to form that close to their stars. However, we think that it is possible for planets to migrate through their disks because of interactions with the gas in the disk.

14. Many of the extrasolar planets have highly eccentric orbits. We think that these eccentricities might be the result of planets passing through resonances with each other or having close encounters. Either type of interaction would allow the orbits of the planets to become more eccentric.
15. While most of the planetary systems around other stars that we have discovered so far do not look much like our solar system, this does not necessarily mean that our solar system is an uncommon case. The techniques we use to search for extrasolar planets tend to pick out certain types of systems more than others. Our system, for example, is one that we would have difficulty finding. It is possible that systems like ours are common and we are just having trouble seeing them.
16. The Kepler mission will stare at a patch of sky in the constellation Cygnus and measure the brightnesses of 100,000 stars. We hope to see the dimming due to an Earth-size planet transiting.
17. *GAIA* and *SIM* will both use astrometry to search for extrasolar planets. *SIM* will use interferometry to improve its resolution and thus make better measurements. It will test nulling interferometry, a technique that puts together the light from the two telescopes so that they cancel out the light from the star, leaving the planet's light intact.
18. There are several technologies that we hope will help us find Earth-like planets. One is nulling interferometry, a technique that uses multiple telescopes as with ordinary interferometry, but combines the light in such a way that the light from the central star cancels out, leaving the light from the planet. The Terrestrial Planet Finder Interferometry mission will use this technique, as will Darwin. We are also looking forward to the launch of the Terrestrial Planet Finder Coronographic mission, which will use a specially designed telescope to block the light of the central star. Before either of these launch, we plan to launch the Kepler mission, which will make sensitive measurements of the brightnesses of 100,000 stars to look for transits of Earth-size planets.
19. *An extraterrestrial astronomer surveying our solar system with the Doppler technique could discover the existence of Jupiter with just a few days of observation.* Surprising. Unless her technology operates on fundamentally different principles, it's necessary to observe one or more orbits before a planet's existence (and period) can be known.
20. *The fact that we have not yet discovered an Earth-mass planet tells us that such planets must be very rare.* False. Our observational techniques are not yet sensitive to planets of such low mass.
21. *Within the next few years, astronomers expect to confirm all the planet detections made with the Doppler technique by observing transits of these same planets.* False. Only planets whose orbits pass directly in front of their stars can be detected by the transit method.
22. *Although "hot Jupiters" are unlikely places to find life, they could be orbited by moons that would have pleasant, Earth-like temperatures.* False. Although some extrasolar planets may lie at distances where moons might be habitable, the planets in the "hot Jupiter" category are generally so close as to be uninhabitable.
23. *Before the discovery of planetary migration, scientists were unable to explain how Saturn could have gotten into its current orbit.* False. Saturn's orbit was easier to explain before the discovery of planetary migration.

24. *It's the year 2007: Astronomers have successfully photographed an Earth-size planet, showing that it has oceans and continents.* Surprising. No such technology will be available so soon.
25. *It's the year 2025: Astronomers have just announced that they have obtained a spectrum showing the presence of oxygen in the atmosphere of an Earth-size planet.* Possible. In fact, this is the mission of the Terrestrial Planet Finder and other proposed spacecraft.
26. *It's the year 2040: Scientists announce that our first spacecraft to reach an extrasolar planet is now orbiting a planet around a star located near the center of the Milky Way Galaxy.* Not possible. This would require faster-than-light travel.
27. *An extrasolar planet is discovered with a year that lasts only 3 days.* Not surprising. In fact, several such planets are discovered every (Earth) year.
28. *Later this year, scientists use the Doppler technique to identify a planet whose mass is equal to Earth's mass.* Surprising. At least for the coming year, the available technology is not expected to allow such a discovery.
29. *Astronomers announce that all the Doppler technique discoveries of extrasolar planets made to date are actually more massive brown dwarfs, and we had thought they were less massive only because we didn't realize that they have nearly face-on orbits.* Surprising. While this hypothesis was plausible when only a few planets were known, there are now far too many to share the suggested orbital coincidence. Furthermore, the transiting planets' orbits do not have such orbits.
30. *The number of extrasolar planets increases from around 150 in 2005 to more than 1,000 by the year 2010.* Not surprising. This is the goal of the Kepler mission, which will use the transit method.
31. c; 32. b; 33. b; 34. b; 35. c; 36. a; 37. b; 38. b; 39. b; 40. c.
41. Extrasolar planets were discovered in large numbers much more quickly than expected because many of them are so close to their stars. This not only increases the sensitivity of the Doppler method to their existence, but also shortens the time necessary to observe an entire orbit.
42. No current telescope—not even Hubble—can reduce the glare of normal stars enough to see a planet orbiting around it. The planet imaged in Figure 13.9 does not orbit a regular star, but a much, much fainter kind of object called a *brown drawf.*
43. This essay question asks students to repeat the explanation of the Doppler method in very simple terms. Answers should be graded based on simplicity and use of analogies such as the Doppler shift of a siren, or alternatively, waves on a pond.
44. The Doppler method has the main advantage of being able to detect planets in a wide range of orbits—as long as the orbit is not face-on. It is most sensitive to large planets close to their stars. Its disadvantage is that it can yield only the planet's mass and orbital properties. The transit method has the advantage of yielding completely different information: a planet's size. This method is even more biased to large planets close to their stars. Its disadvantage is that few planets have the necessary orbital alignment to even be detectable.
45. We think that "hot Jupiters" formed beyond the frost line, as in our solar system, and migrated inward due to interaction with the solar nebula. None formed in our solar system because the nebula must have dispersed shortly after the formation of our jovian planets, either by late formation of these planets, or by early turn on of the solar wind.

46. Orbital resonances between planets can cause one or both orbits to become elliptical, explaining why many extrasolar planets have orbits unlike those of our jovian planets. Such resonances can also lead to migration and even ejection of a planet from its solar system. These resonances are most similar to resonances between the Galilean moons, which makes their orbits elliptical, though the possibility of migration and ejection more resembles Jupiter's influence on the asteroid belt. The resonances of extrasolar planets are different from those in our solar system primarily by occurring between the major planets themselves. In our solar system, only Neptune and Pluto share an orbital resonance.
47. Only Saturn has a low density among our jovian planets, because it is the smallest planet that is still rich in hydrogen and helium. Among extrasolar planets, many are smaller in mass than Saturn, meaning they are less compressed and therefore lower density. A second important factor is how close these objects lie to their stars. The extra heating puffs them up, lowering their densities.
48. Essay question—answers will vary. Answers should include the shortness of the year, the high temperatures and unusual cloud composition, and the large apparent size of the Sun.
49. Because detection methods are most sensitive to massive planets, we should not yet conclude that low-mass planets are rare. We won't know until astronomers are able to detect planets like ours, and determine if they are truly absent from other planetary systems.
50. Observing project. This is probably the most challenging observing project in the textbook, but the reward is so great that students should be rewarded even for attempting it. Keeping copies of successful attempts may inspire others to try, since they'll know their peers could do it.
51. a. We will ignore the effects of varying reflectivities and assume that both planets reflect half of the light that hits them. What we see, then, is the ratio of the amount of light that the Earth reflects to the total amount of light emitted by the Sun. To find this ratio, let us assume that the amount of light emitted by the Sun is *L*. By the time this light has reached Earth's orbit, it has spread out over a sphere of radius *a*, the semimajor axis of Earth's orbit. The amount of light per unit area is $L/4\pi a^2$ since the sphere has an area of $4\pi a^2$. The amount of light that the Earth intercepts is equal to the amount of light per area times Earth's cross-sectional area, πr^2, where *r* is Earth's radius. The amount of light that Earth reflects is:

$$\frac{1}{2}\frac{L}{4\pi a^2}\pi r^2$$

Where the extra $\frac{1}{2}$ is for the reflectivity. Now, taking the ratio of this to the amount of light that the Sun emits, we will get the fraction of the Sun's total light that is reflected by the Earth:

$$\text{fraction reflected} = \frac{\frac{1}{2}\frac{L}{4\pi a^2}\pi r^2}{L}$$

Of course, we can do some algebra and simplify this quite a bit:

$$\text{fraction reflected} = \frac{1}{2}\frac{r^2}{4\,a^2}$$
$$= \frac{1}{2}\left(\frac{r}{2\,a}\right)^2$$

Appendix E tells us that the radius of Earth's orbit is 149.6×10^6 km and that the radius of the Earth is 6,378 kilometers. Putting those numbers in, we get:

$$\text{fraction reflected} = \frac{1}{2}\left(\frac{6378 \text{ km}}{2(149.6 \times 10^6 \text{ km})}\right)^2$$
$$= 2.27 \times 10^{-10}$$

Earth would appear only 2.27×10^{-10} as bright as the Sun.

b. The expression we derived in part (a) will work here as well, since we did not assume anything specific to Earth until we applied the actual numbers. All we need are Jupiter's radius and distance from the Sun. Appendix E stands ready to assist, as usual, with 71,492 kilometers for the radius and 778.3 10^6 km for the distance to the Sun. We can compute away:

$$\text{fraction reflected} = \frac{1}{2}\left(\frac{r}{2\,a}\right)^2$$
$$= \frac{1}{2}\left(\frac{71.492 \text{ km}}{2(778.3 \times 10^6 \text{ km})}\right)^2$$
$$= 1.06 \times 10^{-9}$$

Jupiter would appear 1.06×10^{-9} times as bright as the Sun. In this case, the extra radius beats out the extra distance.

52. a. From Mathematical Insight 13.3, we know that we can relate the fraction of the light that is blocked to the radius of the planet with the equation:

$$r_{\text{planet}} = r_{\text{star}} \times \sqrt{\text{fraction of light blocked}}$$

If the star TrES-1 is 85% of the Sun's radius, we can find its radius in kilometers just by looking up the Sun's radius in Appendix E. The Sun's radius is 696,000 kilometers, so we get TrES-1's radius: $0.85 \times (696{,}000 \text{ km}) = 592{,}000$ km. Also knowing that the planet blocks 2% of the star's light, we can find the planet's radius:

$$r_{\text{planet}} = (592{,}000 \text{ km}) \times \sqrt{0.02}$$
$$= 83{,}700 \text{ km}$$

The planet has a radius of 83,700 kilometers, a bit larger than Jupiter.

b. We need to convert the radius to centimeters and the mass of the planet to grams to get the density in the units we want. First, the radius:

$$83{,}700 \cancel{\text{km}} \times \frac{1000 \cancel{\text{m}}}{1 \cancel{\text{km}}} \times \frac{100 \text{ cm}}{1 \cancel{\text{m}}} = 8.37 \times 10^9 \text{ cm}$$

Now we find the mass. We are told that the planet has a mass 0.75 times that of Jupiter. We get a mass of $0.75 \times (1.90 \times 10^{27}$ kg. Converting to grams:

$$1.43 \times 10^{27} \cancel{\text{kg}} \times \frac{1000 \text{ g}}{1 \cancel{\text{kg}}} = 1.43 \times 10^{30} \text{ g}$$

To find the density, we need mass over volume. The last thing we need to do is to turn the radius of the planet into its volume. Since the planet is spherical, its volume will be:

$$\begin{aligned}\text{volume} &= \frac{4}{3} \pi r^3 \\ &= \frac{4}{3} \pi (8.37 \times 10^9 \text{ cm})^3 \\ &= 2.46 \times 10^{30} \text{ cm}^3\end{aligned}$$

The density is:

$$\begin{aligned}\text{density} &= \frac{\text{mass}}{\text{volume}} \\ &= \frac{1.43 \times 10^{30} \text{ g}}{2.46 \times 10^{30} \text{ cm}^3} \\ &= 0.581 \text{ g/cm}^3\end{aligned}$$

The planet has a density of 0.581 g/cm^3, less than half of Jupiter's density.

53. The planet around 51 Pegasi has an orbital period of 4.23 days. Because the star has about the same mass as the Sun, we can find the planet's distance from its star by using Kepler's third law in its original form. But first we must convert the period of 4.23 days into units of years:

$$p = 4.23 \cancel{\text{days}} \times \frac{1 \text{ yr}}{365 \cancel{\text{days}}} = 0.0116 \text{ yr}$$

Now we can use Kepler's third law to find the planet's distance in AU:

$$p^2 = a^3 \Rightarrow a = \sqrt[3]{p^2} = \sqrt[3]{0.0116^2} = 0.0512 \text{ AU}$$

The planet around 51 Pegasi orbits with an average distance (semimajor axis) of 0.051 AU—much closer to its star than Mercury is to our Sun.

At 0.6 times the mass of Jupiter, the planet in 51 Pegasi seems likely to be a jovian planet. In the nebular theory, jovian planets form outside the "frost line," which in our solar system is outside the orbit of Mars (1.5 AU). Thus, it is surprising to find a jovian planet that is located closer to its star than Mercury is to our Sun.

54. a. First we will find out how many stars are born each year. We are told that there are 100 billion stars in the galaxy and that they were born over 10 billion years ago. So we can find how long we go between births of stars:

$$\frac{10 \text{ billion years}}{100 \text{ billion stars}} = 0.1 \frac{\text{yr}}{\text{star}}$$

We are told to assume that 5%, or 1 in 20, of these stars have planetary systems. All that remains is to multiply this time by 20 stars/planetary system. We find that a planetary system is born about every 2 years.

b. If the planet formation rate in our galaxy is typical and there are 100 billion galaxies in the universe, then we can take the time between births of planetary systems in our galaxy and divide by 100 billion galaxies/universe to find out how often a planetary system is born in our universe. That comes out to 2×10^{-10} yr between star births. We can convert this to seconds and get 0.006 second between planetary systems being born.

c. Students responses will vary.

55. a. If we assume that the star is a solar mass star, we can use the equation in Mathematical Insight 13.1 to find the mass of the planets. First, we need the semimajor axes of the planets, which we can find with Mathematical Insight 13.1:

$$a = \sqrt[3]{\frac{GM_{\text{star}}}{4\pi^2} p^2}$$

Putting in the mass of the Sun (2×10^{30} kg) and the periods of the planets (3 and 300 days, which, converted to seconds, are 2.6×10^5 sec and 2.6×10^7 sec), we find that they have semimajor axes of 6.1×10^9 m and 1.3×10^{11} m. The equation for planet mass is:

$$M_{\text{planet}} = \frac{M_{\text{star}} v_{\text{star}} p_{\text{planet}}}{2\pi a_{\text{planet}}}$$

Putting in the values for the two planets given, we find that the masses are 6.8×10^{26} kg and 3.2×10^{27} kg. The planet with the longer period is the larger one.

b. Repeating the steps from part (a), we find that the semimajor axes of the two planets are now 4.3×10^9 m and 9.2×10^{10} m. Putting these values into the mass formula, we find that the masses are 4.8×10^{26} kg and 2.3×10^{27} kg. The masses are smaller for these planets than for the corresponding ones in part (a).

56. We can use the formula in Mathematical Insight 13.2 to find the masses of the smallest planets detectable at 0.1 AU and 10 AU. First, we need the orbital periods of these planets. We can use Newton's version of Kepler's third law to get this from the mass of the star (2×10^{30} kg) and the semimajor axes:

$$p = \sqrt{\frac{4\pi^2}{GM} a^3}$$

We will need the semimajor axes in meters rather than AU for this. Recalling that 1 AU is 1.496×10^{11} m, the conversion is straightforward and yields semimajor axes of 1.5×10^{10} m and 1.5×10^{12} m. Putting in these values, then, we find that the orbital periods are 1.0×10^5 sec and 1.0×10^8 sec. (*Note*: For this part of the problem, we could have just used Kepler's law rather than Newton's version. For the $0.5M_{\text{Sun}}$ star, we would have needed Newton's version, though.)

Now that we have the period, we have everything we need to find the mass of the smallest planet that we can detect. The smallest planets will cause the smallest speeds in their stars, so since 1 m/s is the smallest speed we can detect, we will use this for the speed in the formula for mass:

$$M_{\text{planet}} = \frac{M_{\text{star}} v_{\text{star}} p_{\text{planet}}}{2\pi a_{\text{planet}}}$$

Putting in the semimajor axes, periods, speed, and mass of the star, we find that the smallest planets that we could detect are 2.1×10^{25} kg for a planet 0.1 AU from the star and 2.1×10^{26} kg for a planet 10 AU from the star. These planets are still a few times larger than Earth.

If we change the mass of the star to $0.5M_{\text{Sun}}$, we will have to recompute both the periods and then the masses of the planets. The periods of the two planets (still at 0.1 and 10 AU) are 1.4×10^6 sec and 1.4×10^9 sec. The masses are 1.5×10^{25} kg for the planet at 0.1 AU and 1.5×10^{26} kg for the planet at 10 AU. Both of these are still a few times larger than Earth is.

57. a. From Appendix E, we know that the mass of the Sun is 2×10^{30} kg, so a 2-solar-mass star has a mass of 4×10^{30} kg. The period of the planet is 5 days. We can convert this to seconds to get 4.32×10^5 sec. Using Newton's version of Kepler's third law:

$$a = \sqrt[3]{\frac{GM_{\text{star}}}{4\pi^2} p^2}$$

we put in the numbers and get the semimajor axis:

$$a = \sqrt[3]{\frac{\left(6.67 \times 10^{-11} \frac{\text{m}^3}{\text{kg} \times \text{s}^2}\right)(4 \times 10^{30} \text{ kg})}{4\pi^2}(4.32 \times 10^5 \text{ s})^2}$$
$$= 1.08 \times 10^{10} \text{ m}$$

This can easily be converted to kilometers to get 1.08×10^7 km. Since 1 AU is 1.496×10^8 km we can also change the value to astronomical units and get 0.0722 AU.

b. We do the same thing as above, except this time we get a star with half a solar mass, or 1×10^{30} kg, and a planet with an orbit of 100 days, or 8.64×10^6 sec. So putting the numbers into the equation:

$$a = \sqrt[3]{\frac{\left(6.67 \times 10^{-11} \frac{\text{m}^3}{\text{kg} \times \text{s}^2}\right)(1 \times 10^{30} \text{ kg})}{4\pi^2}(8.64 \times 10^6 \text{ s})^2}$$
$$= 5.01 \times 10^{10} \text{ m}$$

This is the same as 5.01×10^7 km or 0.335 AU.

58. a. To solve this problem we can use Kepler's third law as Kepler wrote it:

$$a^3 = p^2$$

We are told that the minimum period for a planet that we can detect is 4 days. We will need this in years. Converting gives us 0.0164 year. So we calculate the semimajor axis of this planet:

$$\begin{aligned} a &= p^{2/3} \\ &= (0.0164 \text{ yr})^{2/3} \\ &= 0.0646 \text{ AU} \end{aligned}$$

The maximum period we could detect is 40 days, or 10 times the shortest period. That must be 0.164 year. Calculating the semimajor axis:

$$\begin{aligned} a &= p^{2/3} \\ &= (0.164 \text{ yr})^{2/3} \\ &= 0.300 \text{ AU} \end{aligned}$$

We could find planets with semimajor axes between 0.0646 AU and 0.300 AU.

b. For this problem we will need Newton's version of Kepler's third law:

$$a = \sqrt[3]{\frac{GM_{\text{star}}}{4\pi^2} p^2}$$

In this case, the stars are half and twice the mass of the Sun. Since the mass of the Sun is about 2×10^{30} kg, these two stars have masses of 1×10^{30} kg and 4×10^{30} kg. The periods are still 4 to 40 days, but now we will need those in seconds. Converting is straightforward and we get 3.46×10^5 sec and 3.46×10^6 sec.

Starting with the smaller star, putting in the numbers yields semimajor axes of 5.87×10^9 m and 2.72×10^{10} m. In kilometers, these are 5.87×10^6 km and 2.72×10^7 km. Since 1 AU is 1.496×10^8 km, we can convert the semimajor axes to AU to get 0.039 AU and 0.18 AU.

Moving to the larger star, we repeat all of these calculations. We get semimajor axes between 1.48×10^{10} m and 6.87×10^{10} m (1.48×10^7 km and 6.87×10^7 km). We convert this to astronomical units to find that the range of possible semimajor axes is between 0.099 AU and 0.46 AU.

From these, we see that the combination of most massive star and longest period orbit will result in the largest semimajor axis for the planet.

59. We will use Mathematical Insight 13.2 to find out how fast each planet in our solar system makes our Sun move. The speed that the star moves is given by:

$$v_{\text{star}} = \frac{M_{\text{planet}}}{M_{\text{Sun}}} \frac{2\pi a_{\text{planet}}}{p_{\text{planet}}}$$

We can make a table with the relevant data:

Planet	Mass (kg)	Semimajor Axis (m)	Period (s)	Speed of Sun (m/s)
Mercury	3.30E+23	5.79E+10	7.61E+06	7.89E-03
Venus	4.87E+24	1.08E+11	1.94E+07	8.52E-02
Earth	5.97E+24	1.50E+11	3.16E+07	8.88E-02
Mars	6.42E+23	2.28E+11	5.94E+07	7.73E-03
Jupiter	1.90E+27	7.78E+11	3.75E+08	1.24E+01
Saturn	5.69E+26	1.43E+12	9.30E+08	2.74E+00
Uranus	8.66E+25	2.87E+12	2.65E+09	2.94E-01
Neptune	1.03E+26	4.50E+12	5.17E+09	2.81E-01
Pluto	1.31E+22	5.92E+12	7.84E+09	3.11E-05

We see that Jupiter causes the largest effect on the Sun, making the Sun move at about 12.4 meters per second. This is not surprising since Jupiter is the largest planet in the solar system. The Doppler shift of Jupiter is enough to be seen by our current technologies if the limit is about 1 meter per second, though the alien astronomers would have to observe one or more Jupiter years to be sure.

60. a. Following the method of Mathematical Insight 11.1 and assuming a reflectivity of 0.15, we find the "no greenhouse" temperature of the planet around 51 Pegasi:

$$T = 280\text{ K} \times \sqrt[4]{\frac{1-\text{reflectivity}}{d^2}} = 280\text{ K} \times \sqrt[4]{\frac{(1-0.15)}{0.051^2}} = 1{,}190\text{ K}$$

This is far warmer than the Earth's temperature of under 300 K.

b. With a very high reflectivity of 0.8, the "no greenhouse" temperature of the planet around 51 Pegasi is:

$$T = 280\text{ K} \times \sqrt[4]{\frac{1-\text{reflectivity}}{d^2}} = 280\text{ K} \times \sqrt[4]{\frac{(1-0.8)}{0.051^2}} = 830\text{ K}$$

This is still significantly warmer than the Earth's temperature of under 300 K.

c. Even in the reflective atmosphere case, the planet is probably too hot to be habitable.

61. a. We will start by drawing a picture for ourselves. We seek the angle, θ, that is the inclination of the planet's orbit. If the planet has a semimajor axis of a and the star's radius is r, we get the following picture:

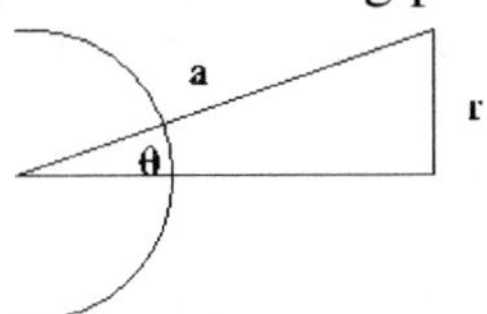

To solve this, recall that by the definition of the sine function:

$$\sin(\theta) = \frac{r}{a}$$

We can solve this for θ:

$$\theta = \arcsin\left(\frac{r}{a}\right)$$

Now all that remains is to find r and a (in the same units) and plug in the numbers. We are told to assume that the radius of the star is the same as the Sun. Appendix E says that the Sun's radius is 696,000 kilometers. We are also told that the semimajor axis of the planet's orbit is 0.05 AU. We can convert from astronomical units to kilometers by recalling that 1 AU is 1.496×10^8 km. So 0.05 AU is 7.48×10^6 km. Now we put those numbers into our formula and find that the maximum inclination for a transit is 5.3°.
Since the inclinations can fall between 0° and 90° and only inclinations between 0° and 5.3° will cause transits, we can find the fraction of possible inclinations that lead to transits:

$$\text{fraction of inclinations} = \frac{5.3°}{90°}$$
$$= 0.059$$

If we observe 10,000 stars and 2% have "hot Jupiters" like this, that's 200 "hot Jupiters." We just calculated what fraction of those will be detectable via transits, so multiplying the fraction of inclinations that lead to transits (0.059) by the number of planets (200), we find that about 12 planets will be detectable.

b. Doing the same thing as above, except for the case where a is 1 AU (1.496×10^8 km), we find that the maximum inclination is 0.27°. Thus, the fraction of such planets that we should be able to see transit is 0.003.
If we look at 10,000 stars and 10% have Earth-like planets, then there are 1,000 stars with Earth-like planets. Using the fraction of those that we should be able to see transit as above, we should be able to see about three Earth-like planets transit out of the 10,000 stars we survey.

c. Doing the same thing as above, except for the case where a is 5 AU (7.48×10^8 km), we find that the maximum inclination is 0.053°. Thus, the fraction of such planets that we should be able to see transit is 5.9×10^{-4}.
If we look at 10,000 stars and 10% have Jupiter-like planets, then there are 1,000 stars with Jupiter-like planets. Using the fraction of those that we should be able to see transit as above, we should be able to see about 0.6 Jupiter-like planet transit out of the 10,000 stars we survey. This means that we have a bit better than a 50-50 chance of seeing any such planets.
Things get worse for the Jupiter-like planets when we consider how long we would have to observe the stars to see three transits. Since Jupiter has an orbital period of 12 years (Appendix E), we would have to wait up to 36 years (three orbital periods) to see the three transits. (Technically, if we got lucky, we would have to wait as little as 24 years.) On the other hand, we would only have to wait at most 3 years to see three transits of Earth-like planets. For the "hot Jupiter" we examined in part (a), we can find the orbital period with Kepler's third law, $a^3 = p^2$. The period is 0.011 year, so would have to wait at most 0.033 year (about 12 days).

Part IV: A Deeper Look at Nature

Chapter S2. Space and Time

General Notes on Part IV (Chapters S2–S4)

The chapters of Part IV (Chapters S2–S4) are labeled "supplementary" because coverage of them is optional. Covering these chapters will give your students a deeper understanding of the topics that follow—on stars, galaxies, and cosmology—but the later chapters are self-contained and may be covered without having covered Chapters S2–S4 at all. We have therefore designed these chapters so that they can be used in any of three ways:

1. You may skip Part IV entirely. We have written the chapters of Parts V through VII so that they are not dependent on the Part IV chapters.
2. If you want to present a brief overview of relativity and quantum mechanics, you can cover only the first section in each of Chapters S2–S4. Each first section presents an overview of the key ideas. Covering only the first section will provide about the amount of background found in most other introductory astronomy texts, in which relativity and quantum mechanics are given just a few pages when relevant topics come up.
3. You can cover Chapters S2–S4 in depth. Our experience shows that each of these three chapters requires a minimum of about a week in class (e.g., 3 hours) to do it justice; it is even better if you can spread the three chapters over 4 weeks. Note that, while this coverage takes significant class time, it should enable you to move more quickly through later chapters of the book, especially Chapter 16 (in which we discuss brown dwarfs), Chapter 18 (on stellar corpses), and the chapters with significant cosmology (especially Chapters 20–23).

The placement of these three supplementary chapters is dictated by the fact that, if you cover them, they will enhance students' understanding of the chapters in Parts V and VI (even though they are not prerequisites to those chapters).

Incidentally, if you have not previously covered relativity in your astronomy classes, we hope you will at least consider incorporating it in the future. Based on having taught it for many years in the way presented in these chapters, we've found not only that students are able to understand the topic, but also that most of our students name relativity as their favorite part of their course in astronomy.

General Notes on Chapter S2

This chapter provides students with a short but solid introduction to the special theory of relativity. The first section provides a concise overview of special relativity, and the remaining sections explain the logic and evidence behind this overview. Thus, it is possible to teach only the first section if you want to give students an overview of special relativity but do not have time to go into any details of the theory.

Because relativity forces students to think in a different way than they are accustomed to, we recommend that you cover the key elements and thought experiments of this chapter essentially verbatim from the text. We have found that repetition of explanations in class and in the text helps solidify student understanding of unfamiliar topics such as relativity.

As always, when you prepare to teach this chapter, be sure you are familiar with the relevant media resources (see the complete, section-by-section resource grid in Appendix 3 of this Instructor's Guide) and the online quizzes and other study resources available on the Mastering Astronomy Web site.

Teaching Notes (By Section)

Section S2.1 Einstein's Revolution

This first section introduces the idea of relativity and summarizes the key findings of relativity that are discussed in the remainder of the chapter.

- The bulleted list in this section lists what we refer to as five key ideas of special relativity: (1) inability to exceed c, (2) time dilation, (3) relativity of simultaneity, (4) length contraction, and (5) mass increase and $E = mc^2$. The rest of the chapter essentially explains these five ideas in detail; if you cover only this first section, use this list as the main element in your overview of relativity.
- We recommend that you bring a large globe (a large, inflatable Earth globe works great) when explaining the example of the supersonic airplane flight from Nairobi to Quito that appears in Figure S2.1. We have found this example to be particularly effective in helping students understand that motion really is relative.
- Please note that, although many people often state that relativity defies common sense, we have found that such statements only serve to make students resistant to learning relativity. Thus, we instead choose the strategy of arguing that relativity does not violate common sense because we have no common experience with speeds at which the effects of relativity become noticeable.

Section S2.2 Relative Motion

This section introduces thought experiments to help students understand the idea of relative motion and then explains why the absoluteness of the speed of light is such a surprising fact—and why this fact implies that we cannot reach or exceed the speed of light.

- The idea of relative motion is counterintuitive for many students even in its low-speed, everyday applications. Thus, we begin with three thought experiments involving ordinary speeds before moving to thought experiments with relative motion at speeds approaching the speed of light. We have found that it helps to "act out" these low-speed thought experiments in class. For example, you might recruit two volunteers—one to play the role of Jackie and one to hold the baseball. You and your volunteers can then demonstrate the motion involved in the thought experiments, which help students see why the different observers see things in different ways. If your students are having difficulty with this concept, you may wish to do several additional thought experiments of a similar nature.
- The key thought experiment of this section is Thought Experiment 5, because this is the one that shows students the counterintuitive nature of the idea that the speed of light is absolute. Unless students understand what is "strange" about the absoluteness of the speed of light, they will not be able to follow the remaining thought experiments.

- Thought Experiment 6 then uses the absoluteness of the speed of light to "prove" that no material object can reach or exceed the speed of light; note that this is the first of the five bullets from the bulleted list in Section S2.1. Students will invariably look for loopholes in the logic establishing the absoluteness of the speed of light, so be prepared for many such questions. (You may wish to refer the students to the Think About It box titled "What if Light Can't Catch You?") In addition, students will question the postulate that the speed of light is absolute. For this latter objection, we recommend emphasizing that the absoluteness of the speed of light is an experimentally verified fact—and reminding the students that this point is discussed further in Section S2.3.
- Note that we have chosen to use the term *free-float frame* in place of the more common *inertial reference frame* because we feel it is much more intuitive. The term *free-float frame* was coined by E. F. Taylor and J. A. Wheeler in their book *Spacetime Physics,* 2d ed. (W.H. Freeman and Company, New York, 1992).
- FYI: We are often asked why we chose a woman named Jackie to be the participant in our thought experiments. The answer is that we wrote the first draft of this chapter during the 1996 Olympics in Atlanta—and decided that if anyone was capable of moving at speeds close to the speed of light, it would be Jackie Joyner Kersee. Thus, the Jackie in our thought experiments was named in her honor.

Section S2.3 The Reality of Space and Time

By this point, students should understand what we mean by the absoluteness of the speed of light. This section therefore uses this idea in a series of thought experiments to establish the ideas listed in the remaining four bullets of the bulleted list in Section S2.1 and to discuss the evidence supporting these ideas.

- Under the subheading "Time Differs in Different Reference Frames," you may be wondering why we begin with the example of the ball being tossed into the train rather than going directly to the light paths in Thought Experiment 7. We've found that, if you ask students what a vertical light path would look like to an observer moving past them, most students initially think the observer would see the light path slanting backward, rather than forward with your perceived motion. By beginning with the example of the ball in the train, we usually can clear up this confusion. In addition, the comparison of the two situations helps students understand why the absoluteness of the speed of light implies that time varies in different reference frames, whereas there is no such obvious implication with the ball in the train (since different observers can disagree on the ball's speed).
- There is a subtlety in the relativity of simultaneity that may bother some thoughtful students. At the instant of the flashes, both you and Jackie agree that she is midway between them (i.e., she is in the middle of the train, and flashes occur on its front and rear ends). You also both agree that the flashes travel at the speed of light. Thus, from your point of view, the green flash reaches Jackie first because its speed relative to her is greater than the speed of light: It is the light's speed plus Jackie's speed toward the position where you saw the light emitted. Similarly, you see the red flash moving relative to her at less than the speed of light. There are no violations of relativity: Everyone still agrees on the speed of the light, and no one sees any individual object moving at greater than the speed of light.

A good way to explain this point in class is with the following thought experiment: Suppose that you see spaceship A going at 0.99 c to your left and spaceship B going at 0.99 c to your right. According to you, they are separating at 1.98 c and therefore will be 1.98 light-years apart after 1 year. But either of the two spaceships will see the other going at less than c and therefore will measure the other to travel less than 1 light-year in 1 year. Of course, the three observers (A, B, and you) will disagree about both the distances and the times between events.

- Note that we have deliberately chosen not to address the fact that Jackie sees your time dilated just as you see hers dilated; this issue is covered in Section S2.4.
- FYI: Note that, while time dilation applies to any reference frame moving relative to you, what you actually *see* may be more complex. If the moving frame has a radial component of motion toward or away from you, what you see is complicated by kinematic effects arising from the fact that the reference frame is moving at a speed close to the speed at which its light moves out ahead of or behind it. In fact, if the reference frame is moving toward you, these kinematic effects will cause it to appear to be moving faster than the speed of light—which is the origin of the apparent superluminal motion often seen in quasars. This issue is discussed at a level accessible to many students (if their math skills are reasonably strong) in Taylor and Wheeler's *Spacetime Physics*, 2d ed.
- FYI: It is also difficult to *see* length contraction, because motion near the speed of light introduces apparent rotations of objects. Again, this point is discussed at a level accessible to many students in Taylor and Wheeler's *Spacetime Physics*, 2d ed.
- Note that we have chosen not to go into detail concerning the Michelson-Morley experiment or to discuss the historical issue of the ether. While both topics are interesting and relevant, they are not necessary to make the case for special relativity. If you have time in your class, you may wish to discuss these topics to give historical context to the subject of relativity.

Section S2.4 Toward a New Common Sense

In this section we point out the fact that Jackie must claim the same things about you (e.g., time dilation, length contraction, and mass increase) that you claim about her, as long as you are moving at a constant relative velocity. We then use this idea to help students begin to establish what we refer to as a "new common sense"—a common sense that can incorporate the ideas of relativity. Note that we do not expect students to develop such a new common sense easily; at this point, we want them only to recognize that it is possible to do so. They can move further toward developing this new common sense when they study Chapter S3, but they will need to study relativity further if they wish to truly get a handle on this new common sense. If they have enjoyed learning about relativity, we hope they will be encouraged to read more about it in the future.

- As discussed above (fourth bullet under Section S2.3), Thought Experiment 12 works as written only if Jackie's motion is transverse to yours. We have chosen not to mention this subtlety in the text.
- The final learning goal discusses how special relativity offers a "ticket to the stars"—a way to make interstellar trips in times that are reasonably short for the traveler, if not for those who stay behind on Earth.

- Note that we do not discuss the twin paradox at this point, except in a footnote. As described in the footnote, it is possible to resolve the twin paradox with special relativity. (Again, a good discussion of this issue can be found in Taylor and Wheeler's *Spacetime Physics*, 2d ed.) However, we have found that it is easier for students at this level to understand the twin paradox in the context of general relativity, so we discuss it in a Special Topic box in Chapter S3.
- The actual distance to Vega is close to 26 light-years; we use "about 25 light-years" to make the mathematics easier for students to follow.

Answers/Discussion Points for Think About It Questions

The Think About It questions are not numbered in the book, so we list them in the order in which they appear, keyed by section number.

Section S2.1

- (p. 413) On the treadmill, your speed is measured relative to the moving tread. Your speed relative to the ground is essentially zero. An observer on the Moon would see you moving with the rotation of the Earth (and also with the motion of Earth relative to the Moon). Other viewpoints on your speed might include an observer on the Sun, who would see you moving with the Earth in its orbit at some 100,000 km/hr, or an observer in a distant galaxy, who might see you moving away with the expansion of the universe at a substantial fraction of the speed oflight.
- (p. 413) This question is designed to help students recognize that maps and globes are made with the Northern Hemisphere bias of most map makers. (*Note:* If you can get hold of one, you should show students a map in which the Southern Hemisphere is placed "up"—ask an Australian friend to buy one for you.)

Section S2.2

- (p. 416) If you throw the ball in Jackie's direction at 80 km/hr while Jackie is moving away from you at 90 km/hr (as in Thought Experiment 3), Jackie will see the ball moving away from her at 10 km/hr. If you throw it toward her at 80 km/hr while she is moving toward you at 90 km/hr, she will see the ball coming toward her at 170 km/hr.
- (p. 417) If Jackie is moving away from you at a speed short of the speed of light by only 3 km/s, you'll see a light beam traveling faster than Jackie by only 3 km/s. She'll measure the same light beam to be traveling past her at the full speed of light.

Section S2.3

- (p. 426) The point of this question is to make students think about the fact that you can't simply discount an idea without considering the consequences. In this case, many students would like to make the strange ideas of relativity "go away" by saying that the speed of light is not truly absolute. Here, they must consider the fact that a nonabsolute speed of light would introduce other consequences in terms of how people perceive events.

Solutions to End-of-Chapter Problems (Chapter S2)

1. The theory of relativity refers to both Einstein's theory of special relativity and his theory of general relativity. The theory of special relativity deals with the special case when there is no gravity, while general relativity deals with all situations, with or without gravity.
2. The predictions made by special relativity include:
   - No information can travel faster than light and no material object can ever reach the speed of light.
   - Time will run more slowly for a person traveling at speeds close to light.
   - Observers may disagree whether two events were simultaneous if they are moving relative to each other.
   - Objects moving at speeds close to light will appear shorter in their directions of motion.
   - The apparent masses of objects moving at close to the speed of light are larger than when they are not moving.
3. A paradox is a situation that seems to violate common sense or to contradict itself. Paradoxes can lead us to a deeper understanding of an issue by forcing us to accept that our commonsense ideas are incorrect.
4. The absolutes in the universe, according to special relativity, are that the laws of nature are the same for everyone and that the speed of light is constant for everyone.
5. Two objects share a reference frame if they are not moving relative to each other. A free-floating reference frame is a reference frame where everything is weightless.
6. If we each performed experiments of any kind, my friend and I would get the same results. Thus, there is no way we can tell which reference frame is moving and which one is not.
7. Answers will vary.
8. Time dilation refers to the idea that if you observe someone moving relative to you, you will observe time to be running more slowly for them than for you.
9. Observers in two different reference frames can disagree about the order in which events happen because each measure time and space differently—they agree only on the speed of the light traveling from events. Since we cannot tell which observer is moving, both views are equally valid. (Thought Experiment 8 explains the idea in more detail.)
10. Length contraction is the shortening of objects in their directions of motion when they are observed in another reference frame. So from your point of view, if a spaceship flies past you it will be shorter than if you were in the same reference frame as the spaceship.
11. Mass increase refers to the effect whereby objects gain mass when they are moving relative to an observer. So any object you observe that is moving relative to you has a greater mass than it would if it were at rest in your reference frame.
12. Student answers will vary.
13. Special relativity has been tested many times. For example, mass increase and time dilation have been observed with particles that move near the speed of light in particle accelerators. Accelerators have also shown that we cannot get particles to move faster than the speed of light. The Michelson-Morley experiment showed that the speed of light really is absolute. Other experiments include observations of time

dilation with precise clocks on moving airplanes and nuclear energy, demonstrating Einstein's equation $E = mc^2$.

14. Because $E = mc^2$ is a direct consequence of special relativity, testing it tests relativity. If $E = mc^2$ did not hold up, there would have to be something wrong with the theory. We can see examples of this equation in action when we feel the Sun's rays, use nuclear power plants, or see footage of an atomic bomb.
15. While I will see my friend's mass increased, her time dilated, and her length contracted, she will not feel any of those effects. To her, her own mass, length, and time will be normal. However, since I am moving from her perspective, she will see *me* as having a mass increase, a length contraction, and a time dilation.
16. Relativity can allow us to make long trips to distant stars in a reasonably short amount of time by taking advantage of time dilation. While to an observer on Earth a journey could take many years, to a person in the spaceship it might take much less time. (From the traveler's point of view, the distance to the star will be contracted, so the trip will be shorter.) Unfortunately, when the traveler returned to Earth, everyone there would have aged many years.
17. *Einstein proved that everything is relative.* This is a false statement that reflects a common misconception. He showed that motion is relative, while other things—such as the laws of nature and the speed of light—are absolute.
18. *An object moving by you at very high speed will appear to have a higher density than it has at rest.* (*Hint:* Think about the effects on both length and mass.) This is true, because its mass will be increased while its length will be decreased. A greater mass in a smaller volume means a higher density.
19. *Suppose you and a friend are standing at opposite sides of a room, and you each pop a peanut into your mouth at precisely the same instant. According to the theory of relativity, it is possible for a person moving past you at high speed to observe that you ate your peanut before your friend ate hers.* This is true. If two events appear simultaneous in one reference frame, they may not appear simultaneous in a different reference frame.
20. *Suppose you and a friend are standing at opposite sides of a room, and you each pop a peanut into your mouth at precisely the same instant. According to the theory of relativity, it is possible for a person moving past you at high speed to observe that you ate cashews rather than peanuts.* This is false. While people in different reference frames may disagree about the timing of events, they cannot disagree on the specifics of a particular event.
21. *Relativity is "only a theory," and we really have no way to know whether any of its predictions would really occur at speeds close to the speed of light.* This statement is false. We have tested relativity at high speeds for subatomic particles in particle accelerators.
22. *The detonation of a nuclear bomb is a test of the special theory of relativity.* This is true, because the bomb releases energy from its mass, thereby verifying a prediction of Einstein's special theory of relativity.
23. *If you could travel away from Earth at a speed close to the speed of light, you would find yourself feeling uncomfortably heavy because of your increased mass.* This is false: Motion is relative, so you would consider yourself to be at rest and therefore would not notice any changes to your own time, length, or mass.
24. *We do not have rockets that can reach the speed of light today, but someday we will be able to build more powerful rockets that will allow us to travel much faster than*

the speed of light. This is false: More power can get us closer to the speed of light, but we still can't reach it.

25. *If you see someone's time running slowly in a different reference frame, that person must see your time running fast.* This statement is false. As long as you are both in free-float reference frames, you must both get the same experimental results, meaning that the person will see your time running slow, not fast.
26. *If you had a sufficiently fast spaceship, you could leave today, make a round-trip to a star 500 light-years away, and return home to Earth in the year 2020.* This is false. As the traveler on the fast ship, you could potentially make the 1,000 light-year round-trip in just a few years of ship time. However, more than 1,000 years would have to pass on Earth, so you would return in the distant future.
27. c; 28. c; 29. a; 30. b; 31. b;
32. b; 33. c; 34. a; 35. a; 36. c.
37. On a stationary bike, the speed is supposed to represent how fast the bike would be moving if it had wheels in contact with the ground. Clearly, it is a good example of the idea of relative motion, since the actual speed of the bike through the room is zero.
38. According to an astronaut on the Moon, you would be moving with the rotating Earth. The precise speed of rotation depends on your latitude, but for most latitudes where people live it is more than 1,000 km/hr.
39. a. If Bob is coming toward you at 75 km/hr and you throw a baseball in his direction at 75 km/hr, he'll see the ball coming at him at 75 + 75 = 150 km/hr.
    b. If Marie is traveling away from you at 120 km/hr when she throws a baseball in your direction at 100 km/hr, you'll see the ball traveling away from you at 20 km/hr.
    c. José's speed doesn't matter in this case: You will always measure the light beam to be traveling at the speed of light.
40. a. If Carol is traveling away from you at 75 km/hr and Sam is going away from you in the opposite direction at 90 km/hr, Carol will see Sam going away from her at 165 km/hr.
    b. If you now throw a ball toward Sam at 120 km/hr, he'll see the ball coming at him at 120 – 75 = 45 km/hr. Carol will see the ball going away from her at 75 + 120 = 195 km/hr.
    c. Cameron's speed doesn't matter in this case: You will always measure the light beam to be traveling at the speed of light.
41. a. As you observe a spaceship moving past, you will see their clocks running slow. That is, everything on the spaceship would appear to be taking place in slow motion.
    b. The spaceship would be shortened in the direction of its motion. Its height and width would be unchanged.
    c. The mass of the spaceship would be increased compared to its rest mass.
    d. A passenger on the spaceship would say that your clocks are slow, your length is contracted, and your mass is increased. Because all inertial frames are equivalent, the situations seen by you and by passengers on the spaceship must be symmetric.
42. a. All observers, including Bob, will agree that Jackie is illuminated first by the green flash and then by the red flash.

b. Because Bob is traveling in the opposite direction of Jackie, he will be illuminated first by the red flash and then by the green flash. Again, all observers will agree on this point.

c. Bob will conclude that the red flash occurred before the green flash—the opposite of the order seen by Jackie, and not simultaneously as seen by you.

43. We will let t' represent the time for the traveler, and t is the time according to those who stay at "rest" on the Earth. The traveler's time will be slowed by the time dilation factor so that:

$$t' = t\sqrt{1-\left(\frac{v}{c}\right)^2}$$

In this problem v = 89 km/hr, which we must convert to a fraction of the speed of light:

$$\frac{v}{c} = \frac{89\frac{\cancel{\text{km}}}{\cancel{\text{hr}}} \times \frac{1\ \cancel{\text{hr}}}{3600\ \cancel{\text{s}}}}{3\times 10^5 \frac{\cancel{\text{km}}}{\cancel{\text{s}}}} = 8.2\times 10^{-8} \Rightarrow \left(\frac{v}{c}\right)^2 = 6.8\times 10^{-15}$$

In this case, $(v/c)^2$ is so small that, when you subtract it from 1, your calculator still gives an answer of 1. Thus, to the accuracy of your calculator, the time that has passed for the traveling student is the same as that for the people at "rest" on the Earth, or 70 years. We can get a more accurate answer by using the *binomial expansion*, which has the generalform:

$$(a+x)^n = a^n + na^{n-1}x + \frac{n(n-1)}{2!}a^{n-2}x^2 + \frac{n(n-1)(n-2)}{3!}a^{n-3}x^3 + \cdots$$

In this case, we set a = 1 and $x = -(v/c)^2$ to find:

$$\sqrt{1-\left(\frac{v}{c}\right)^2} = \underbrace{\left(1+\underbrace{-\left(\frac{v}{c}\right)^2}_{x}\right)^{1/2}}_{a=1,\,x=-\left(\frac{v}{c}\right)^2,\,n=\frac{1}{2}} = 1-\frac{1}{2}\left(\frac{v}{c}\right)^2+\frac{3}{8}\left(\frac{v}{c}\right)^4-+\cdots$$

For $(v/c)^2 < 1$, the series converges rapidly and we may neglect all but the first two terms:

$$\begin{aligned}\sqrt{1-\left(\frac{v}{c}\right)^2} &\approx 1-\frac{1}{2}\left(\frac{v}{c}\right)^2\\ &= 1-\frac{1}{2}(6.9\times 10^{-15})\\ &= 1-(3.5\times 10^{-15})\end{aligned}$$

The time for the traveling student can then be computed as:

$$t' = (70 \text{ yr})(1 - (3.5 \times 10^{-15})) = 70 \text{ yr} - (2.5 \times 10^{-13}) = 70 \text{ yr} - (7.7 \times 10^{-6} \text{ s})$$

That is, the time for the traveler at 55 mph is less than the time for those at "rest" on Earth by only 7.7 microseconds!

44. We use the same formula as in the previous problem to figure out the time passage for our second student. Because she is traveling at 95% of the speed of light, we have $(v/c) = 0.95$ or $(v/c)^2 = 0.9025$. Plugging this value into the time dilation formula, we find:

$$t' = t\sqrt{1 - \left(\frac{v}{c}\right)^2} = (70 \text{ yr})\sqrt{1 - (0.9025)} = (70 \text{ yr})\sqrt{0.0975} = (70 \text{ yr})(0.312) \approx 21.9 \text{ yr}$$

For the traveler moving at 95% of the speed of light, only a little less than 22 years passes while 70 years pass for those of us left behind on Earth.

45. To solve this problem we will use the length contraction formula:

$$\underset{(\text{moving})}{\text{length}} = (\text{rest length}) \times \sqrt{1 - \left(\frac{v}{c}\right)^2}$$

We are told that the speed is 75% the speed of light, so v/c is 0.75. The rest length was 50 meters, making it a simple matter to compute the measured length: 33.1 meters. We would measure the spaceship as being 33.1 meters in length even though Marta would measure it at 50 meters.

46. To solve this problem we will use the length contraction formula:

$$\underset{(\text{moving})}{\text{length}} = (\text{rest length}) \times \sqrt{1 - \left(\frac{v}{c}\right)^2}$$

The reset length in this case is 8.6 light-years. The speed is 92% the speed of light, so v/c is 0.92. Calculating the measured length, we get the 3.4 light-years. So from our point of view on our trip, the length is 3.4 light-years.

We see the star coming toward us at 92% the speed of light. This means that it is traveling at 0.92 light-year per year. The time it takes to travel the 3.4 light-years to us is therefore:

$$\frac{3.4 \ \cancel{\text{light-years}}}{0.92 \ \cancel{\text{light-year}}/\text{yr}} = 3.7 \text{ yr}$$

It will take 3.7 years for us to reach Sirius from our perspective.

47. For this problem we will use the mass increase formula:

$$\underset{(\text{moving})}{\text{mass}} = \frac{(\text{rest mass})}{\sqrt{1 - \left(\frac{v}{c}\right)^2}}$$

The rest mass is 500,000 tons and the speed is half the speed of light so that v/c is 0.5. Putting in the numbers, we calculate the apparent mass to be 577,000 tons.

48. For this problem we will use the mass increase formula:

$$\underset{(\text{moving})}{\text{mass}} = \frac{(\text{rest mass})}{\sqrt{1-\left(\frac{v}{c}\right)^2}}$$

However, we are given the moving mass and the rest mass and are asked for the speed. We need to solve the equation for v/c, then:

$$\frac{v}{c} = \sqrt{1-\left(\frac{(\text{rest mass})}{\underset{(\text{moving})}{\text{mass}}}\right)^2}$$

Now we can use the number we were given to solve this. The rest mass is 1 gram while the moving mass is 3,000 kilograms or 3 million grams. Putting in the numbers, we discover that to make the fly have the mass of an SUV, v/c must be very, very close to 1—most calculators cannot distinguish it from 1. The fly must travel at a speed within a hair of the speed of light.

49. The π^+ meson will always "think" it is at rest and therefore decay after about 18-billionths of a second (1.8×10^{-8} s) in its *own* reference frame. However, from the point of view of the scientists conducting the experiment, the π^+ meson represents the *moving* reference frame. Hence, the π^+ meson's lifetime of 1.8×10^{-8} s represents t' in the time dilation formula, or $t' = 1.8 \times 10^{-8}$ s. To solve for the lifetime observed by the scientists, t, divide both sides of the time dilation formula by the square root term:

$$t' = t\sqrt{1-\left(\frac{v}{c}\right)^2} \Rightarrow t = \frac{t'}{\sqrt{1-\left(\frac{v}{c}\right)^2}} = \frac{1.8 \times 10^{-8}\text{ s}}{\sqrt{1-(0.998)^2}} = \frac{1.8 \times 10^{-8}\text{ s}}{0.0632} = 2.8 \times 10^{-7}\text{ s}$$

When produced at $0.998c$, the π^+ meson is expected to last about 280-billionths of a second, rather than its "normal" lifetime of 18-billionths of a second. That is, its moving lifetime is more than 15 times as long as its lifetime at rest.

Actual experiments in particle accelerators allow scientists to measure the lifetime of π^+ mesons (and other particles) produced at high speeds. The formulas of special relativity allow us to calculate the expected lifetimes. The fact that predicted lifetimes match observed experimental results provides strong evidence in support of the special theory of relativity.

50. To solve this problem we will use the time dilation formula:

$$\underset{(\text{moving})}{\text{time}} = \underset{(\text{moving})}{\text{time}} \times \sqrt{1-\left(\frac{v}{c}\right)^2}$$

We will first have to convert the Shuttle's speed to kilometers per second in order to take a ratio with the speed of light. The conversion goes as follows:

$$30{,}000\frac{\text{km}}{\cancel{\text{hr}}}\times\frac{1\ \cancel{\text{hr}}}{60\ \cancel{\text{min}}}\times\frac{1\ \cancel{\text{min}}}{60\ \text{s}}=8.33\frac{\text{km}}{\text{s}}$$

Now remembering that the speed of light is 3.00×10^5 km/s, we can use the time dilation formula. The resulting time in the moving frame is 0.9999999996 hour. This time is so close to an hour that we could hardly measure the difference. However, we can take the difference from 1 full hour and convert into seconds. Doing this shows us that the time that passes on the Shuttle is 1.39×10^{-6} shorter than the time that passes on the ground.

51. Your sister's speed is $v = 0.99\ c$, or $v/c = 0.99$. The time that passes for you is $t = 20$ yr (you age from 25 to 45). The time that passes for your sister is:

$$t' = t\sqrt{1-\left(\frac{v}{c}\right)^2} = (20\ \text{yr})\sqrt{1-(0.99)^2} = (20\ \text{yr})\sqrt{1-0.9801} = (20\ \text{yr})\sqrt{0.0199} = 2.8\ \text{yr}$$

While 20 years pass for you, only a little less than 3 years passes for your traveling sister. This is a real difference: Although 20 years will have passed on Earth, only 3 years will have passed for your sister—she will return only 3 years older, at age 28. Although you were the same age before she left, you now are 17 years older than she.

52. This problem is simply velocity addition. Let u be the speed of the baseball according to those on the Earth. Let u' be the speed of the baseball according to Santana ($u' = 0.8\ c$), and let v be the speed of the train carrying Santana ($v = 0.9\ c$). To then find the velocity, we measure u by use of the addition formula:

$$u=\frac{u'+v}{1+\frac{u'v}{c^2}}=\frac{0.8\ c+0.95\ c}{1+\frac{(0.8\ c)(0.95\ c)}{c^2}}=\frac{1.75\ c}{1+0.76}=\frac{1.75}{1.76}c=0.994\ c$$

We will see the baseball traveling at 99.4% of the speed of light—less than the speed of light, but faster than either the train or the baseball independently.

53. The light beam travels at the speed of light and therefore covers a distance of 100 meters in a time of:

$$\text{time}=\frac{\text{distance}}{\text{speed}}=\frac{100\ \cancel{\text{m}}}{3\times10^8\ \cancel{\text{m}}/\text{s}}=3.3\times10^{-7}\ \text{s}$$

The light beam covers the 100-meter course in less than a millionth of a second, much faster than the sprinter's 8.7 seconds. Thus, the light beam wins the race easily.

54. a. As seen from the stands, Jo is going at a speed of 0.999 c while the light beam's speed is c. Thus, the light beam is only 0.001 c faster than Jo and wins a very close race.
    b. As seen by Jo, the light beam goes the full speed of light ahead of him because everyone always measures the same absolute value for c.
    c. From Jo's point of view, the light is going faster than he is by the same 300,000 km/s as before; thus, he has gained nothing in terms of catching up with the light. In contrast, the spectators see Jo staying much closer to the light beam in the second race than in the first.

d. Jo sees a shorter course, because of length contraction due to his high speed relative to the Earth. We simply plug values into the length contraction formula:

$$\begin{aligned}\text{length according to Jo} &= (\text{rest length})\sqrt{1-\left(\frac{v}{c}\right)^2} \\ &= (100\text{ m})\sqrt{1-(0.999)^2} \\ &= (100\text{ m})(0.0447) \\ &\approx 4.5\text{ m}\end{aligned}$$

Jo finds the course to be only about 4.5 meters long, instead of the usual 100 meters.

Chapter S3. Spacetime and Gravity

Please see the general notes about Chapters S2–S4 that appear on page 231.

This chapter provides students with a short but solid introduction to the general theory of relativity. As with Chapter S2, this chapter begins with a section that provides a concise overview of the topic; the remaining sections explain the logic and evidence behind this overview. Thus, it is possible to teach only the first section if you want to give students an overview of general relativity but do not have time to go into any details of the theory.

As always, when you prepare to teach this chapter, be sure you are familiar with the relevant media resources (see the complete, section-by-section resource grid in Appendix 3 of this Instructor's Guide) and the online quizzes and other study resources available on the Mastering Astronomy Web site.

Teaching Notes (By Section)

Section S3.1 Einstein's Second Revolution

This first section introduces the ideas of general relativity, including a bulleted list that highlights the ideas that we will encounter in the rest of the book and that are discussed in further detail in the rest of this chapter. After introducing the basic ideas, we introduce the equivalence principle and discuss how it plays a role in general relativity similar to the role played by the absoluteness of the speed of light in special relativity.

- Note that, in stating the equivalence principle in the main body of the text, we do not explicitly state that it holds true only on a highly localized scale—although we do mention this fact in footnote 1. We have found that, unless you have a great deal of time to discuss this point, it tends to confuse all but the best students. Thus, we have chosen to be technically correct by addressing it in the footnote, but otherwise to treat it as a subtlety that is beyond the scope of this book.
- Historical note: In the "Special Topic: Einstein's Leap" box, we state that someone was bound to come up with special relativity around the time that Einstein did. In fact, French mathematician Jules Henri Poincaré gave a speech in 1904 in which he predicted the development of "an entirely new mechanics" in which the speed of light became an absolute limit.

Section S3.2 Understanding Spacetime

This section discusses the concept of spacetime, including the construction of spacetime diagrams and a discussion of what it means for spacetime to be curved.

- Note that we do not make much use of spacetime diagrams elsewhere in the chapter; we introduce them because we have found that they help students understand the idea of spacetime.
- Confronted with the idea of history as being "viewable" in spacetime, many students will wonder whether the future is similarly "viewable" in spacetime and whether this issue has any implications for the idea of fate or a predetermined future. This issue is the topic of Discussion Question 61 at the end of the chapter and can make for a great in-class discussion if time permits. You may wish to tell students about the "many worlds" hypothesis in which each possible future has an independent reality, as one of many possible ways that spacetime might have a fixed structure and yet not imply a predetermined future.
- We would like to acknowledge and thank Martin Gardner, whose book *The Relativity Explosion* gave us the idea for the analogy used in this section to show how a three-dimensional book can look different in two-dimensional views. Gardner's book has many other wonderful ideas and analogies that also helped shape our approach to writing Chapters S2 and S3.

Section S3.3 A New View of Gravity

This section builds on the previous sections to explain why general relativity provides a new view of gravity in which gravity arises from the curvature of spacetime. It also explains gravitational time dilation and discusses the possibilities of an overall geometry of the universe.

- We suggest emphasizing the bulleted list of caveats about rubber sheet analogies; without these caveats, these analogies often lead to misconceptions instead of helping students understand spacetime curvature.
- Note that we discuss black holes only very briefly in this section; more detailed discussion of black holes appears in Chapter 18.

Section S3.4 Testing General Relativity

By this point we have discussed all the key ideas of general relativity laid out in the bulleted list of Section S3.1 except gravitational waves. In this section, we turn to the evidence supporting these ideas and also introduce gravitational waves in the context of discussing evidence of their existence.

- Note that we keep the discussion of gravitational lensing brief here; it is discussed further in Chapter 22.

Section S3.5 Hyperspace, Wormholes, and Warp Drive

This short section discusses the science fiction devices of hyperspace, wormholes, and warp drive, explaining how these ideas arise from speculation about the implications of relativity.

- It is important to emphasize the speculative nature of these ideas so that students do not confuse them with established science. Nevertheless, students usually enjoy this discussion.

Section S3.6 The Last Word

This very short section (which does not contain any formal Learning Goals) essentially consists of a quotation from Einstein that summarizes his thoughts about the meaning of relativity. It is a great quotation with which to end a series of classes on relativity and can spawn great discussions.

Answers/Discussion Points for Think About It Questions

The Think About It questions are not numbered in the book, so we list them in the order in which they appear, keyed by section number.

Section S3.1

- (p. 436) After 1 second of 1 *g* acceleration, Jackie will be traveling relative to you at about 10 m/s. After 10 seconds she will be going about 100 m/s. After a minute, she will be going about 600 m/s.

Section S3.2

- (p. 439) The idea in this question is that your body would look like its normal three-dimensional self, but with an added dimension stretching through time. Bumping into someone on a bus would be represented by a short stretch along the time axis at which your four-dimensional self would be in contact with his or hers.
- (p. 440) On a spacetime diagram in which the speed of light is represented by a 45° line, ordinary speeds are so slow that their angles are nearly indistinguishable from the vertical.
- (p. 442) This question asks students to examine great circle routes for themselves with the aid of a globe. Airplanes try to follow great circle routes because they are the shortest distances between points on Earth. (However, weather, topography, legal restraints [such as not flying over certain sensitive areas] and political constraints [such as not having permission to cross a nation's air space] can force significant deviations from great-circle routes.)
- (p. 444) When you are standing on a scale—or for that matter standing anywhere on Earth—your worldline is not following the straightest possible path through spacetime because you are feeling weight.

Section S3.3

- (p. 447) You would age more slowly on the Earth than on the Moon because of the Earth's stronger gravitational field. However, the Earth's gravitational field is still so weak that the effects of gravitational time dilation are barely measurable, and the difference between your aging rate on the Earth and your aging rate on the Moon would be virtually unnoticeable.

Section S3.4

- (p. 448) If Mercury's perihelion were closer to the Sun, it would be in a region of spacetime with greater curvature and hence greater gravitational time dilation. Hence, the discrepancy between the actual orbit and the orbit predicted by Newton's laws would be greater than it actually is.
- (p. 450) This question is designed simply to bring students up-to-date with evidence from the GP-B mission that was not available at the time this book went to press.

Solutions to End-of-Chapter Problems (Chapter S3)

1. The straightest possible path on the Earth's surface is the shortest distance between two points if we travel along the Earth's surface. It is always part of a great circle, a circle whose center is the center of the Earth.
2. Spacetime is the four-dimensional combination of time and the three ordinary dimensions of space (e.g., length, width, depth).
3. Major ideas of general relativity include:
   - Gravity arises from distortions of spacetime.
   - Time runs more slowly in gravitational fields.
   - Black holes can exist in spacetime, and falling into a black hole means leaving the observable universe.
   - The universe has no boundaries and no center.
   - A large mass that undergoes rapid changes in motion or structure will emit gravitational waves.
4. The equivalence principle states that the effects of gravity are equivalent to the effects of uniform acceleration. This means, for example, that a person in an enclosed box cannot tell if she is sitting on the surface of the Earth or accelerating at 1 *g* in space.
5. By dimension we mean the number of independent directions in which movement is possible. A point is zero-dimensional; a being confined to a point could not go anywhere. A line is one-dimensional; a being who can move in a line is free to move in only one direction. A plane is two-dimensional; a being who can move on a plane can move lengthwise or widthwise. A three-dimensional space allows a being to move in three directions: length, width, and depth. A four-dimensional space allows a fourth direction of motion, although it is impossible to visualize. We call any space with more than three dimensions a hyperspace.
6. Just as different two-dimensional pictures of the same three-dimensional object can appear quite different, different three-dimensional views of spacetime can look different. But the underlying spacetime is really the same thing, just seen from different perspectives.
7. A spacetime diagram is a plot showing position on one axis and time on the other. An event on such a plot is represented by a point. A worldline is a path through four-dimensional spacetime.
8. A vertical worldline represents an object that is not moving at all since the space coordinate is staying the same. A slanted worldline corresponds to an object that is moving at constant velocity. A curved worldline corresponds to an object that is accelerating.
9. In a flat geometry, the straightest possible paths are straight lines, the circumference of a circle is $2\pi r$, parallel lines remain the same distance apart, and the interior angles of triangles add up to 180°. For a spherical geometry, the straightest possible paths are pieces of great circles, a circle's circumference is less than $2\pi r$, parallel lines always meet, and the interior angles of a triangle add up to more than 180°. In a saddle-shaped geometry, the straightest possible paths are hyperbolas, a circle's circumferences is greater than $2\pi r$, parallel lines diverge, and the interior angles of triangles add to less than 180°.

10. Flat or saddle-shaped geometries go on forever, so there are no edges and no centers to them. Spherical geometries curve back on themselves, so while they are finite, there are still no edges and no centers.
11. You can tell if you are following the straightest possible path in your spacetime if you feel no weight. If you feel weight, you are not on the straightest possible path.
12. According to general relativity, gravity is the curvature of spacetime. In this view, Earth orbits the Sun because Earth is following the straightest possible path through the curved spacetime around the Sun.
13. The "rubber sheet" analogy to spacetime represents spacetime as a two-dimensional rubber sheet. Massive objects on the sheet create depressions in the sheet that distort the local spacetime, resulting in gravity. The analogy, while useful, suffers from a few limitations. For example, it represents only two dimensions of space and does not represent the time dimension at all. Another limitation is that objects in spacetime do not sit *on* top of spacetime; rather, they are embedded within it.

    The rubber sheet analogy gives us two ways to increase how much an object curves spacetime. One is to increase the mass, which will make the amount of curvature at any distance from the object greater. The other way to increase the curvature is to use the same mass and make the object smaller in size. While the curvature far away from the object would be the same, close to the object the curvature would be much higher.
14. A black hole is a bottomless pit in spacetime. Nothing can escape the black hole once it is within the event horizon, the point of no return around the black hole.
15. Gravitational time dilation refers to the slowing of time due to gravity. The amount that time is slowed depends on the strength of the gravitational field: A stronger field results in more slowing.
16. Tests of general relativity include:
    - Mercury's orbit—Mercury's orbit precesses around the Sun more than it should from just the effects of Newtonian gravity. General relativity predicts the precession because of gravitational time dilation.
    - Gravitational lensing—When light passes near a massive object like the Sun, its path is bent. This makes the source appear to be in a slightly different position than it should be. This was first observed in 1919, making Einstein famous.
    - Gravitational time dilation—Clocks at different altitudes move at slightly different speeds as predicted by gravitational time dilation. Also, spectral lines on stars appear redshifted slightly because the gases that are creating the lines are in a stronger gravitational field than we are on Earth. And the accuracy of GPS devices depends on calculations that take the gravitational time dilation into account.
    - Gravitational waves—General relativity predicts that gravitational waves will be emitted when the curvature of spacetime suddenly changes. These waves have not yet been directly detected, but we have observed their effects on the orbits of neutron stars.
17. Because a massive object curved the local spacetime, the path of light passing near the object bends, making the object appear to be in a different position than it would otherwise appear. This effect is called "gravitational lensing."
18. Because time travels more slowly on the surface of the Sun (where the acceleration of gravity is higher), the time it takes for a trillion cycles to occur for a photon is longer. So on Earth, we will see fewer cycles in each second, making the light appear to have a lower frequency or a longer wavelength. This shifts the photon's wavelength in the "red" direction, and we call the effect "gravitational redshift."

19. Gravitational waves are ripples of curvature in spacetime caused by a change in the curvature of spacetime (due to a massive object changing its motion or its structure). We have not yet detected these waves directly, but we have detected changes in the orbits of binary neutron stars that can be precisely predicted by assuming that the orbits are losing energy by radiating gravitational waves.
20. Wormholes could allow us to travel between two points faster than light can make the trip by making a tunnel through hyperspace that would be shorter than the distance through normal space. They are plausible according to our current understanding of general relativity, but there is no evidence that they actually exist. Moreover, many scientists think that they will eventually be shown to be impossible, because their existence would seem to allow for time-travel paradoxes.
21. *The equivalence principle tells us that experiments performed on a spaceship accelerating through space at 1g will give the same results as experiments performed on Earth.* This is true, and basically defines the equivalence principle. (*Technical note:* In reality, you could measure a slight tidal effect that occurs on Earth but not in the spaceship. That is why the equivalence principle is technically valid only in small regions where tidal effects are negligible.)
22. *The equivalence principle tells us that there's no difference at all between a planet and a human-made spaceship.* This is false. There is an obvious difference. The equivalence principle says that on an accelerating spaceship you will get the same experimental results that you would get on a planet with the same acceleration of gravity, but it doesn't mean that the two objects (planet and spaceship) are the same.
23. *A person moving by you at high speed will measure time and space differently from you, but you will both agree that there is just a single spacetime reality.* This is true: Spacetime is the same for everyone.
24. *With a sufficiently powerful telescope, we could search for black holes by looking for funnel-shaped objects in space.* This is false. The funnel shape that we see in rubber sheet diagrams is an artifact of the way we must portray curved three-dimensional space on a curved two-dimensional rubber sheet. A black hole would actually appear spherical if we could see it.
25. *Time runs slightly slower on the surface of the Sun than it does here on Earth.* This is true: Time runs slower in stronger gravitational fields, and gravity is stronger on the Sun's surface than it is here on Earth.
26. *Telescopes sometimes see multiple images of a single object, just as we should expect from the general theory of relativity.* This is true, and represents the phenomenon of gravitational lensing.
27. *When I walk in circles, I am causing curvature of spacetime.* This is false; the structure of spacetime is determined by the masses within it.
28. *Although special relativity deals only with relativity of motion, general relativity tells us that everything is relative.* This is false. In fact, general relativity emphasizes that there is a single spacetime reality, even though different observers may view it differently depending on their relative motion.
29. *The shortest distance between two points is always a straight line.* This is true only in a flat (Euclidean) geometry; it is not true in other geometries.

30. *According to the general theory of relativity, it is impossible to travel through hyperspace or to use anything like* Star Trek's *"warp drive."* This is false. General relativity allows such travel in principle, though other laws of physics (not yet known) may prohibit it in reality.
31. b; 32. b; 33. b; 34. c; 35. c;
36. b; 37. c; 38. b; 39. b; 40. b.
41. This is a subjective question, but you might look in particular to see if students recognize that Einstein considered time to be like an illusion because spacetime has a fixed reality.
42. The two paths are both parallel to begin with, because they are aimed due north from the same latitude. However, they will not remain parallel as the people continue their journeys, and the people will meet at the North Pole. This convergence of parallel lines tells us that Earth's surface has a spherical geometry.
43. Sample answers: A table has flat geometry; a ball has spherical geometry; a saddle has saddle-shaped geometry. You can determine the geometry by checking the geometrical rules that apply.
44. A black hole would actually be spherical. To explain the misconception, students should focus on the way the rubber sheet analogies differ from reality.
45. a. Spacetime would be much more highly curved in the region that used to be within the Sun, with the curvature becoming more extreme near the center.
    b. Because the mass of the black hole is the same as the mass of the Sun, there would be *no change* to the curvature of spacetime at distances far from the object, such as at the orbit of Earth.
    c. Because there is no change to the curvature of spacetime at Earth's orbit, Earth's orbit would not be affected.
46. Essay question; answers will vary, but the key point should be that, from the point of view of general relativity, masses move in response to the curvature of spacetime. Because the curvature is well defined by any large mass, small masses will all follow the same trajectories through spacetime regardless of differences in mass between them.
47. If we have a constant acceleration through space of 1 g, we will feel gravity in the same way we feel it on Earth. Thus, the trip will be very comfortable. Moreover, we will not notice anything unusual about our own length, mass, or time. Thus, from our point of view, we will not notice anything different from what we would notice in a closed room on Earth (aside from the view out the windows).
48. Answers will vary; this is a fun question in which students evaluate the science in a science fiction movie that involves interstellar travel.
49. Answers will vary; this question requires research to learn about the Eötvös experiment.
50. Answers will vary; this question requires students to read a book or article about wormholes.

51. The following diagram shows the worldlines for the five situations described:

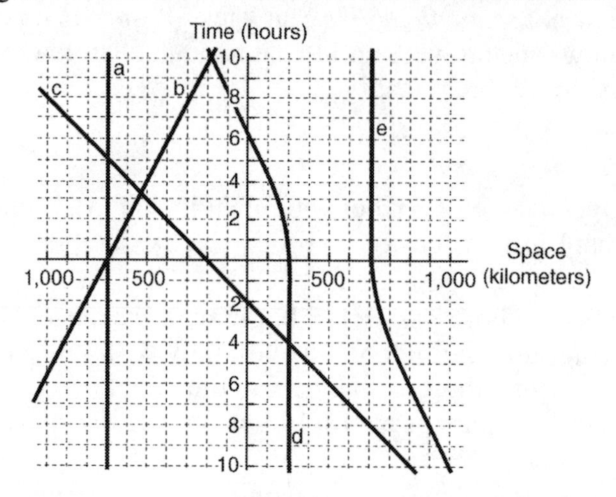

52. The following diagram shows the worldlines for the three cases described:

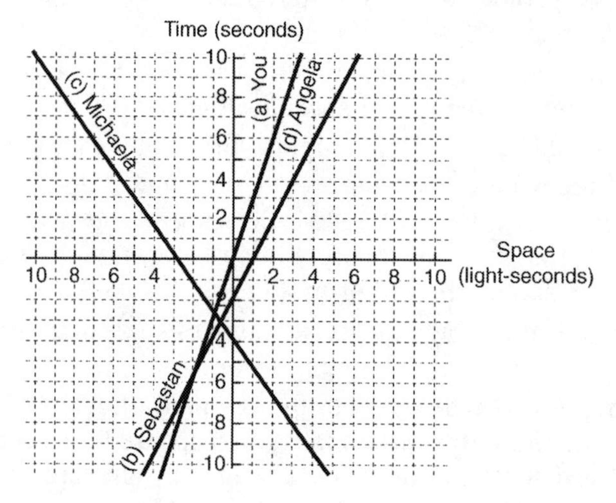

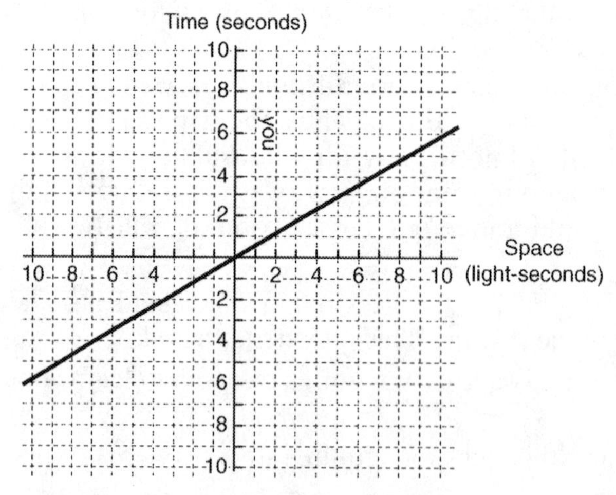

53. To determine the speed, recall the definition of speed:

$$\text{speed} = \frac{\text{distance traveled}}{\text{time}}$$

On the other hand, if we locate a point on the worldline, we can form a right triangle where the hypotenuse is the worldline, the vertical leg is the time, and the horizontal leg is the distance traveled. In this case, we know that the definition of the tangent function means that:

$$\tan(30°) = \frac{\text{time}}{\text{distance traveled}}$$

Looking at the right-hand sides of these two equations, we can see that they are the reciprocals of each other. This means that:

$$\text{speed} = \frac{1}{\tan(30°)}$$

So we run this through our calculators to find out that the speed of travel is 1.73 light-seconds per second—in other words, 1.73 times the speed of light. This is clearly impossible since no object can travel faster than light.

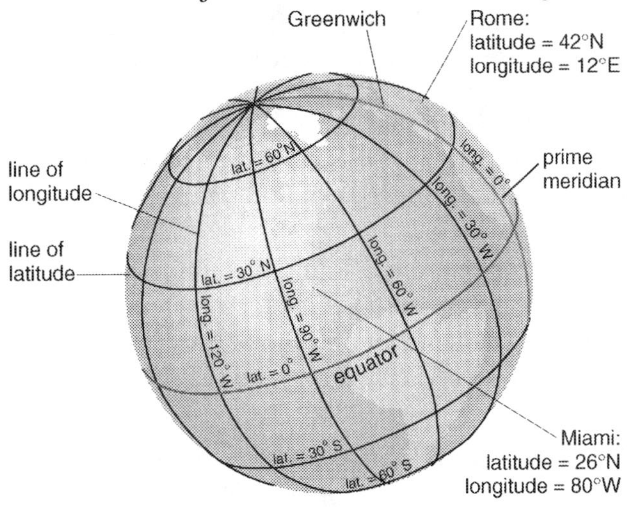

54. The angle made at each vertex of this triangle is 90°. We can see this because every line of longitude intersects the equator at a 90° angle, and two lines of longitude that are 90° apart also intersect each other at right angles. If all three angles are 90°, then the triangle's interior angles must add to 270°. If Earth's surface were a flat geometry, this would be surprising. (It would also be impossible.) However, we know that if the interior angles of a triangle add up to more than 180°, we are not looking at a flat geometry, but a spherical one. So our result tells us that the Earth's surface is spherical.

55. a. As long as time dilation is not a major factor, we can determine the velocity after some time with a given acceleration from the formula:

$$\text{velocity} = \text{acceleration} \times \text{time}$$

Although time dilation is noticeable at half the speed of light (1.5×10^8 m/s), it is still a relatively small factor until much higher speeds are reached. Thus, we can get a reasonable approximation of how long it would take the ship to reach half the speed of light at a constant acceleration of 1 *g* by neglecting time dilation. We therefore just solve the above formula for the time:

$$\text{time} = \frac{\text{velocity}}{\text{acceleration}} = \frac{1.5 \times 10^8 \, \cancel{\text{m}}/\text{s}}{9.8 \, \cancel{\text{m}}/\text{s}^2} = 1.5 \times 10^7 \text{ s}$$

Converting to days gives a time of:

$$1.5 \times 10^7 \cancel{s} \times \frac{1 \cancel{hr}}{3600 \cancel{s}} \times \frac{1 \text{ day}}{24 \cancel{hr}} \approx 174 \text{ days}$$

At a constant acceleration of 1 g, the ship will reach a speed of about half the speed of light in about 174 days, or almost 6 months.

b. As the ship continues to accelerate away from Earth at 1 g, you (on Earth) will not continue to see its speed increase by 9.8 m/s with each passing second. From your point of view on Earth, time dilation will begin to noticeably affect the rate at which time is passing on the accelerating ship. Thus, each second on the ship will become much longer than a second on Earth, which means the ship's speed will increase by much less than 9.8 m/s during an Earth second. This effect will become more and more pronounced as the ship's speed approaches the speed of light away from Earth.

c. Because the ship will reach half the speed of light in just a few months, we can conclude that on a long journey it will be traveling very close to the speed of light for most of its journey. Thus, the ship will take only slightly longer to make a long trip than it would take a light beam from the point of view of observers on Earth. Thus, from your point of view on Earth, it will take the ship only a little more than 500 years to reach a star 500 light-years away.

56. As mentioned in the problem, these solutions require putting D in meters, with $g = 9.8 \text{ m/s}^2$ and $c = 3 \times 10^8$ m/s, to get an answer in units of seconds. A spaceship is traveling to a star at a distance of 500 light-years, with a constant acceleration of 1 g. To find the amount of time that passes on the ship, we set:

$$D = 500 \cancel{\text{light-years}} \times \frac{9.5 \times 10^{15} \text{m}}{\cancel{\text{light-year}}} = 4.75 \times 10^{18} \text{ m}$$

We now use this value for D in the formula to find the ship time:

$$T_{\text{ship}} = \frac{2c}{g} \ln\left(\frac{g \times D}{c^2}\right)$$

$$= \frac{2 \times 3 \times 10^8 \text{ m/s}}{9.8 \text{ m/s}^2} \times \ln\left(\frac{9.8 \text{ m/s}^2 \times 4.75 \times 10^{18} \text{ m}}{\left[3 \times 10^8 \text{ m/s}\right]^2}\right)$$

$$= (6.1 \times 10^7 \text{ s}) \times \ln(517.2)$$

$$= (6.1 \times 10^7 \text{ s}) \times 6.25 = 3.8 \times 10^8 \text{ s}$$

Finally, we convert this answer to years:

$$3.8 \times 10^8 \cancel{s} \times \frac{1 \cancel{hr}}{3600 \cancel{s}} \times \frac{1 \cancel{day}}{24 \cancel{hr}} \times \frac{1 \text{ yr}}{365 \cancel{days}} = 12 \text{ yr}$$

The ship accelerating at 1 g will reach the star at a distance of 500 light-years in a time of about 12 years, ship time. Meanwhile, on Earth, about 500 years will go by:

The ship will be traveling *at* very close to the speed of light for most of its journey, so we can assume that it takes the ship only slightly longer than it would take a light beam from the point of view of observers on Earth.

57. This time your spaceship travels 28,000 light-years to the center of the Milky Way Galaxy. Thus, we set:

$$D = 28{,}000 \text{ light-years} \times \frac{9.5 \times 10^{15} \text{ m}}{\text{light-year}} \approx 2.7 \times 10^{20} \text{ m}$$

We now use this value for D in the formula to find the ship time:

$$T_{\text{ship}} = \frac{2\,c}{g} \ln\left(\frac{g \times D}{c^2}\right)$$

$$= \frac{2 \times 3 \times 10^8 \text{ m/s}}{9.8 \text{ m/s}^2} \times \ln\left(\frac{9.8 \text{ m/s}^2 \times 2.7 \times 10^{20} \text{ m}}{\left[3 \times 10^8 \text{ m/s}\right]^2}\right)$$

$$= (6.1 \times 10^7 \text{ s}) \times \ln(29{,}400)$$

$$= (6.1 \times 10^7 \text{ s}) \times 10.3 = 6.3 \times 10^8 \text{ s}$$

Finally, we convert this answer to years:

$$6.3 \times 10^8 \cancel{\text{s}} \times \frac{1 \cancel{\text{hr}}}{3600 \cancel{\text{s}}} \times \frac{1 \cancel{\text{day}}}{24 \cancel{\text{hr}}} \times \frac{1 \text{ yr}}{365 \cancel{\text{days}}} = 20 \text{ yr}$$

The ship accelerating at 1 g will reach the center of the galaxy in a time of about 20 years, ship time. Note that, even though this distance is more than 50 times as far as the 500-light-year distance from Problem 56, the travel time is less than twice as long—this is because the continuing acceleration makes the average speed much faster for the longer trip. During this trip of 20 years, ship time, about 28,000 years will pass on Earth because the ship's average speed will be very close to the speed of light, and thus the travel time will be only slightly longer than the time required for a light beam to make the trip.

58. The Andromeda Galaxy is about 2.5 million light-years away, or:

$$D = 2.5 \times 10^6 \text{ light-years} \times \frac{9.5 \times 10^{15} \text{ m}}{\text{light-year}} = 2.4 \times 10^{22} \text{ m}$$

If you traveled in a spaceship accelerating at 1 g, the time it would take for the trip is given by the formula in the text, from which we find:

$$T_{\text{ship}} = \frac{2\,c}{g} \ln\left(\frac{g \times D}{c^2}\right)$$

$$= \frac{2 \times 3 \times 10^8 \text{ m/s}}{9.8 \text{ m/s}^2} \times \ln\left(\frac{9.8 \text{ m/s}^2 \times 2.4 \times 10^{22} \text{ m}}{\left[3 \times 10^8 \text{ m/s}\right]^2}\right)$$

$$= (6.1 \times 10^7 \text{ s}) \times \ln(2{,}613{,}333)$$

$$= (6.1 \times 10^7 \text{ s}) \times 14.8 = 9.0 \times 10^8 \text{ s}$$

Converting this answer to years, we find:

$$9.0\times10^{8}\,\cancel{s}\times\frac{1\,\cancel{hr}}{3600\,\cancel{s}}\times\frac{1\,\cancel{day}}{24\,\cancel{hr}}\times\frac{1\text{ yr}}{365\,\cancel{days}}=285\text{ yr}$$

The round-trip travel time is twice as long, or 57 years. In other words, the ship time for the round-trip to Andromeda would be about 57 years, so you could easily make this trip in your lifetime. However, when you returned, you would find that 5 million years would have passed on Earth!

59. The mass of the Earth is 5.98×10^{24} kg and its radius is 6,378 kilometers or 6.378×10^{6} m. We are told that the fractional time dilation factor at a distance r from an object of mass M is:

$$\frac{1}{c^2}\times\frac{GM_{\text{object}}}{r}$$

For Earth's surface, this becomes:

$$\frac{1}{(3\times10^{8}\text{ m/s})^2}\times\frac{\left(6.67\times10^{-11}\dfrac{\text{m}^3}{\text{kg}\times\text{s}^2}\right)(5.98\times10^{24}\text{ kg})}{6.378\times10^{6}\text{ m}}=6.95\times10^{-10}$$

Thus, while 1 hour passes in deep space, the time that passes on Earth is an extra amount of this fraction above multiplied by 1 hour, or 6.95×10^{-10} hr. The answer is easier to interpret if we convert it to seconds by multiplying by 3,600 s/hr; it becomes 2.50×10^{-6} s. An extra 2.5 microseconds would pass on Earth while an hour passed in deep space. This is a measurable amount of time, but it is not something we on Earth would notice.

60. We use the same formula as in Problem 59. This time we save a step by realizing that the extra amount of time on the Sun will be the time in deep space times the fractional amount given in the formula. The mass of the Sun is 2×10^{30} kg and its radius is 6.96×10^{8} m. We can calculate how much more time the clocks on the surface of the Sun will run in time t in deep space:

$$(t)\times\frac{1}{(3\times10^{8}\text{ m/s})^2}\times\frac{\left(6.67\times10^{-11}\dfrac{\text{m}^3}{\text{kg}\times\text{s}^2}\right)(2\times10^{30}\text{ kg})}{6.96\times10^{8}\text{ m}}=(2.13\times10^{-6})\times t$$

To find how much slower the clocks on the Sun are running compared to the clocks in deep space, we take a ratio of the extra time on the Sun to the time in deep space and multiply by 100%:

$$\frac{(2.13\times10^{-6})t}{t}\times100\%=2.13\times10^{-4}\%$$

Clocks on the Sun run 2.13×10^{-4}% slower than clocks in deep space. We expect to see the wavelengths of spectral lines longer by this same percentage. For example, a line with a wavelength of 121.6 nanometers will be 121.6 nm $\times$ $(2.13 \times 10^{-6}) = 2.59 \times 10^{-4}$ nm longer when viewed on the Sun from deep space.

Chapter S4. Building Blocks of the Universe

Please see the general notes about Chapters S2–S4 that appear on page 231.

This chapter provides students with a brief overview of the standard model of physics and of the two key principles of quantum mechanics (uncertainty and exclusion). Like the chapters on relativity, it is designed to enhance student appreciation of the remaining chapters in the book, but it is not prerequisite to those chapters. Also like the relativity chapters, its first section provides a concise overview of the chapter ideas, while the remaining sections discuss these ideas in greater detail. Thus, it is possible to teach only the first section if you want to give your students an overview of modern particle physics without the details.

As always, when you prepare to teach this chapter, be sure you are familiar with the relevant media resources (see the complete, section-by-section resource grid in Appendix 3 of this Instructor's Guide) and the online quizzes and other study resources available on the Mastering Astronomy Web site.

Teaching Notes (By Section)

Section S4.1 The Quantum Revolution

This first section introduces the history of the quantum revolution and provides a bulleted list of key quantum and particle physics ideas that are important in astronomy.

Section S4.2 Fundamental Particles and Forces

This section essentially provides a summary of the standard model of physics, explaining what we mean by fundamental particles and forces.

- We have included the terminology of the standard model (e.g., fermions and bosons, names of the quarks and leptons) because students are likely to encounter it in articles they read about cosmology. However, only the terms *quarks* and *leptons* arise again in this text, primarily in Chapter 23 on cosmology.

Section S4.3 Uncertainty and Exclusion in the Quantum Realm

This section serves to introduce and explain the meaning of the uncertainty and exclusion principles.

- Because the uncertainty principle is often misused by those who don't really understand it (and especially in the pseudosciences), we recommend focusing not only on what the uncertainty principle says, but also on what it doesn't say—e.g., it does not say that uncertainty is inevitable on macroscopic scales. (And, contrary to many news articles at the time, it did not play a role in the uncertainty surrounding the 2000 presidential election results in Florida.)
- As physicists, we use the term *state* so often that its meaning seems obvious. It is important to remember that this meaning of *state* in physics is not obvious to most students. Hopefully, the discussion of state in this section will clarify the point for students, but you may need to use additional examples of your own.

Section S4.4 Key Quantum Effects in Astronomy

Having established the key ideas of quantum mechanics, in this section we describe several important quantum effects that arise in astronomical contexts: degeneracy pressure, quantum tunneling, and virtual particles and Hawking radiation.

- The effects described here will come up again in context in later chapters. You should suggest that students return to this section to review the physics behind the effects when they arise later in context.

Answers/Discussion Points for Think About It Questions

The Think About It questions are not numbered in the book, so we list them in the order in which they appear, keyed by section number.

Section S4.2

- (p. 460) Quarks and leptons are more fundamental than protons, neutrons, and electrons, because protons and neutrons are made of quarks, while electrons represent just one type of lepton.

Section S4.3

- (p. 464) This question is meant to encourage thought about how a precise quantum statement such as the uncertainty principle relates to the way it is often stated colloquially. You can use this question to get into a discussion of how the uncertainty principle is often misused.
- (p. 466) The quantum uncertainties in the position and momentum of a baseball are so small in comparison to the baseball's size and momentum that they are unnoticeable. (In class, it can be fun to discuss what the world would be like if Planck's constant were much larger.)
- (p. 467) The velocity of 5 km/hr is consistent with walking or swimming, but not with driving or riding a bike down a hill. The high heart rate of 160 and high metabolic rate of 1,200 calories per hour suggest that the person is swimming vigorously, not walking slowly.

Section S4.4

- (p. 471) This question is designed to point out to students that the concept of something being "virtual" is not as unfamiliar as it may at first seem.

Solutions to End-of-Chapter Problems (Chapter S4)

1. The quantum realm is the realm of the very small. Quantum mechanics is the science of this realm.
2. Five major ideas to come from the laws of quantum mechanics include:
   - The ultimate building blocks of matter are quarks and leptons.
   - Antimatter is real. When a particle and antiparticle meet, they annihilate each other.
   - Four forces govern all interactions between particles: gravity, electromagnetism, the strong nuclear force, and the weak nuclear force. These forces in turn may be combined, perhaps into a single force.
   - Tiny particles exhibit a wave-particle duality.
   - Quantum effects are important to astrophysics with such effects as tunneling, degeneracy pressure, and virtual particles.

3. Fundamental particles are the most basic units of matter, impossible to divide further. Particle colliders let us examine these particles by accelerating them to high speeds and then smashing them together. By watching the particles that come out of the collision, we can learn what the original particles were made of. The results from the particle colliders have led us to the standard model, which builds all of the particles we see from a few fundamental components.
4. Spin is a measure of the inherent angular momentum of a particle. Based on spin, we break particles into two categories: bosons and fermions.
5. Quarks make up neutrons and protons. Leptons include electrons and neutrinos. Quarks and leptons both have spins that qualify them as fermions.
6. The six quarks are the up, down, charm, strange, top, and bottom. The six leptons are the electron, muon, tauon, electron neutrino, mu neutron, and tau neutrino. Protons are made of two ups and a down quark, while neutrons are made of two downs and an up quark.
7. Neutrinos are very small, fast particles with no electrical charge. But while their masses are tiny, the particles appear to be extremely common so that the total mass of neutrinos is probably significant (though still a tiny fraction of the total mass of the universe).
8. Antimatter refers to particles that are the exact opposite of ordinary matter particles, except that they have the same masses. When a particle and its antiparticle meet, they annihilate each other and convert to energy according to $E = mc^2$.
9. Antimatter is formed during pair production, when two particles pop into existence. The particles are always a particle and its antiparticle, so the process is the exact reverse of annihilation. Pair production satisfies conservation laws such as the conservation of charge.
10. The four basic forces in nature are: the electromagnetic force (carried by photons), the strong force (carried by gluons), the weak force (carried by weak bosons), and gravity (carried by gravitons).
11. In atomic nuclei, the strong and weak forces are the most powerful. Outside the nucleus, these two forces are unimportant and electromagnetism dominates gravity. However, when considering large objects like stars, galaxies, and planets, the electrical charges almost always very nearly cancel out. So gravity, for which there is no known way to make it cancel, wins out and becomes the most important force on large scales.
12. Many scientists feel that the standard model will one day be replaced by an even simpler model of nature because there are 24 fundamental particles in the standard model. This seems like too many to be the most basic set of components.
13. The uncertainty principle states that the more we know about a particle's position, the less we can know about its momentum and vice versa. This suggests that particles have a double nature. When we measure the momentum, we are measuring a wavelike property. When we measure the position, we are measuring a particle-like property. So we say that these particles have a wave-particle duality.
14. We can quantify the uncertainty principle in two ways. The first is:

$$\text{(uncertainy in location)} \times \text{(uncertainty in momentum)} \approx \text{Planck's constant}$$

This says that if we determine the location of a particle to within a certain amount, say, to within 500 nanometers, we can calculate how precisely we can determine the momentum. We can also express the uncertainty principle as:

$$(\text{uncertainy in energy}) \times (\text{uncertainty in time}) \approx \text{Planck's constant}$$

This version tells us that we can determine the energy of a particle to some precision and that we can then calculate how precisely we can know when it has that energy.

15. The quantum state tells us the conditions for a subatomic particle.
16. The exclusion principle tells us that two fermions of the same type cannot occupy the same quantum state at the same time. This means that multiple electrons in an atom cannot all be at the lowest energy state, so as we add more electrons, they must populate higher energy states.
17. Degeneracy pressure is a resistance to compression due to the uncertainty and exclusion principles. Since two fermions cannot be in the same state at the same time, they are forced apart, creating a kind of pressure. This differs from thermal pressure in that it does not depend on the temperature of the object. Degeneracy pressure is important in several kinds of astronomical objects including brown dwarfs, white dwarfs, and neutron stars.
18. Because each particle in a highly compressed plasma has little space to exist, the uncertainty principle says that its momentum must be very large. A large momentum suggests a large speed. However, the speed of light limits how fast the particle can move, so if we compress the plasma far enough, the particles will have to move faster than this limit to compensate for their small space, and degeneracy pressure will no longer be able to resist the collapse. The point where this happens depends on the particles exerting the degeneracy pressure, though: Neutrons, being much more massive than electrons, can resist to much higher pressures because their momentum at the same speed is much higher.
19. Quantum tunneling occurs when a particle moves through an energy barrier. This can happen because of the uncertainty principle's allowance that the particle might be across the barrier at any given time. This effect allows us to control the current in modern electronics since we know that we can reduce the probability of tunneling occurring by making the barrier "higher." This effect is also important to nuclear fusion in the Sun because it allows protons to get close enough together to bind to each other even though they do not have the energy to overcome their electrical repulsions.
20. Virtual particles are particles that pop in and out of existence constantly. They exist for such short periods of time that they are impossible to detect. Virtual particles may be able to oppose the force of gravity and drive objects in space farther apart. This might help explain why the expansion of the universe is accelerating. Virtual particles can also explain the prediction (as yet untested) that black holes should lose mass through Hawking radiation: If a particle and antiparticle pop into existence near an event horizon and one of the particles falls in while the other does not, energy has been added to the universe. To conserve energy, the black hole must give up some of its mass to compensate, shrinking slightly. Over time, this effect will cause black holes to evaporate.
21. *Although there are six known types of quarks, ordinary atoms contain only two of these types.* This is true. Ordinary atomic nuclei are made only from up and down quarks.

22. *If you put a quark and a lepton close together, they'll annihilate each other.* This is false. Quarks and leptons are different types of particles, so one can never be the antiparticle of the other.
23. *There's no such thing as antimatter, except in science fiction.* This is false. Antimatter is routinely produced in laboratories and in the cosmos.
24. *Some particle accelerators have been known to build up a huge electrical charge because of the electrons produced inside them.* This is false. An antielectron (positron) with opposite charge is always produced along with an electron, so the total charge remains zero.
25. *According to the uncertainty principle, we can never be certain whether one theory is really better than another.* This is false. The uncertainty principle concerns only particular measurements that cannot be made to perfect precision. It does not apply to scientific theories or to anything else.
26. *The exclusion principle describes the cases in which the uncertainty principle is excluded from being true.* This is false, and in fact the statement really makes no sense at all.
27. *No known astronomical objects exhibit any type of degeneracy pressure.* This is false. Degeneracy pressure is important in many astronomical objects, including brown dwarfs, white dwarfs, and neutron stars.
28. *Although we speak of four fundamental forces—gravity, electromagnetic, strong, and weak—it is likely that these forces are different manifestations of a smaller number of truly fundamental forces.* This is true. A well-established theory already links the electromagnetic and weak forces (into the electroweak force), and physicists suspect that the other forces are also linked.
29. *Imagine that, somewhere in deep space, you meet a person made entirely of antimatter. Shaking that person's hand would be very dangerous.* This is true. you and a person made of antimatter would undergo mutual annihilation, with nearly all your mass turning into energy.
30. *Someday, we may detect radiation coming from an evaporating black hole.* This statement makes sense, because evaporating black holes should in principle give off detectable radiation.
31. b; 32. a; 33. c; 34. b; 35. a;
36. c; 37. b; 38. b; 39. a; 40. b.
41. An atomic nucleus is made of protons and neutrons. The neutrons have no charge, but protons are positively charged. Thus, if it were just up to the electromagnetic force, a nucleus would fall apart due to the repulsion between the positive protons. Because the strong force is holding the nucleus together despite the electromagnetic repulsion, it must be the stronger (per particle) force *within* the nucleus. Note that this strength holds only over distances roughly the size of an atomic nucleus. Over larger distances, the strong force cannot be felt at all.
42. Gravity is far too weak to play a role in creating and breaking bonds between atoms or molecules; in fact, its only role in life is keeping us "stuck" to the ground. The nuclear forces are of such short range that they have no effects outside the nucleus itself. The only force that remains is the electromagnetic force, which influences interactions between the charged electrons and nuclei. Thus, all events in our ordinary lives—all chemistry and biology—are dominated by the electromagnetic force.

43. (i) When we say that gravity is the weakest of the four forces, we mean that it is weaker than the other forces in circumstances in which the other forces act. For example, the gravitational attraction of two particles within an atomic nucleus is far weaker than the strong or weak force between those particles, and the electromagnetic force between two charged subatomic particles is always stronger than the gravitational force between the same two particles. (ii) Despite its far greater strength per particle, the electromagnetic force is unable to attain very large values because it is impossible to accumulate a very large charge. This is because large objects tend to have *equal* amounts of positive (protons) and negative (electrons) charge, and the electromagnetic force thus "cancels" itself out. Gravity, on the other hand, always attracts: As objects get more and more massive, gravity continues to gain strength. Thus, for very large objects, there will be a great deal of gravitational attraction but virtually no electromagnetic force, because the overall object is neutral.
44. Short essay question; answers will vary, but the key point should be that quantum tunneling plays a major role in nuclear fusion in stars and thus is crucial to the creation of the elements by stars. *Note:* In fact, quantum tunneling is most important in low-mass stars, not the high-mass stars that make most heavy elements. However, low-mass carbon stars are probably the source of most of the carbon in the universe.
45. Short essay question; answers will vary, but the key point should be that matter-antimatter annihilation is the only way to convert 100% of the mass-energy of matter into other forms of energy. Practical problems in developing matter-antimatter engines include finding a way to produce all the antimatter and finding a way to store it safely so that it does not come into contact with any ordinary matter until we are ready to use it in the engines.
46. Short essay question; answers will vary, but the key point should be that we learn about the electron's motion because it generates an electromagnetic force that is transmitted through space by photons—the exchange particle for the electromagnetic force. Because the photons travel at the speed of light, the time it takes for us to learn of the electron's motion is the light travel time from the electron to us. This idea is very important to astronomy, because it explains how we are able to learn about distant objects by studying their light.
47. Short essay question; answers will vary, but the key point should be that the uncertainties predicted by the uncertainty principle are far smaller than the sizes or momenta of everyday objects like people and baseballs.
48. Short essay question; answers will vary, but the key point should be that an electron does not have a simple analog in our macroscopic world, because it sometimes acts like a tiny particle but other times acts like a wave.
49. a. We calculate the gravitational force, in *newtons*, simply by plugging the given numbers into the formula:

$$F_g = G\frac{M_1M_2}{d^2} = \left(6.67\times10^{-11}\frac{\text{N}\times\cancel{\text{m}^2}}{\cancel{\text{kg}^2}}\right)\frac{(9.1\times10^{-31}\ \cancel{\text{kg}})\times(9.1\times10^{-31}\ \cancel{\text{kg}})}{(10^{-10}\ \cancel{\text{m}})^2}$$

$$= 5.52\times10^{-51}\ \text{N}$$

The gravitational force of attraction between the two electrons is 5.52×10^{-51} newton.

b. We calculate the electromagnetic force by plugging the charges of the two electrons into the formula for the electromagnetic force:

$$F_{EM} = k\frac{q_1 q_2}{d^2} = \left(9.0\times10^{9}\frac{\text{N}\times\cancel{\text{m}^2}}{\cancel{\text{Coul}^2}}\right)\frac{(-1.6\times10^{-19}\,\cancel{\text{Coul}})\times(-1.6\times10^{-19}\,\cancel{\text{Coul}})}{(10^{-10}\,\cancel{\text{m}})^{\cancel{2}}}$$

$$= 2.3\times10^{-8}\text{ N}$$

The electromagnetic force of repulsion between the two electrons is 2.3×10^{-8} newton.

c. The ratio of the two forces is:

$$\frac{F_{EM}}{F_g} = \frac{2.3\times10^{-8}\text{ N}}{5.52\times10^{-51}\text{ N}} = 4.2\times10^{42}$$

The electromagnetic repulsion between the electrons is stronger than their gravitational attraction by a factor of over 10^{42}! Clearly, gravity will not play an important role until collections of very large amounts of mass are put together.

50. First, we compute the gravitational force between the Earth and Sun:

$$F_g = -G\frac{M_{\text{Sun}}M_{\text{Earth}}}{d^2}$$

Appendix E tells us that $M_{\text{sun}} = 2 \times 10^{30}$ kg, $M_{\text{earth}} = 5.98 \times 10^{24}$ kg, and $d = 1.496 \times 10^{11}$ m. So the gravitational force between the Sun and the Earth is 3.56×10^{22} newtons.

We can also compute the electromagnetic force between the two bodies:

$$F_{\text{EM}} = k\frac{q_{\text{Sun}}\,q_{\text{Earth}}}{d^2}$$

The Sun and Earth both have charges of 1 electron in this problem, so they each have a charge of -1.6×10^{-19} Coul. The force due to electromagnetism therefore comes out to 1.03×10^{-50} newtons. It is clear that at these scales, gravity is much, much more important than the electromagnetic force.

51. Note that the lifetime of a black hole depends on the *cube* of its mass. Thus, lower-mass black holes have much shorter lifetimes than more massive ones. For example, the lifetime of a 2-solar-mass black hole is $2^3 = 8$ times longer than that of a 1-solar-mass black hole. This also means that the evaporation process accelerates as the black hole loses mass. For example, suppose you calculate the lifetime of some black hole with a mass M. Some time in the future, the mass of the black hole will have decreased by a factor of 2, to 0.5 M. Its remaining lifetime at that point will be $0.5^3 = 1/8$ of its original lifetime. Therefore, each successive "half-life" for a black hole requires only one-eighth of the time it previously took to reduce the mass in half. The evaporation process thus begins slowly, while the final evaporation is a runaway process emitting a violent burst of energy. If such bursts occur, they might be observable with gamma-ray detectors.

52. To calculate the lifetime of a black hole with the mass of the Sun (2.0×10^{30} kg), we simply plug values into the formula given:

$$t = 10{,}240\ \pi^2 \frac{G^2 M^3}{hc^4} = 10{,}240\ \pi^2 \frac{\left(6.67 \times 10^{-11} \dfrac{\text{m}^3}{\text{kg} \times \text{s}^2}\right)^2 (2 \times 10^{30}\ \text{kg})^3}{\left(6.63 \times 10^{-34} \dfrac{\text{kg} \times \text{m}^2}{\text{s}}\right)\left(3 \times 10^8 \dfrac{\text{m}}{\text{s}}\right)^4} = 6.7 \times 10^{74}\ \text{s}$$

Converting this answer to years, we find:

$$t = 6.7 \times 10^{74}\ \cancel{\text{s}} \times \frac{1\ \cancel{\text{hr}}}{3600\ \cancel{\text{s}}} \times \frac{1\ \cancel{\text{day}}}{24\ \cancel{\text{hr}}} \times \frac{1\ \text{yr}}{365\ \cancel{\text{days}}} \approx 2 \times 10^{67}\ \text{yr}$$

The lifetime of a black hole with the mass of the Sun is some 10^{67} years—which is some 10^{57} times the current age of the universe (which is about 10^{10} years).

53. One trillion (10^{12}) solar masses is about 2.0×10^{42} kg. To calculate the lifetime of a black hole with this mass, we simply plug values into the formula given:

$$t = 10{,}240\ \pi^2 \frac{G^2 M^3}{hc^4} = 10{,}240\ \pi^2 \frac{\left(6.67 \times 10^{-11} \dfrac{\text{m}^3}{\text{kg} \times \text{s}^2}\right)^2 \left(2 \times 10^{42}\ \text{kg}\right)^3}{\left(6.63 \times 10^{-34} \dfrac{\text{kg} \times \text{m}^2}{\text{s}}\right)\left(3 \times 10^8 \dfrac{\text{m}}{\text{s}}\right)^4}$$

Because most calculators do not register numbers over 10^{100}, it is necessary to rearrange the calculation so that you can do the powers of 10 in your head:

$$t = 10{,}240\ \pi^2 \frac{\left(6.67 \times 10^{-11} \dfrac{\text{m}^3}{\text{kg} \times \text{s}^2}\right)^2 (2 \times 10^{42}\,\text{kg})^3}{\left(6.63 \times 10^{-34} \dfrac{\text{kg} \times \text{m}^2}{\text{s}}\right)\left(3 \times 10^8 \dfrac{\text{m}}{\text{s}}\right)^4}$$

$$= 10{,}240\ \pi^2 \frac{\left(6.67 \times 10^{-22} \dfrac{\text{m}^6}{\text{kg}^2 \times \text{s}^4}\right)(8 \times 10^{126}\,\text{kg}^3)}{\left(6.63 \times 10^{-34} \dfrac{\text{kg} \times \text{m}^2}{\text{s}}\right)\left(81 \times 10^{32} \dfrac{\text{m}}{\text{s}}\right)}$$

$$= 10{,}240\ \pi^2 \frac{6.67^2 \times 8}{6.63 \times 81} \times (10^{-22+126-(-34)-32}) \left(\frac{\text{m}^6 \times \text{kg}^3}{\text{kg}^2 \times \text{s}^4} \times \frac{\text{s}^2}{\text{kg} \times \text{m}^3}\right)$$

$$= 66{,}979 \times 10^{106}$$

$$\approx 7 \times 10^{110}$$

Thus, the lifetime of the trillion-solar-mass black hole is about 7×10^{110} seconds, which is equivalent to about 2×10^{103} yr—a long time, but still a lot shorter than infinite time!

54. To calculate the lifetime of a black hole with the mass of the Earth (6×10^{24} kg), we plug values into the formula given:

$$t = 10{,}240\ \pi^2 \frac{G^2 M^3}{hc^4} = 10{,}240\ \pi^2 \frac{\left(6.67 \times 10^{-11} \frac{\text{m}^3}{\text{kg} \times \text{s}^2}\right)^2 (6 \times 10^{24}\ \text{kg})^3}{\left(6.63 \times 10^{-34} \frac{\text{kg} \times \text{m}^2}{\text{s}}\right)\left(3 \times 10^8 \frac{\text{m}}{\text{s}}\right)^4} = 1.8 \times 10^{58}\ \text{s}$$

Converting this answer to years, we find:

$$t = 1.8 \times 10^{58}\ \cancel{\text{s}} \times \frac{1\ \cancel{\text{hr}}}{3600\ \cancel{\text{s}}} \times \frac{1\ \cancel{\text{day}}}{24\ \cancel{\text{hr}}} \times \frac{1\ \text{yr}}{365\ \cancel{\text{days}}} \approx 6 \times 10^{50}\ \text{yr}$$

The lifetime of a black hole with the mass of the Earth is some 10^{50} years—which is some 10^{40} times the current age of the universe (which is about 10^{10} years).

55. To solve the lifetime formula for the mass, we first divide both sides by all of the constants to isolate the mass term, then take the cube root of both sides. We find:

$$t = 10{,}240\pi^2 \frac{G^2 M^3}{hc^4} \Rightarrow M = \sqrt[3]{\frac{t \times hc^4}{10{,}240\ \pi^2 \times G^2}}$$

Before plugging in the lifetime of 12 billion years, we must convert it to units of seconds; you should find that 12 billion years is equivalent to about 3.8×10^{17} s. Plugging in all the values, we find that the mass of the black hole is:

$$M = \sqrt[3]{\frac{(3.8 \times 10^{17}\ \text{s}) \times \left(6.63 \times 10^{-34} \frac{\text{kg} \times \text{m}^2}{\text{s}}\right)\left(3 \times 10^8 \frac{\text{m}}{\text{s}}\right)^4}{10{,}240\ \pi^2 \times \left(6.67 \times 10^{-11} \frac{\text{m}^3}{\text{kg} \times \text{s}^2}\right)}} = 1.7 \times 10^{11}\ \text{kg}$$

Dividing by 1,000 kilograms per metric ton, we see that a mini–black hole with a mass of about 170 million tons would be evaporating during the present epoch.

56. Say I have a mass of 50 kilograms. The uncertainty in my momentum is the uncertainty in my speed times my mass. The uncertainty in my speed is 0.5 kms/hr. We will want this in meters per second, so we convert it to get 0.139 m/s. The uncertainty in my momentum is therefore 6.95 kg $\times$ m/s. Now we recall that the uncertainty principle states that:

$$(\text{uncertainty in location}) \times (\text{uncertainty in momentum}) \approx \text{Planck's constant}$$

We can solve for the uncertainty in location:

$$(\text{uncertainty in location}) \approx \frac{\text{Planck's constant}}{(\text{uncertainty in momentum})}$$

Planck's constant is 6.626×10^{-34} joules $\times$ second, giving us an uncertainty in the position of 9.54×10^{-35} m. This is a very small uncertainty compared to my size, so this effect is unimportant.

57. We will use the uncertainty principle to find the minimum uncertainty in the electron's momentum. The uncertainty principle states:

$$(\text{uncertainty in location}) \times (\text{uncertainty in momentum}) \approx \text{Planck's constant}$$

We can solve this to find the uncertainty in the momentum:

$$(\text{uncertainty in momentum}) \approx \frac{\text{Planck's constant}}{(\text{uncertainty in position})}$$

We know Planck's constant: 6.626×10^{-34} joules $\times$ seconds. We also know the uncertainty in the position, 10^{-10} meters. We can calculate the uncertainty in the momentum: 6.626×10^{-24} kg $\times$ m/s. Knowing that the mass of the electron is 9.1×10^{-31} kg, we can find the uncertainty in the velocity since momentum is mass times velocity. The uncertainty in the velocity comes out to 7.28×10^{6} m/s.

Part V: Stellar Alchemy

Chapter 14. Our Star

This chapter on the Sun describes how the Sun works, laying the groundwork for the study of stars in general by focusing on this all-important example.

As always, when you prepare to teach this chapter, be sure you are familiar with the relevant media resources (see the complete, section-by-section resource grid in Appendix 3 of this Instructor's Guide) and the online quizzes and other study resources available on the Mastering Astronomy Web site. As with all chapters in this edition, the end-of-chapter questions and problems have been updated and augmented.

What's New in the Fourth Edition That Will Affect My Lecture Notes?

Those who have taught from previous editions of *The Cosmic Perspective* should be aware of the following organizational or pedagogical changes to this chapter (i.e., changes that will influence the way you teach) from the Third edition:

- We have reorganized the chapter into just three sections. Section 14.1 gives an overview of the Sun and its properties. Section 14.2 covers energy generation in the Sun. Section 14.3 covers solar activity and the Sun-Earth connection. The headings are different, but the content is basically the same as in the Third edition.
- The discussion of the solar neutrino problem has been updated to include new results from the Sudbury Neutrino Observatory.
- Since solar vibrations are continuous, we decided that the term *star quakes* gives the wrong idea, so we don't call them that anymore. They are henceforth called *solar vibrations*.

Teaching Notes (By Section)

Section 14.1 A Closer Look at the Sun

We have already introduced students to the Sun briefly in Chapter 7, during our tour of the solar system in Section 7.1. Here, we take a more in-depth look at the Sun. We begin by addressing one of the most basic of all astronomical questions: Why does the Sun shine? This question immediately engages students because it relates directly to their personal experience of the Sun and perhaps even to questions they began asking as children. They are often surprised to find out how recently we learned the answer, and the process of elimination that led to the right answer provides a good example of scientific reasoning. We then use an imaginary journey into the Sun to discuss basic properties of the Sun and its structure.

- Relating the history of ideas about how the Sun shines provides an early opportunity to discuss gravitational contraction, a mechanism that will arise repeatedly throughout the rest of the book. Introducing this idea early allows students to digest it somewhat before they encounter it again in star formation and stellar evolution.
- This section also introduces the very important idea of the balance between pressure and gravity within a star. Among astronomers, this kind of equilibrium is known as *hydrostatic equilibrium.* We have found that the word "hydrostatic" is so foreign to students that they often have trouble remembering what it describes. Thus, we have elected to use the term *gravitational equilibrium* in this book so that the link between the term and the concept is easier for students to remember. Note that this term allows the following very simple contrast: Gravitational contraction occurs when gravity overwhelms pressure, and gravitational equilibrium occurs when gravity is in balance with pressure. (In using the term *gravitational equilibrium,* we are following the lead of Mitch Begelman and Martin Rees in their book *Gravity's Fatal Attraction,* Scientific American Library, 1996.)
- This section also raises a potential dilemma for instructors: whether to describe the Sun as yellow or white. Because students already *know* that the Sun is yellow, they can have trouble with the idea that the Sun would look nearly white if they were not seeing it through the Earth's atmosphere. However, when astronomers speak of "white stars," they are usually talking about stars considerably hotter than the Sun. For this reason, we depict the Sun as yellow in our figures. (One way to demonstrate that the Sun is whiter than it appears is to point out that clouds look white because they are scattering both the blue light from the sky, which is indirect solar radiation, and the direct yellow light from the Sun. Combining these two colors of light more closely approximates the original color of the Sun.)

Section 14.2 Nuclear Fusion in the Sun

This section focuses on the process of nuclear fusion in the Sun and describes how the balance between gravity and pressure acts as a thermostat in the solar core, maintaining a constant fusionrate.

- The topic of fusion in the solar core presents an opportunity to discuss the role of mathematical modeling in science and to explain how we can be so certain of what's going on when we can't see the core and can't send in a probe to observe what's going on there. Up to this point in the course, we've been discussing planets, which some students regard as "more real" because we have landed probes there. The Sun provides these students with their first challenge to understand how we might be able to learn about a place without actually visiting it.
- In this section we also discuss solar neutrinos as a way to "observe" the solar core, and we present the solar neutrino problem as a nearly solved problem in our understanding of the Sun.
- Based on recent experimental results confirming the fact that neutrinos oscillate, we presume that neutrinos do have mass (e.g., Fukuda et al., 1998, *Physical Review Letters,* 81, 1562), though the amount of mass remains unknown. (Ironically, the best *upper* limits on the neutrino mass at about 1 eV come not from detector experiments but from cosmological measurements of the largest gravitationally bound structures in the universe; e.g., Crotty, Lesgourgues, & Pastor, 2004, *Physical Review D*, 69, 123007.) Again, be sure to stay current in class with the latest results on neutrino mass.

Section 14.3 The Sun-Earth Connection

This section discusses solar activity, including sunspots, and the ways in which solar activity can affect the Earth.

- Solar activity is one of the most direct ways in which cosmic events affect human activity. Don't miss this opportunity to engage your more practically minded students with some examples of how solar weather affects the Earth.

Answers/Discussion Points for Think About It Questions

The Think About It questions are not numbered in the book, so we list them in the order in which they appear, keyed by section number.

Section 14.1

- (p. 479) At higher altitudes there is less overlying air pressing down. Because of the lower pressure, the atmosphere is less compressed.
- (p. 481) We need to know Earth's period and distance from the Sun. We measure the mass of the Sun by measuring the periods and semimajor axes of planetary orbits and then plugging these measurements into Newton's version of Kepler's third law. Because the Sun is so much heavier than anything else in the solar system, the sum of the masses in the equation is essentially equal to the Sun's mass.

Section 14.2

- (p. 484) Heavier elements have more protons in their nuclei. Their greater positive charge means they repel each other more strongly than do hydrogen nuclei, which contain only one proton. The stronger repulsion means that higher temperatures are required to overcome the repulsion and make the nuclei fuse.
- (p. 487) Some other examples of diffusive processes are the ways smells propagate through air or pollution leaks into a river from a tiny pipe. Sociological examples might be the way a particularly vicious rumor spreads through a middle school or the way an e-mail computer virus propagates to many PCs.

Section 14.3

- (p. 495) Because the Sun rotates faster at the equator, after a few days the sunspot near the equator would have moved ahead (east) of the higher latitude sunspot. By measuring the rate at which we see the sunspots move at different latitudes, we can determine the Sun's rotation period at each latitude.

Solutions to End-of-Chapter Problems (Chapter 14)

1. As something contracts, its gravitational potential energy is converted into thermal energy. This energy source was important for the Sun when the Sun was forming billions of years ago because it provided the energy needed to start the fusion in the Sun's core.
2. Gravitational equilibrium is a balance between the force of gravity pulling inward and pressure pushing outward. In the Sun's core, the weight of the layers above is very large so that the pressure needed to balance it is also large. Pressure is the product of temperature and density, so high pressure means high densities and high temperatures in the core.

3. The Sun's radius is about 700,000 kilometers, more than 100 times the radius of the Earth. The Sun's mass is 2×10^{30} kg, more than 1,000 times the mass of all of the planets combined. The Sun's luminosity is 3.8×10^{26} watts and just 1 second's worth of that would be enough to meet humanity's current energy demands for the next 500,000 years. The Sun's surface temperature is 5,800 K. A blast furnace, by comparison, is about 1,500 K. (The fireball temperature of an atmospheric detonation of a nuclear weapon can achieve and maintain 5,000 K, but hopefully this is not an "everyday experience" of anybody.)
4. The corona is a region several million kilometers above the surface of the Sun that is at a temperature of about 1 million K. It is from here that the Sun's X rays are emitted. Beneath that layer is the chromosphere, where the temperature drops to 10,000 K and where the Sun's ultraviolet light is emitted. The lowest layer of the atmosphere is the photosphere, where the temperature is 6,000 K and where the visible light from the Sun is emitted. (The photosphere is what we see as the surface of the Sun.)

   The topmost layer inside the Sun is the convective zone, where the energy generated in the core is transported upward by hot gas rising and cooler gas sinking. About a third of the way to the middle of the Sun, the convective zone ends and the radiative zone begins. In the radiative zone, the Sun's energy is carried outward by photons of light. Finally, the innermost layer of the Sun is the core. The Sun is able to produce its energy here through nuclear fusion, thanks to the temperature of 15 million K and a density 100 times that of water.
5. Nuclear fission is the process of splitting an atomic nucleus into two. Nuclear fusion is the process of combining two nuclei. Nuclear power planets use fission to generate energy here on Earth, but the Sun uses fusion.
6. Nuclear fusion requires high temperatures to keep the protons colliding at high enough speeds that they can get close enough to stick together rather than be deflected by the electromagnetic force. The high pressure, generated by the weight of all the Sun's layers above the core, is required to keep the hot gas in the Sun's core from exploding into space, shutting off the nuclear reactions. A high particle density is required to sustain a high rate of fusion.
7. The Sun's overall nuclear reaction is to combine four protons to form a single helium nucleus. The actual process, called the proton-proton chain, requires three steps. In the first, two protons collide to form an isotope of hydrogen called deuterium. (This is done twice for each reaction.) In the second step, a deuterium nucleus is struck by a proton to become helium-3. (This step also occurs twice per reaction.) Finally, when two helium-3 nuclei collide, the formed helium-4 and two protons are released.
8. The Sun's energy output is steady in time. This steadiness results because the rate of fusion is sensitive to temperature. If the Sun's core were a bit hotter, the fusion rate would increase. This would produce more energy, which would cause the core to expand slightly and cool. The cooling would cause the fusion rate to slow back down until the Sun was back to the original size and temperature and fusion occurred at the same rate.
9. The Sun brightens gradually with time because as fusion converts protons into helium nuclei in the core, the number of particles must decrease. This means a lower pressure so that the core contracts and heats up slightly. The hotter core results in a higher rate of fusion, making the Sun brighter.

10. Photons take hundreds of thousands of years to get out of the Sun because their paths zigzag repeatedly. Because the plasma is so dense in the Sun's interior, photons can only travel a fraction of a millimeter before "colliding" with an electron and deflecting into a new direction. So photons bounce around at random and only slowly make their way out of the Sun.
11. Mathematical models use the observed composition and mass of the Sun along with the laws of physics to derive equations that describe the gravitational equilibrium, solar thermostat, and rate at which energy moves from the core to the photosphere. Computers let us calculate the Sun's temperature, pressure, and density at any depth. We can check the models by comparing their predictions for the radius, surface temperature, and luminosity, as well as other observable parameters generated by studying helioseismology. Helioseismology allows us to probe the conditions in the solar interior. The models do indeed make predictions that allow us to explain these observations; we are on the right track.
12. Neutrinos are subatomic particles produced in nuclear reactions. They move at nearly the speed of light and almost never interact with matter. Detectors on Earth found only about $\frac{1}{3}$ of the neutrinos predicted by models of nuclear fusion in the Sun. This disagreement between theory and data was called "the solar neutrino problem." Today, we think we have solved the problem and that the electron neutrinos we were measuring had changed into other kinds of neutrinos. Recent experiments support this idea by showing that neutrinos can change their type. Other experiments are showing that the total number of neutrinos of all types is about what the models for fusion in the Sun predict, indicating that the problem is solved.
13. Solar activity refers to the changing features of the Sun such as sunspots, flares, prominences, and coronal mass ejections. The sunspots are regions of the photosphere that are cooler than the surrounding plasma so that they appear darker. The sunspots occur in pairs with the magnetic field lines arcing from one sunspot to the other. Where gas from the chromosphere or corona becomes trapped in the magnetic field, we see prominences. Flares are intense storms that result in bursts of X rays and fast-moving charged particles being shot off into space. Coronal mass ejections are huge bubbles of energetic, charged particles that are released from the Sun's corona.
14. The Sun's photosphere is at a temperature of 5,800 K. It looks mottled because it is churning constantly with rising and falling gas. However, in some areas the surface is cooler and therefore less bright. These areas are sunspots, where the temperature is "only" 4,000 K.
15. Magnetic fields in sunspots keep the sunspots cooler than the surrounding plasma by keeping the other plasma out. Since new, hot plasma cannot enter the sunspot to warm them, the spots can cool.
16. The chromosphere is best viewed with ultraviolet telescopes because it emits a lot of light in that part of the spectrum, being 10,000 K, and because we cannot usually see the chromosphere in the visible wavelengths because the photosphere drowns the light out. Similarly, the 1 million K corona is so hot that it emits strongly in the X-ray part of the spectrum, making it best viewed in X rays. These two parts of the Sun's atmosphere are heated by magnetic fields carrying energy upward from the surface.
17. The sunspot cycle is a cycle in which the average number of sunspots on the Sun gradually rises and falls over a period of about 11 years. As the cycle progresses, more and more sunspots appear on the Sun's surface and the spots start to move to lower latitudes. As the sunspots approach the solar equator, solar activity reaches

a peak with prominences, flares, and coronal mass ejections. Eventually, the activity decreases and the number of sunspots diminishes as the Sun approaches solar minimum. At this time, the sun's magnetic field reverses orientation so that magnetic north becomes magnetic south and vice versa.

There are also long-term changes in solar activity. Astronomers have detected a period of about 70 years when virtually no sunspots were visible.

18. The leading model to explain the sunspot cycle suggests that as the magnetic fields in the Sun are dredged upward by convection, they are amplified. Because the Sun rotates faster at the equator than at the poles, the field lines are also stretched and shaped. As this goes on, the field lines become wound around the Sun more and more tightly, producing contorted field lines that generate sunspots and other solar activity.

    The sunspot cycle may influence Earth's climate. There are data suggesting that the period from 1645 to 1715 when solar activity seems to have almost ceased also corresponded to low temperatures in Europe and North America. Other data have suggested that droughts or frequencies of storms may be connected to the sunspot cycle. However, the connections are weak, so we do not yet know how much effect the sunspot cycle has on Earth's weather.

19. *Before Einstein, gravitational contraction appeared to be a perfectly plausible mechanism for solar energy generation.* This statement is not sensible. Gravitational contraction has been known to be an insufficient source of stellar power since the 1800s, when geologists realized that the age of rocks on the Earth numbered in the billions of years. Gravitational contraction of the Sun, if the only source of the Sun's energy, would have powered the Sun only for some 25 million years, a time much less than its geological age.

20. *A sudden temperature rise in the Sun's core is nothing to worry about, because conditions in the core will soon return to normal.* This statement is sensible. The Sun's core acts as a thermostat, and thus if the inner core temperature increases, the reaction rate increases too, but the extra energy expands the core and cools it, thus reducing the nuclear reaction rates.

21. *If fusion in the solar core ceased today, worldwide panic would break out tomorrow as the Sun began to grow dimmer.* This statement is not sensible. If fusion in the core ceased, the photons would continue to percolate out of the Sun at about the same rate for many thousands of years. No dimming would be possible to measure the day after such an event. It is a debatable point as to when such an event might be noticed—certainly neutrino fluxes would drop noticeably, and eventually the structure of the Sun would be affected.

22. *Astronomers have recently photographed magnetic fields churning deep beneath the solar photosphere.* This statement is not sensible. The photosphere is as far in toward the Sun as we can "see" in photographs. Conditions beneath the photosphere must be inferred from other types of observations and theoretical models.

23. *Neutrinos probably can't harm me, but just to be safe I think I'll wear a lead vest.* This statement is not sensible. Neutrinos pass right through the Earth and would not be diminished at all by a lead vest.

24. *There haven't been many sunspots this year, but there ought to be many more in about 5 years.* This statement is not sensible in 2006/2007. An inspection of Figure 14.22 shows that we are currently at a time of major activity, and that extension of the trend predicts the sunspot numbers will go down over the next 5 years. [This statement of course is time-dependent—the next minimum in activity will be around 2009/2010, at which time this statement might well be sensible.]

25. *News of a major solar flare today caused concern among professionals in the fields of communications and electrical power generation.* This statement is sensible. Solar flares can cause havoc with satellites, communications, and power grids because of the energetic charged particles.
26. *By observing solar neutrinos, we can learn about nuclear fusion deep in the Sun's core.* This statement is sensible. Neutrinos are produced by fusion in the Sun's core and can travel directly from the Sun's core to neutrino detectors on Earth.
27. *If the Sun's magnetic field somehow disappeared, there would be no more sunspots on the Sun.* This statement is sensible. Sunspots are regions of strong magnetic fields, so with no magnetic field there would be no sunspots.
28. *Scientists are currently building an infrared telescope designed to observe fusion reactions in the Sun's core.* This statement does not make sense. Infrared telescopes cannot see through the Sun.
29. c. Four individual protons
30. a. the photosphere
31. c. the photosphere
32. b. mathematical models of the Sun
33. a. They are cooler than their surroundings
34. c. helium, energy, and neutrinos
35. a. photons
36. a. about the same
37. b. electrons (Because the charge must be neutral, all electrons have a single negative charge; the helium nuclei are +2 positively charged, so the electron must be the most common particle in a net charge neutral solar wind.)
38. c. protons (charged particles)
39. If fusion reactions suddenly shut off in the Sun, photons would continue to diffuse away from the core and radiate from the Sun's surface for about a million years. (The main point is that the Sun would not "turn off" suddenly—we're not looking for students to invent red giant phases.)
40. We would be able to tell if fusion reactions had shut off because we would no longer detect neutrinos from the Sun. Neutrinos, a by-product of nuclear fusion, exit the Sun immediately.
41. A "stronger" strong force would mean that nuclear reactions would not require such high temperatures in order to proceed, so the core temperature of the Sun would be cooler.
42. An observer on Pluto would measure the same solar rotation period with respect to the stars—that is, if your friend takes into account Pluto's orbit around the Sun, and you take into account the Earth's orbit around the Sun. But that's not quite what the question says—it asks for you to wait until the sunspot returns to the same position it had in the first picture. During that time the Earth has moved farther around the Sun than Pluto, so by this definition your friend on Pluto would measure a DIFFERENT (fewer) number of days for this same configuration to happen from your friend's perspective.
43. The Sun would look yellow and bright. The sunspots appear dark only in contrast to their surroundings, but they are actually quite luminous.

44. Scientists use mathematical models, and observations of solar vibrations and neutrinos to infer what is happening in the center of the Sun. A probe would not survive the trip.
45. X rays represent a tiny fraction of the Sun's total energetic output. Therefore, huge fluctuations in X-ray production represent only tiny variations in the Sun's energetic output. X rays in general are produced by either extremely hot or extremely violent phenomena; in the Sun, this violent behavior is time-variable.
46. Essay question, but the key point should be that we are *not* capable of affecting the Sun in any meaningful way, at least with current and any foreseeable future technology.
47. The Sun's mass is 2.0×10^{30} kg, and it is radiating energy at the rate of 3.8×10^{26} W. (1 watt = 1 joule/sec). So its lifetime based on chemical burning is simply its chemical energy content divided by the rate it is losing energy. The total energy content is 2.0×10^{30} kg $\times$ (10^8 joule/kg) = 2.0×10^{38} joules. Divide this quantity by the Sun's luminosity to get the lifetime in seconds, or 2.0×10^{38} joules/3.8×10^{26} W = 5.26×10^{11} seconds, or about 17,000 years (see the 48 b. solution for an example of the time conversion).
48. a. The total amount of mass in the Sun is 2.0×10^{30} kg, 75% of which is hydrogen and 13% of which becomes available for fusion. Thus, the total mass of hydrogen available for fusion over the Sun's lifetime is simply 13% of 75% of the total mass of the Sun or:

$$2.0 \times 10^{30} \text{ kg} \times 0.75 \times 0.13 = 1.95 \times 10^{29} \text{ kg}$$

b. The Sun fuses 6×10^{11} kg per second and has 2×10^{29} kg available for fusion, so the Sun's lifetime is:

$$\text{lifetime} = \frac{\text{mass available}}{\text{rate mass burned}} = \frac{1.95 \times 10^{29} \not{\text{kg}}}{6 \times 10^{11} \not{\text{kg}}/\text{s}} = 3.25 \times 10^{17} \text{ s}$$

or:

$$3.25 \times 10^{17} \not{\text{s}} \times \frac{1 \not{\text{hr}}}{3600 \not{\text{s}}} \times \frac{1 \not{\text{day}}}{24 \not{\text{hr}}} \times \frac{1 \text{ yr}}{365 \not{\text{days}}} = 10.3 \text{ billion yr}$$

c. Subtracting the current age of the Sun from the lifetime found in part (b), we find:

$$10.3 \text{ billion yr} - 4.6 \text{ billion yr} = 5.7 \text{ billion yr}$$

The Sun will run out of fuel in approximately 6 billion years.
49. a. The surface area of a sphere 1 AU (=1.5×10^{11} m) in radius is:

$$4\,\pi r^2 = 4\,\pi(1.5 \times 10^{11} \text{ m})^2 = 2.83 \times 10^{23} \text{ m}^2$$

b. The flux of solar radiation at the surface of this imaginary sphere is the luminosity of the Sun divided by the surface area of the sphere:

$$\frac{3.8 \times 10^{26} \text{ watts}}{2.83 \times 10^{23} \text{ m}^2} = 1{,}344 \text{ watts per square meter}$$

c. The average power per square meter collected by a solar collector on the ground will always be less because of absorption by the atmosphere, the angle of incidence not being 90°, the weather (cloud cover), nighttime, and varying amounts of daylight.

d. To optimize the amount of power collected, a solar collector should be aimed up and south in the Northern Hemisphere, and up and north in the Southern Hemisphere (and up toward the celestial equator). To achieve even more optimization, one might rotate the face of the collectors east to west to follow the Sun's daily path across the sky, and north to south to match the Sun's changing path with the seasons.

50. a. The power requirement in watts (or joules/second) in the United States is:

$$\frac{2\times 10^{20}\ \text{joules/yr}}{3.1\times 10^{7}\ \text{s/yr}} = 6.45\times 10^{12}\ \text{watts}$$

b. If we can achieve a power conversion efficiency for 200 watts per square meter, we require:

$$\frac{6.45\times 10^{12}\ \text{watts}}{200\ \text{watts/m}^2} = 3.22\times 10^{10}\ \text{m}^2 \times \underbrace{\left(\frac{1\ \text{km}}{1000\ \text{m}}\right)^2}_{\text{convert to km}^2} = 3.22\times 10^{4}\ \text{km}^2$$

of solar detector area to supply the entire United States.

c. The total surface area of the United States is 2×10^7 km^2, so this enterprise requires:

$$\frac{3.22\times 10^{4}\ \text{km}^2}{2\times 10^{7}\ \text{km}^2} = 0.0016, \quad \text{or} \quad 0.16\%$$

of the surface area of the United States. The solar collectors would probably be optimally placed in the sunnier, emptier parts of the country, where there aren't many cloudy days and the land is cheap and relatively unoccupied. The energy generated is renewable, so if the technology required to produce the solar cells, service the arrays, and distribute the power were also eco-friendly, this would be a reasonable way to supply power from an ecological standpoint. However, alternative power may be expensive to implement and might be opposed by some political constituencies. (For example, there will always be constituencies that would be less than thrilled to see this land paved over like a gigantic parking lot in Nevada.)

51. Wien's law states:

$$\lambda_{\text{max}} = \frac{2,900,000}{T(\text{Kelvin})}\text{nm}$$

Plugging in the average temperature of the Sun's photosphere, 5,800 K, gives:

$$\lambda_{\text{max}} = \frac{2,900,000}{5800}\text{nm} = 500\ \text{nm}$$

Thus, the Sun's thermal spectrum peaks at a wavelength of 500 nanometers, which is in the green part of the visible spectrum. However, because the Sun also radiates other colors of the visible spectrum and the atmosphere scatters the bluer light, it appears white or yellow to our eyes.

52. To solve this problem, we will use Wien's law to find the peak wavelength for a sunspot. Wien's law states that:

$$\lambda_{max} = \frac{2,900,000}{T(\text{Kelvin})} \text{nm}$$

Since sunspots have a typical temperature of 4,000 K, we can plug the value in to get:

$$\lambda_{max} = \frac{2,900,000}{4000} \text{nm} = 725 \text{ nm}$$

So the peak wavelength from a sunspot is 725 nanometers. This is a longer wavelength than we get for the rest of the Sun's surface, corresponding to the reddest end of the visible spectrum.

53. Mathematical Insight 14.1 told us that the Sun loses 4.2×10^9 kg of mass every second due to fusion. So we need to multiply this rate by the time that the Sun will live (10 billion years) to get how much mass it will lose over its lifetime. First, we convert 10 billion years to seconds to get 3.16×10^{17} s. Then, we multiply this lifetime in seconds by the rate of mass lost to get 1.33×10^{27} kg lost over the lifetime of the Sun. We recall that the Earth's mass is 5.28×10^{24} kg, so the mass that the Sun will lose due to fusion over its lifetime greatly exceeds the mass of the Earth.

54. We will use the relationship between temperature and pressure:

$$P = nkT$$

where P is the pressure; n is the number density of gas molecules; k is Boltzmann's constant, 1.38×10^{-23} joules/Kelvin; and T is the temperature. We can express the results in a table:

Layer	**Temperature (Kelvin)**	**Density (particles/cm^3)**	**Pressure (J/cm^3)**	**Ratio to Earth Pressure**
Top	4,500	1.60E+16	9.94E–04	1.00E–06
Middle	5,800	1.00E+17	8.00E–03	8.06E–06
Bottom	7,000	1.50E+17	1.45E–02	1.46E–05

For the last column, we have used the fact that the Earth's atmosphere at sea level is about 300 K and has a density of about 2.4×10^{19} particle/cm^3, leading to a pressure of 994 J/cm^3. We see from this table that the pressure of the photosphere increases as we get lower. This makes sense as each deeper layer of the photosphere has to exert enough pressure to support the layers above it.

55. We can assume that the temperatures of our tires and the air are the same. In this case, we can find the ratio of the densities by using the relationship between pressure, temperature, and number density:

$$P = nkT$$

We can solve this for number density:

$$n = \frac{P}{kT}$$

Taking the ratio of the two densities:

$$\frac{n_{\text{tire}}}{n_{\text{outside}}} = \frac{\dfrac{P_{\text{tire}}}{kT_{\text{tire}}}}{\dfrac{P_{\text{outside}}}{kT_{\text{outside}}}}$$

Since the temperatures are the same and Boltzmann's constant is the same, we can cancel and get the simple relationship:

$$\frac{n_{\text{tire}}}{n_{\text{outside}}} = \frac{P_{\text{tire}}}{P_{\text{outside}}}$$

The tire pressure is 30 pounds per square inch and the air pressure is 15 pounds per square inch. Thus, the number density of particles inside the tire must be twice as high as the density outside the tire.

If we puncture the tire, the number density of particles decreases. Eventually, the tire's pressure becomes the same as the surrounding air, and the number densities of particles are equal inside and outside the tire, and the tire goes flat (it shrinks). This behavior is superficially like the core of the Sun in that the Sun's core shrinks as the effective number density decreases. However, in the case of the Sun, the energy generation by fusion continues (as long as the hydrogen is still available), the gravitational pressure increases (smaller radius, same mass), so the core temperature increases. The Sun's luminosity thus increases gradually over time. The tire's temperature is the same as its surroundings because of rapid energy exchange, so it goes flat because it has no source of energy generation, and its source of pressure is not gravitational equilibrium.

56. I have a mass of about 60 kilograms. We can convert 0.7% of that mass, 0.42 kilogram, into energy. Using Einstein's famous equation, $E = mc^2$, we can convert this into energy. Putting in the numbers we find that 3.78×10^{16} joules would be available to be used via fusion. Since the rate at which I use energy is 100 watts, we can determine how long I could operate with this energy supply:

$$\begin{aligned} \text{time} &= \frac{\text{energy supply}}{\text{rate of use}} \\ &= \frac{3.78 \times 10^{16}\ \text{J}}{100\ \text{W}} \\ &= 3.78 \times 10^{14}\ \text{s} \end{aligned}$$

We had better convert from seconds into years! Running through the usual conversion we find that the energy supply would last me 12 million years.

Chapter 15. Surveying the Stars

This chapter outlines how we measure and classify stars. It introduces many important ideas such as the relationship between luminosity, brightness, and distance. It is also where we introduce the H-R diagram.

- Whenever possible, we have used real stellar data from Hipparcos and other sources to construct the H-R diagrams in Chapters 15 and 16.

As always, when you prepare to teach this chapter, be sure you are familiar with the relevant media resources (see the complete, section-by-section resource grid in Appendix 3 of this Instructor's Guide) and the online quizzes and other study resources available on the Mastering Astronomy Web site.

What's New in the Fourth Edition That Will Affect My Lecture Notes?

Those who have taught from previous editions of *The Cosmic Perspective* should be aware of the following organizational or pedagogical changes to this chapter (i.e., changes that will influence the way you teach) from the Third edition:

- We have reorganized the chapter into three sections. Section 15.1 covers basic properties of stars. Section 15.2 covers stellar classification and the H-R diagram. Section 15.3 focuses on star clusters and how we determine their ages.
- We now refer to the relationship between apparent brightness and luminosity as the *inverse square law for light* (we formerly called it the *luminosity-distance formula*).

Teaching Notes (By Section)

Section 15.1 Properties of Stars

This section explains how we determine basic properties of stars: luminosity, surface temperature, and mass.

- We avoid using the term *flux* in this book because it's an unfamiliar word that students do not find particularly descriptive. Instead we use the term *apparent brightness*, which refers explicitly to the concept of brightness. Likewise, we avoid the term *bolometric luminosity* and use the term *total luminosity* instead.
- This section introduces the term *parsecs* to provide a reference for that term, but for this book, we now consistently use the term *light-year* for distance. We also define *solar luminosity* here.
- We introduce the magnitude system in this section, including a Mathematical Insight involving magnitudes, but we do not use magnitudes in our own general education courses at all. We find that magnitudes require great effort for students to master yet add little to students' understanding of other astronomical concepts. Very gradually, astronomers themselves are moving away from the magnitude system as astronomy grows to include more regions of the electromagnetic spectrum for which there are no standard filter sets. Magnitudes are included here largely because of their historical importance and because they are often indicated

on star charts and some astronomical tables. You can skip them if you choose, because we do not use magnitudes elsewhere in the book.

- The discussion of star colors, temperatures, and absorption lines assumes that students have covered the material on light and color in Chapter 5.
- For simplicity, we limit the discussion of stellar classes to OBAFGKM, which covers the standard main sequence and omits special cases like R, N, and S stars and Wolf-Rayet stars. Our intention is to make sure students clearly understand the main sequence, the bedrock of stellar studies, before branching off into more specialized topics. We have found that hosting a student contest to identify a modern mnemonic always stimulates some witty and timely mnemonics involving current events and next weekend's major sports event.
- Students, particularly women students, enjoy the stories of Annie Jump Cannon and Cecilia Payne-Gaposchkin. For many of them, these are the first stories about women scientists that they have heard, beyond (perhaps) Madame Curie.
- This section mentions several different types of binary systems but does not provide an exhaustive nomenclature. We describe astrometric binaries without using the formal term, because the word *astrometric* is uninformative to students. Visual, spectroscopic, and eclipsing binaries are mentioned explicitly, because these terms are usefully descriptive.
- Teaching students about measuring masses is one of the main themes in our course. We found that the most effective way to communicate this difficult but pervasive concept in a large class is to use a combination of frequent conceptual questions in class and peer instruction. (See Eric Mazur's Peer Instruction, or his Web site at *http://mazur-www.harvard.edu/education/educationmenu.php* for some ideas.)
- We refer to the fundamental law for measuring stellar masses as *Newton's version of Kepler's third law* in order to give credit to both individuals.

Section 15.2 Patterns among Stars

This section summarizes stellar classification and introduces the H-R diagram. The previous section laid the groundwork for this section on the H-R diagram. While students can understand the observational H-R diagram before covering stellar masses, we find it prudent to discuss the H-R diagram after covering masses, because we can immediately point out that the main sequence is fundamentally a sequence of stellar masses. This approach helps counteract the tendency of students to think of the main sequence as a temporal evolution of stellar properties.

- In small class sections, it can be illustrative to have students plot stellar temperatures and luminosities on their own before you cover the H-R diagram, enabling them to discover the main sequence for themselves.
- Appendix F lists the 20 brightest stars as well as the stars within 12 light-years. One way to get students to think more deeply about the H-R diagram is to ask them why the spectral types of the stars in the two tables are so different.
- This section includes a discussion of main-sequence lifetimes, showing how some simple order of magnitude estimates can reveal the vast differences between stellar lifetimes.
- Because this text focuses on the "big picture" of astronomy, we limit any specific coverage of variable stars to Cepheid variables, which will resurface in the chapters on galaxies. (RR Lyrae variables are pointed out on the instability strip in Figure 15.14.)

Section 15.3 Star Clusters

The chapter concludes with this section on star clusters, emphasizing that they are excellent laboratories for comparing the properties of stars and establishing how stars evolve.

- This chapter emphasizes that a star's color depends primarily on its mass and age but does not mention that a star's heavy-element content also affects its color. Heavy elements tend to hinder the flow of energy in the outer layers of a star, making it slightly larger and redder than it would be without these heavy elements. We find that downplaying this fact does not hinder students' comprehension of the "big picture," but it does complicate the issue of the main-sequence turnoff. Because globular clusters are poor in heavy elements, their main sequences are displaced to the left in the H-R diagram. Their turnoff points thus lie blueward of the Sun but at lower luminosities. We allude to this effect in our figure caption for Figure 15.18.

Answers/Discussion Points for Think About It Questions

The Think About It questions are not numbered in the book, so we list them in the order in which they appear, keyed by section number.

Section 15.1

- (p. 503) Apparent brightness is directly proportional to luminosity, so at the same distance Star A would be four times as bright in our sky as Star B. If Star A were twice as far away, the inverse square law tells us it would appear four times dimmer—which means it would then look the same in apparent brightness as Star B.
- (p. 509) Inventing an original OBAFGKM mnemonic is a fun way for students to burn this sequence into their memory.

Section 15.2

- (p. 513) The colors of the stars are similar to the star colors determined by thermal radiation at the given surface temperature. The colors of stars are not necessarily related to their interior temperatures. For example, a red supergiant and a red main-sequence star have very different core temperatures.
- (p. 515) Bellatrix: spectral type B, luminosity class V, radius ~$7R_{Sun}$, mass ~$8M_{Sun}$, lifetime ~3×10^7 yr. Vega: spectral type A, luminosity class V, radius ~$3R_{Sun}$, mass ~$2.5M_{Sun}$, lifetime ~5×10^8 yr. Antares: spectral type M, luminosity class I, radius ~$400R_{Sun}$, mass and lifetime uncertain. Pollux: spectral type K, luminosity class III, radius ~$10R_{Sun}$, mass and lifetime uncertain. Proxima Centauri: spectral type M, luminosity class V, radius ~$0.1R_{Sun}$, mass ~$0.09M_{Sun}$, lifetime ~10^{12} years.
- (p. 519) DX Cancri is the star in Figure 15.10 with the longest main-sequence lifetime. It is the coolest, least luminous star pictured on this main sequence, so it will putter along for well over 100 billion years. Its tiny luminosity means that it is burning its fuel at a very conservative rate.

Section 15.3

- (p. 520) The main-sequence turnoff point in a 10-billion-year-old star cluster should be around $1L_{Sun}$, because the Sun itself leaves the main sequence at an age of around 10 billion years. The turnoff should therefore be around spectral type G (in a cluster with solar proportions of heavy elements). All of the original M stars should still remain, but none of the original F stars will. (Beyond the scope of the book: A few stars called *blue stragglers*, which are bluer than the stars at the turnoff point, can be found in old clusters. These stars probably resulted from relatively recent mergers of two smaller stars to form a single, more massive star.)

Solutions to End-of-Chapter Problems (Chapter 15)

1. A star's luminosity measures how much energy it radiates into space. The apparent brightness tells us how bright it seems in our sky. The two concepts differ because objects that are farther away appear dimmer. This relationship is described by the inverse square law for light, which says that the brightness of a star follows an inverse square law with distance, getting four times dimmer every time we move twice as far away.
2. Stellar parallax is the tiny movement of stars in our sky due to Earth's motion around the Sun. Since more distant stars show smaller parallaxes than closer ones, we can measure the amount that stars move over 6 months (half of an Earth orbit) and find the distance to the stars. Once we know this, we can use the apparent brightness of the star along with the inverse square law for light to determine the star's luminosity.
3. The apparent magnitude of a star is how bright it appears from here on Earth, measured on a special scale where a factor-of-100 increase in brightness corresponds to five magnitudes lower in value. The absolute magnitude is what the apparent magnitude would be if the star were at a distance of 10 parsecs. These two quantities are similar to the apparent brightness and luminosity; they are just measured on different scales, are computed using formulas with base 10 logarithms, and scale so that the smaller the magnitude, the brighter the star.
4. Spectral types are a way of classifying stars according to their color or what spectral lines we see in their light. The spectral types run OBAFGKM, where O stars are the hottest and M are the coolest.
5. The spectral sequence was discovered from work done by a large number of people. It began with wealthy astronomer Henry Draper, who was a pioneer in stellar spectroscopy. When he died, his widow donated money to the Harvard College Observatory, where Edward Pickering used the money to hire a group of "computers," mostly women, who studied the stellar spectra. One of them, Williamina Fleming, found that she could classify the spectra by the strength of their hydrogen lines. She ordered them A, B, C, . . . , O. However, this scheme proved inadequate to shedding any understanding on the nature of stars. Annie Jump Cannon realized that she could reorganize the system into a more natural order, in the process eliminating or combining some of the original classes. This left us with OBAFGKM.
6. There are three kinds of binary star systems. The first is visual binaries, ones in which we can see both stars distinctly as they orbit each other. The second type of binary system is the eclipsing binary, which we see by examining the light curve. Light curves of eclipsing binaries show periodic dimming, corresponding to when one of the stars passes behind the other and its light is blocked. The final type of binary is the spectroscopic binary. For these systems we detect the presence of two stars (rather than one) by the Doppler shifts in the spectral lines.

   Eclipsing binaries are particularly important for finding stellar masses because we can measure the orbital periods of the stars and the velocities. (We can get the velocities in this case because we know that these systems orbit in the plane of our line of sight.) With this information, we can determine the orbital separation and then the masses via Newton's version of Kepler's third law.

7. The sketch should look like Figure 15.10. Stars that are cool and dim are located in the lower right of the plot. Cool and luminous stars are in the upper right, hot and dim stars are in the lower left, and hot and luminous stars are in the upper left.
8. Luminosity classes of stars are designated by Roman numerals and tell us what region of the H-R diagram the star falls in. We use both spectral type and luminosity class to completely classify stars since the spectral type tells us the temperature while the luminosity class tells us the radius. So, for example, our Sun is a G2 V, where G2 is the spectral class (it says that it's a yellow-white star) and V is the luminosity class (it tells us that it is a main-sequence star).
9. The defining characteristic of a main-sequence star is that it falls along a specific line on the H-R diagram and so it follows a particular relationship between luminosity and surface temperature. This relationship occurs because the more luminous stars have larger masses and therefore have higher rates of fusion in their cores. Because of the particular relationship between luminosity and radius along the main sequence, the more massive stars must also be much hotter than the less massive ones in order to emit their energy from their surfaces. (Hotter surfaces emit more light per unit area.)
10. Smaller stars have longer lifetimes than larger stars. This is because the larger stars are much more luminous than the smaller ones. While the larger stars have more fuel to use up, their luminosities are so great that they consume their fuel supply faster and end their main-sequence lives sooner. If stellar luminosities were simply proportional to stars' masses, all stars would have the same lifetimes. But massive stars are much more luminous compared to their mass than are low-mass stars.
11. A star's birth mass determines its quantity of hydrogen fuel, its central pressure and temperature, and therefore its luminosity. These in turn set the star's main-sequence lifetime and surface temperature. These relationships mean that the birth mass of the star sets most of its other properties.
12. Giant and supergiant stars are stars that have left the main-sequence after exhausting their supplies of hydrogen fuel in their central cores. They release fusion energy so furiously that they have to expand in order to radiate it away. When a star that is similar in mass to our Sun runs out of fuel completely, it forms a dead core in which the nuclear fusion has ceased. This core is called a "white dwarf." They are small (about the size of Earth) and very dense. Because they are basically exposed stellar cores, they are also very hot.
13. Pulsating variable stars vary in brightness because they cannot achieve a steady equilibrium. Their upper layers are too opaque, trapping photons beneath them, so they expand. But the expansion results in upper layers that are too transparent, so the photons escape, and the layers contract again. This cycle of expanding and contracting cause the luminosity to vary.
14. Open clusters are located in the galactic disk. They contain up to several thousand stars, are typically about 30 light-years across, and tend to be young. Globular clusters are quite old, are found in the galactic halo, and contain more than a million stars.
15. The H-R diagram looks different for clusters of stars because the largest stars have left the main sequence. This is because all of the stars were born together, but the largest stars have shorter main-sequence lifetimes. We can use the main-sequence turnoff point to find the age of the cluster because stars at that turnoff point have lifetimes that are equal to the age of the cluster.

16. *Two stars that look very different must be made of different kinds of elements.* This statement does not make sense because most stars have very similar proportions of elements. Differences among the appearances of stars arise primarily because of differences in age and mass, not element content.
17. *Two stars that have the same apparent brightness in the sky must also have the same luminosity.* This statement does not make sense. Apparent brightness depends on both luminosity and distance.
18. *Sirius looks brighter than Alpha Centauri, but we know that Alpha Centauri is closer because its apparent position in the sky shifts by a larger amount as Earth orbits the Sun.* This statement makes sense. The parallax for Alpha Centauri is larger than that for Sirius because Alpha Centauri is closer to us.
19. *Stars that look red-hot have hotter surfaces than stars that look blue.* This statement does not make sense. Blue stars are hotter than red stars.
20. *Some of the stars on the main sequence of the H-R diagram are not converting hydrogen intohelium.* This statement does not make sense. All main-sequence stars are converting hydrogen to helium.
21. *The smallest, hottest stars are plotted in the lower left-hand portion of the H-R diagram.* This statement makes sense. Temperature on the H-R diagram increases from right to left, and stellar radii on the same diagram increase diagonally from lower left to upper right. So the smallest, hottest stars are in the lower left-hand corner of the H-R diagram.
22. *Stars that begin their lives with the most mass live longer than less massive stars because it takes them a lot longer to use up their hydrogen fuel.* This statement is false. The most massive stars burn their fuel a lot faster than the conservative, low-mass stars. They burn fuel at a profligate rate that negates their size/mass advantage.
23. *Star clusters with lots of bright, blue stars are generally younger than clusters that don't have any such stars.* This statement makes sense. Clusters with no blue stars probably had some blue stars in the past, but as the clusters aged the blue stars rapidly died off.
24. *All giants, supergiants, and white dwarfs were once main-sequence stars.* This statement is true. Giants, supergiants, and white dwarfs are later stages in the evolution of stars that began life as main-sequence stars.
25. *Most of the stars in the sky are more massive than the Sun.* This statement is false. Most of the stars are less massive than the Sun. Many more low-mass stars are formed than are high-mass stars. The high-mass stars burn out sooner, too, while the low-mass stars persist for billions of years.
26. b. get smaller
27. c. apparent brightness and distance
28. a. the time between eclipses and the average distance between the stars
29. c. A K star has the coolest surface temperature.
30. a. A main-sequence A star is the most massive.
31. c. A main-sequence M star has the longest lifetime.
32. a. A giant K star has the largest radius (you may have to consult Figure 15.10).
33. a. A 30-solar-mass main-sequence star has the hottest surface temperature.
34. c. A cluster containing stars of all colors is the youngest (because the blue stars are still around).

35. b. A cluster whose brightest main-sequence stars are yellow is the oldest cluster.

36. a. Sirius appears brightest in our sky because it has the smallest (most negative) apparent magnitude.
    b. Regulus appears faintest of the stars on the list because it has the largest apparent magnitude.
    c. Antares has the greatest luminosity of the stars on the list because it has the smallest (most negative) absolute magnitude.
    d. Alpha Centauri A has the smallest luminosity of the stars on the list because it has the largest absolute magnitude.
    e. Sirius has the highest surface temperature of the stars on the list because its spectral type, A1, is hotter than any other spectral type on the list.
    f. Antares has the lowest surface temperature of the stars on the list because its spectral type, M1, is cooler than any other spectral type on the list.
    g. Alpha Centauri A is most similar to the Sun because it has the same spectral type and luminosity class, G2 V.
    h. Antares is a red supergiant; its spectral type M means it is red, and its luminosity class I indicates a supergiant.
    i. Antares has the largest radius because it is the only supergiant on the list.
    j. Aldebaran, Antares, and Canopus have luminosity classes other than V, which means that they have left the main sequence and are no longer burning hydrogen in their cores.
    k. Spica is the most massive of the main-sequence stars listed because it has the hottest spectral type of the main-sequence stars; thus, it appears higher on the main sequence of an H-R diagram, where masses are larger.
    l. Alpha Centauri A, with spectral type G2, is the coolest and therefore the longest-lived main-sequence star in the table.

37. The list of the brightest stars will include the very luminous hot stars from distances greater than 12 light-years, while the list of the fainter yet closer low-mass stars will not. The list of stars within 12 light-years is "volume-limited," which means that nearly all of the stars within that distance are listed regardless of luminosity. Such a list is more representative of the total population of stars and is more likely to be dominated by low-mass stars. The list of brightest stars contains only those stars that are above a certain apparent brightness threshold. Therefore, the faintest nearby stars are left out, but the brightest and rarer hot stars are included in higher proportion than they are in a volume-limited list.

38. In Figure 15.10 Proxima Centauri is redder and has a cooler surface temperature than Sirius. Proxima Centauri is at least 10 times smaller in radius than Sirius. Both stars are on the main sequence, but Proxima Centauri is less massive and will have a longer life than Sirius. We can't tell from this plot how old each star is or how bright they are in our sky compared to each other.

39. The parallax of stars, as viewed from the orbit of Jupiter, would be about five times larger, since Jupiter's orbit around the Sun is about five times larger than that of Earth's. Parallactic distances would be easier to measure from Jupiter's orbit—for the same accuracy experiment, one could measure distances of about five times farther than we can from Earth.

40. If a star doubled in size with no change in luminosity, its surface temperature would go down, because the surface area of the star would increase but its energy output would stay the same, so a lower temperature would be required to maintain its energy output. Mathematically, the surface area would go up by a factor of four, so the temperature would go down by a factor of x, where $(x)^4 = \frac{1}{4}$ or $x = 1/\sqrt{2} \sim 0.71$ to maintain a constant luminosity.

41. The blue star is smaller than the red star, so when the blue star is eclipsed, none of its light reaches us. When the red star is eclipsed by the blue star, some of its light still gets to us because it is only partially eclipsed.

42. Both stars are at the same distance from Earth, so the fact that one has an orbit that is large enough to be seen means that the other's orbit is physically smaller. Since smaller orbits for a given mass imply larger velocities, the binary star that is not in a visual binary might be expected to have the greater Doppler shifts in its spectra. However, a visual binary with very massive stars could potentially have larger Doppler shifts than a spectroscopic binary with low-mass stars at the same distance.

43. This is an essay question. The key points are that stars of all colors form (with colors and luminosities described by the main sequence) as the cluster is born, and the bluest, most massive ones evolve away from the main sequence first. The cluster starts life with a complete main sequence, and after 13 billion years only stars somewhat less massive than the Sun remain on the main sequence, while the more massive stars have gone on to become giants and supergiants, and many of these will have either exploded or left behind white dwarfs.

44. Inverse square law. We don't need the value of the AU to do this problem, since the apparent brightness (flux) of the Sun scales with the distance.

Distance from Sun (d_{new})	$\dfrac{d_{new}}{1\text{ AU}}$	$\dfrac{1}{(d_{new}/1\text{ AU})^2}$	**New Apparent Brightness (watts/m^2)**
a. 1/2 AU	1/2	4	$4 \times 1{,}300 = 5{,}200$
b. 2 AU	2	1/4	$0.25 \times 1{,}300 = 325$
c. 5 AU	5	1/25	$0.04 \times 1{,}300 = 52$

45. a. We rearrange the inverse square law for light formula to solve for the luminosity:

$$\text{apparent brightness} = \frac{L}{4\pi \times d^2} \Rightarrow L = (\text{apparent brightness}) \times 4\pi \times d^2$$

Now we need to convert the values for distance and apparent brightness into standard units. A light-year is about 9.5 trillion kilometers, or 9.5×10^{15} m, so Alpha Centauri's distance of 4.4 light-years = 4.2×10^{16} m. Combining this with its apparent brightness in our night sky of 2.7×10^{-8} watt/m^2, we find:

$$L = \left(2.7 \times 10^{-8} \frac{\text{watt}}{\text{m}^2}\right) \times 4\pi \times (4.2 \times 10^{16}\text{ m})^2 = 6.0 \times 10^{26}\text{ watts}$$

Note that the luminosity of Alpha Centauri A, about 6.0×10^{26} watts, is similar to that of our Sun.

b. In this problem we must solve for the distance of a light bulb with a luminosity of 100 watts and an apparent brightness of 2.7×10^{-8} watt. First, we solve the formula for the distance.

Starting formula:

$$\text{apparent brightness} = \frac{\text{luminosity}}{4\ \pi \times (\text{distance})^2}$$

Multiply both sides by distance^2 and divide by (apparent brightness):

$$(\text{distance})^2 = \frac{\text{luminosity}}{4\ \pi \times (\text{apparent brightness})}$$

Take the square root of both sides:

$$\text{distance} = \sqrt{\frac{\text{luminosity}}{4\ \pi \times (\text{apparent brightness})}}$$

Now we plug in the numbers to find the distance of the light bulb:

$$\text{distance} = \sqrt{\frac{100 \text{ watts}}{4\ \pi \times 2.7 \times 10^{-8} \text{ watt/m}^2}} = \sqrt{2.9 \times 10^{8} \text{ m}^2} = 1.7 \times 10^{4} \text{ m}$$

The light bulb must be located at a distance of 17,000 meters, or 17 kilometers, to have the same apparent brightness as Alpha Centauri A.

46. a. A star with the same luminosity as our Sun but at a distance of 10 light-years would have an apparent brightness of:

$$\frac{L}{4\ \pi r^2} = \frac{3.8 \times 10^{26} \text{ watts}}{4\ \pi \left[10 \not{\text{ly}} \times \left(9.5 \times 10^{15} \frac{\text{m}}{\not{\text{ly}}}\right)\right]^2} = 3.35 \times 10^{-9} \frac{\text{watt}}{\text{m}^2}$$

b. A star with the same apparent brightness as Alpha Centauri (see Problem 45), but located at a distance of 200 light-years, has an intrinsic luminosity of:

$$L = 4\ \pi \times \left(200 \text{ ly} \times 9.5 \times 10^{15} \frac{\text{m}}{\text{ly}}\right)^2 \times \left(2.7 \times 10^{-8} \frac{\text{watt}}{\text{m}^2}\right) = 1.2 \times 10^{30} \text{ watts}$$

or 3,200 solar luminosities.

c. If a star has a luminosity of 8×10^{26} watts and an apparent brightness of 3.5×10^{-12} watt/m^2, its distance is (using F for apparent brightness):

$$d = \sqrt{\frac{L}{4\ \pi F}} = \sqrt{\frac{8 \times 10^{26} \text{ watts}}{4\ \pi \left(3.5 \times 10^{-12} \frac{\text{watt}}{\text{m}^2}\right)}} = 4.3 \times 10^{18} \text{ m} = 4.3 \times 10^{15} \text{ km}$$

$$= 450 \text{ light-years}$$

d. If a star has a luminosity of 5×10^{29} watts and an apparent brightness of 9×10^{-15} watt/m^2, its distance is (using F for apparent brightness):

$$d = \sqrt{\frac{L}{4\,\pi F}} = \sqrt{\frac{5 \times 10^{29}\ \text{watts}}{4\,\pi\left(9 \times 10^{-15}\ \frac{\text{watt}}{\text{m}^2}\right)}} = 2.1 \times 10^{21}\ \text{m} = 2.1 \times 10^{18}\ \text{km}$$

$$= 220{,}000\ \text{light-years}$$

Note that this star lies outside the Milky Way Galaxy.

47. a. Alpha Centauri: parallax angle of 0.742″. Using the parallax formula, we find that the distance to Alpha Centauri is:

$$d = \frac{1}{0.742''} = 1.35\ \text{pc} = 4.39\ \text{light-years}$$

Because 1 parsec is 3.26 light-years, this is the same as 4.39 light-years.

b. Procyon: parallax angle of 0.286″. Using the parallax formula, we find that the distance to Procyon is:

$$d = \frac{1}{0.286''} = 3.5\ \text{pc} = 11.4\ \text{light-years}$$

Because 1 parsec is 3.26 light-years, this is the same as 11.4 light-years.

48. a. A star with apparent magnitude 2 is 100 times brighter than a star with apparent magnitude 7. (Five magnitudes indicate a factor-of-100 difference; larger apparent magnitude stars are always fainter.)

b. A star with absolute magnitude –6 is intrinsically more luminous than a star of magnitude +4. The difference is 10 magnitudes, so the difference in luminosity is a factor of 100 for the first five magnitudes and a factor of 100 for the second five magnitudes, making an overall difference of a factor of $100^2 = 10{,}000$.

49. Each of the stars in the binary completes an orbit once every 6 months (or 0.5 year), at a velocity of 80,000 m/s. The circumference of that orbit is thus the time it takes to complete an orbit multiplied by the speed at which the stars move through the orbit, or:

$$\text{orbital distance} = \text{velocity} \times \text{time} = 80{,}000\frac{\text{m}}{\text{s}} \times 0.5\ \text{yr} \times 3.1 \times 10^{7}\frac{\text{s}}{\text{yr}} = 1.26 \times 10^{12}\ \text{m}$$

The average distance a of that orbit is thus:

$$r = \frac{\text{circumference}}{2\,\pi} = \frac{1.26 \times 10^{12}\ \text{m}}{2\,\pi} = 2.01 \times 10^{11}\ \text{m} = 1.3\ \text{AU}$$

We can now plug the period ($p = 0.5$ yr) and average distance ($a = 1.3$ AU) into the formula for Kepler's third law, normalized to solar values, to get the $M_1 + M_2$ value in solar masses:

$$M_1 + M_2 = \frac{a^3}{p^2} = \frac{(1.3)^3}{(0.5)^2} = 8.8\ \text{solar masses}$$

Because the stars have equivalent orbits, they must have equal masses, so each one has a mass of $4.4M_{\text{Sun}}$.

50. Sirius has a luminosity of $26L_{Sun}$ and a surface temperature of 9,400 K, so its radius is:

$$r = \sqrt{\frac{L}{4\pi\sigma T^4}} = \sqrt{\frac{26 \times 3.8 \times 10^{26}\ \cancel{\text{watts}}}{4\pi \times \left(5.7 \times 10^{-8}\ \frac{\cancel{\text{watt}}}{\text{m}^2 \cancel{\text{K}}^4}\right) \times (9400\ \cancel{\text{K}})^4}} = 1.3 \times 10^9\ \text{m}$$

or about twice the radius of the Sun.

51. Lifetime of a red giant.
    a. The masses of the stars left on the main sequence of this H-R diagram with a turnoff of around 13 billion years, when compared to other H-R diagrams with masses listed on it (like 15.10), are one solar mass or less.
    b. The luminosity of the most luminous stars in this cluster (the red giant stars) is around 1,000 solar luminosities.
    c. The ratio of luminosity to mass is therefore about 1,000 solar luminosities/solar mass, approximately. That ratio is about 1,000 times the same ratio for the Sun, which is, of course, 1 solar luminosity/solar mass.
    d. Since these stars are burning fuel about 1,000 times more rapidly than the Sun is, but are about the same mass, one might estimate that their lives as red giants at maximum luminosities of 1,000 solar luminosities are about 1,000 times shorter than their main-sequence lifetimes, or about 10 million years.

Chapter 16. Star Birth

This brand-new chapter covers star formation. Why an entire chapter on star birth? In the earlier editions of *The Cosmic Perspective*, star formation was covered in the chapter on Star Stuff, where we followed the story of stellar evolution from birth to death. The number of concepts in that single chapter was daunting for students. Furthermore, the research field of star formation is large and active, with many new and fundamental results coming from both space- and ground-based observatories. So by writing a new chapter on star formation, we lightened the conceptual load in Chapter 17 on star stuff and we can provide more depth on a vital area of astronomical research. We emphasize the competition between *pressure* and *gravity* as the conflict that provides the drama for the life story of stars, even as they are being born.

As always, when you prepare to teach this chapter, be sure you are familiar with the relevant media resources (see the complete, section-by-section resource grid in Appendix 3 of this Instructor's Guide) and the online quizzes and other study resources available on the Mastering Astronomy Web site.

What's New in the Fourth Edition That Will Affect My Lecture Notes?

Those who have taught from previous editions of *The Cosmic Perspective* should be aware of the following organizational or pedagogical changes to this chapter (i.e., changes that will influence the way you teach) from the Third edition:

- This chapter is brand-new. The star formation topic that was formerly covered in Chapter 17.2 has been expanded into Chapter 16; also, some of the material about the interstellar medium that was formerly in Chapter 19.1 and 19.2 (Our Galaxy) was also moved to this chapter.
- The chapter is divided into three sections: Stellar Nurseries, Stages of Star Birth, and Masses of Newborn Stars.

Teaching Notes (By Section)

Section 16.1 Stellar Nurseries

This section explains where and why stars form.

- The introduction to the interstellar medium now occurs here. The first section of old Section 17.2 was expanded to include material on the interstellar medium formerly covered in the chapter on the Milky Way.
- This chapter is the first chapter on stars where we call out the term *thermal pressure* instead of pressure. We will need to distinguish between the various sources of pressure to talk about star birth and stellar evolution in the next chapter.
- Mathematical Insight 16.1 provides order of magnitude reasoning to estimate the pressure in a star due to thermal pressure, compared to the pressure induced by self-gravity in order to illustrate how those forces depend on a cloud's mass, temperature, and density. We also provide an expression for the mass of a molecular cloud that is in balance. This mass insight sets us up to explain why stars form in clusters. The *Jeans Mass* is not identified by name, however.

Section 16.2 Stages of Star Birth

This section explains how stars form. Some of this material overlaps the discussions of solar system formation in Chapter 8.

- The material formerly in the star stuff Chapter (old 17.2) in subsections "From Cloud to Protostar," "Disks and Jets," and "From Protostar to Main Sequence" is now here.
- We have added material on the H^- opacity to explain why a protostar's surface remains around 3,000 K during most of its contraction. This is the magic temperature at which hydrogen begins to collisionally ionize. Some of the now-free electrons make negatively charged hydrogen ions, which strongly trap photons. Gas with trapped photons convects; the convecting gas rises until it cools to 3,000 K, at which point the photons can escape. So the temperature of a protostar is thereby tuned to be about 3,000 K. We have found that the explanation of this process in some other introductory books is actually incorrect. For more reinforcement, read the Hayashi paper: Hayashi, C. 1966, *Annual Reviews of Astronomy and Astrophysics*, 4, 171.
- We use the term *life track* instead of *evolutionary track* to describe a star's path through the H-R diagram. This is not because we are avoiding the term *evolution*. We find that students understand "life track" more quickly than "evolutionary track," since the life track is the path traced by an individual star through the H-R diagram, while students associate evolution to something that happens during multiple generations of individuals.

Section 16.3 Masses of Newborn Stars

This section explains the minimum and maximum masses of stars. The saga of outward pressure resisting gravity continues.

- This chapter is the first chapter on stars where we call out the term *degeneracy pressure*. Degeneracy pressure is covered in more detail in Chapter S4.4; here it is described in terms of its function in preventing brown dwarfs from becoming main-sequence stars.
- Similarly, here is where we identify *radiation pressure* as the reason that stars have a maximum mass. We show a picture of one of the most massive stars known, the Pistol Star, at about 150 solar masses.
- We discuss how we know the ratio of low-mass stars to high-mass stars, but we admit that we do not know exactly how this ratio is determined, beyond that turbulent gas motions probably have some role in the explanation.

Answers/Discussion Points for Think About It Questions

The Think About It questions are not numbered in the book, so we list them in the order in which they appear, keyed by section number.

Section 16.2

- (p. 540) This question encourages students to put the meaning of life track into their own words. A life track is the luminosity and temperature of a star at different stages of its life, plotted as points sequentially connected by lines on an H-R diagram. It also asks students to recognize that a life track can be carried through from protostar to death of the resulting star, and that the life tracks discussed here are limited to the time before the star begins burning hydrogen on the main sequence. It also asks students to extrapolate beyond the life tracks plotted in Figure 16.16 to higher mass and lower mass.

Solutions to End-of-Chapter Problems (Chapter 16)

1. The interstellar medium refers to the gas and dust that fill the spaces between stars within a galaxy. The interstellar medium is made up mostly of hydrogen (70%) and helium (28%) with a small amount of heavier elements (2%). We have learned this by the light from stars lying behind a cloud of gas and studying which absorption lines the cloud leaves in the spectrum.
2. A molecular cloud is an interstellar cloud that is cold and dense enough to allow atoms to combine into molecules. Molecular clouds are much colder and denser than the rest of the interstellar medium.
3. Interstellar dust refers to tiny, solid grains of dust that make up about 1% of the mass of molecular clouds. These grains scatter or absorb virtually all of the light that enters molecular clouds so that the clouds appear black in our sky. At the edges of clouds, the grains cause the stars to appear redder than normal. Infrared light passes through the dust more easily so that we can see into and through the molecular clouds. We can also see the grains glow in the infrared because they are heated by stars in and near the cloud.

4. Molecular clouds are favorable locations for star formation for two reasons: low temperature and high density. Their low temperature keeps their pressures about the same as other interstellar clouds, despite the higher density. But the higher density means that gravity is stronger in molecular clouds, so it is able to overcome the pressure in molecular clouds. This increased gravitational attraction allows collapse, leading to star formation.
5. The heat generated as the clouds contract due to gravity is lost as photons. The photons are generated by the molecules' rotational and vibrational energy levels, which are excited by collisions between molecules. Since the heat can be radiated away effectively, there is no build up and star formation is not stopped.
6. Stars tend to form in clusters because more massive clouds are better able to overcome pressure and collapse. As these larger clouds collapse, the density increases and gravity gets an increasingly large advantage over pressure. So smaller and smaller parts of the cloud are able to collapse on their own, leading to fragmentation of the cloud. Each fragment becomes a star or a system of stars.
7. A small molecular cloud can give birth to a single star as long as it is unusually cold and dense. This means star forming clouds have a density above a few thousand molecules per cubic centimeter and a temperature of about 10 K.
8. We think that the first generation of stars must have been more massive than the Sun because there were no elements heavier than hydrogen and helium to form the molecules that cool molecular clouds. Since the clouds would have been warmer, the clouds would have had to form larger fragments to collapse. The larger fragments would have gone on to form larger stars.
9. When the cloud's thermal energy can no longer escape in the form of photons, the cloud heats up and the pressure rises. This resists the pull of gravity and slows the contraction.
10. A protostar is a clump of gas that will become a new star. It forms when a collapsed cloud fragment can no longer emit its thermal energy via photons so that the core heats up, raising the pressure and resisting the gravitational contraction. Protostars grow in mass because material around the protostar, which feels a weaker pull of gravity and does not collapse as quickly as the protostar, continues to contract. This material rains down on the protostar, gradually increasing the mass.
11. A protostellar disk is a spinning disk of gas around a protostar. Because of friction in the disk, material gradually spirals inward and eventually falls onto the protostar.
12. Jets are streams of gas that are fired out into interstellar space at high speeds. We think that they are related to the protostar's rotation because they are fired out along the protostar's rotation axis. The jets help clear away the cocoon of gas that surrounds a forming star. They also pump significant amount of kinetic energy into the surrounding molecular cloud, causing some of the turbulent motion that we see in the clouds.
13. Because protostars radiate away energy from their surfaces, they are able to contract further. Since only half of the energy is radiated away, this allows their central temperatures to rise. So as the stars continue to contract, the central temperature rises. Eventually, the temperature reaches 10 million K, making it hot enough for hydrogen fusion to operate efficiently through the proton-proton chain. We can represent these stages on a life track, an H-R diagram that tracks the changes of a single star in relation to the standard main sequence.

14. Degeneracy pressure is a quantum mechanical effect that halts the contraction of protostars with masses less than 0.08 solar mass. Unlike thermal pressure, degeneracy pressure depends only on the density and not on the temperature. Since the degeneracy pressure does not weaken as the core cools, it will continue to support a core even when the core becomes cold.
15. The minimum mass for a star is 0.08 solar mass. Below this mass, degeneracy pressure halts the collapse of the core before it gets hot enough to start fusion. A brown dwarf is an object in which the degeneracy pressure halted the collapse of the protostar before fusion began, making it a "failed star."
16. The maximum mass for stars is around 150 solar masses. This limit is set by radiation pressure, the pressure exerted by light. For stars larger than 150 solar masses, energy is generated so furiously that gravity cannot resist the force of radiation pressure and the extra mass is blown away into space.
17. Low-mass stars greatly outnumber the high-mass stars, with a few hundred stars less than 0.5 solar mass for every star greater than 10 solar masses.
18. *If you want to get a more accurate count of the number of stars in our galaxy, use an infrared telescope to observe them instead of a visible-light telescope.* This statement is sensible. An infrared telescope can see both the visible stars and the stars shrouded in dusty molecular clouds.
19. *A molecular cloud needs to trap all the energy released by gravitational contraction in order for its center to become hot enough for fusion.* This statement is not sensible. If a protostar trapped all of its energy, no photons would escape and we wouldn't be able to see it. This situation is physically impossible because even if the cloud is opaque it will emit thermal radiation.
20. *Low-mass stars form more easily in clouds that are unusually cold and dense.* This statement is sensible. A colder cloud requires less mass to overcome thermal pressure; a denser cloud has more gravitational pull in a smaller volume.
21. *The current mass of any star is the same as the mass it had when it first became a protostar.* This statement is not sensible, strictly speaking. A protostar gains mass as matter continues to rain in on it during the protostar phase. A star can also lose mass in various ways during its lifetime.
22. *The rotation of a protostar always speeds up with time because it is surrounded by a spinning disk.* This statement is not sensible. The rotation of a protostar is determined by the amount of angular momentum it has. A disk forms because the angular momentum of the gas particles prevents them from raining directly onto the protostar. A protostar's rotation will speed up as it collapses because angular momentum is conserved, and it can slow only if the protostar can shed its angular momentum somehow (jets, winds, etc.)
23. *Some of the stars in a star cluster live their entire lives and then die off before many of the cluster's stars initiate fusion.* This statement is sensible. The most massive stars have main-sequence lifetimes of less than 10 million years, which is a shorter time than stars like our Sun spend on the pre-main sequence life track.
24. *Protostars are generally best observed in ultraviolet light because their surfaces have to get very hot before fusion can begin.* This statement is not sensible. Ultraviolet light is very easily absorbed by the molecular gas clouds surrounding protostars; also, the surface of protostars tends to be maintained at a constant 3,000 K because of the H^- process.

25. *Degeneracy pressure exists only in objects that are very cold.* This statement is not sensible. Degeneracy pressure is insensitive to the temperature of a material. It is important when the electron density is so high that the laws of quantum physics prevent the electrons from getting any closer together.
26. *If Jupiter were 10 times more massive, we would consider it a brown dwarf, and if it were 100 times more massive we would consider it a star.* This statement is sensible. 0.08 solar mass is 80 times the mass of Jupiter, so if Jupiter were 100 times more massive, it could ignite hydrogen in its core. If it were 10 times more massive, we'd probably consider it to be a brown dwarf companion of the Sun, although this classification could be debated.
27. *Most of the stars that formed from the same cloud as the Sun have already died off.* This statement is not sensible, since (a) the main-sequence turnoff for the stars born at the same time as the Sun is more massive than the Sun, and (b) many more low-mass stars form in any cloud than do high-mass stars. Therefore the majority of the stars born from the same cloud as the Sun are still alive and well.
28. a. yellow light
29. b. CO (Most of the hydrogen is too cold to produce much molecular hydrogen emission.)
30. c. More information is required.
31. a. It breaks into smaller fragments.
32. b. The cloud temperatures were higher because the clouds were pure H and He.
33. c. The cloud first slows down its contraction when it starts to trap its thermal energy.
34. a. They carry away angular momentum and allow the star to contract and accrete matter.
35. b. radiation pressure
36. c. M stars are more common.
37. The key point is that dust scatters blue light preferentially to red light, and it completely blocks our view of some parts of the Milky Way. The Milky Way would be brighter and bluer.
38. If the temperature suddenly dropped, the pressure would drop and gravity would have the upper hand and the cloud would start to collapse. If the temperature rose, pressure would increase and the cloud would begin to expand. If it gained a little mass, gravity would win out and the cloud would start to collapse.
39. This essay describes the life history of a protostar. The key points are in the text. Some conceptual understanding is demonstrated if the student can explain what phenomena would not be present if there were exactly zero rotation in the star: no protostellar disk, no jets, probably no strong winds. Close binaries would be much less likely.
40. The protostar at 100 solar luminosities is probably evolving more quickly than a lower luminosity protostar of the same mass at 10 solar luminosities because it is shedding thermal energy 10 times faster. For a star of a given mass, one would predict it would spend 1/10 of the time as a 100-solar-luminosity protostar than it does as a 10-solar-luminosity protostar.
41. This question asks the student to interpret Figure 16.16. The luminosity of the highest mass protostars changes the least, according to Figure 16.16. The luminosity of the lowest mass protostars changes the most. The temperature of the highest mass protostars changes the most. The temperature of the lowest mass protostars changes the least.

42. If there were no such thing as degeneracy pressure to stop the contraction of Jupiter, Jupiter could have in principle contracted enough to achieve the central temperatures necessary to ignite hydrogen fusion in its core.
43. Both brown dwarfs and jovian planets are formed of hydrogen and helium gas yet are not massive enough to ignite hydrogen fusion in their cores. Both types of objects could be a companion to a regular main-sequence star. Brown dwarfs are like stars in the sense that they are both hydrogen and helium gas, they are both more massive than a jovian planet, and they are both thought to start out as protostars. Planets are thought to form in protoplanetary disks around stars. It is nontrivial to tell the difference between a massive jovian planet and a low-mass brown dwarf orbiting a star.
44. This question asks the student to draw a series of H-R diagrams, where the first diagram has all of the protostars on the far right-hand side of the diagram. The next H-R diagram, at 100,000 years later, would be a line that crosses all the life tracks in Figure 16.16, but the line crosses the low-mass life tracks on the far right of the diagram and the high-mass life tracks on the main sequence. So the line starts midway up, far right, travels somewhat to the left and up, but then bends precipitously to the left and connects to the main sequence somewhere between 9 and 15 solar masses. At 1 million years, the H-R diagram would have a similar line but this time intersect the main sequence at somewhat more massive than 3 solar masses, then curve up the main sequence. At 10 million years, some of the most massive stars have begun to evolve away from the main sequence (and explode), while stars of about 2 solar masses are just starting to reach the main sequence. So the H-R diagram again starts at the midway up, far right, goes nearly horizontally to the main sequence, angles up and to the left along the main sequence, and then, at the O star level, turns off to the right. Finally, at 100 million years, almost all of the stars more massive than about 0.5 solar mass are now on the main sequence, except for the O and B stars, which have either begun their red giant phases or ended their lives in supernovae. So there would be a little line curving through the low-mass protostars around the K and M division, continuing up and to the left along the main sequence to the B stars, and then (for those stars in their giant phase) jumping away from the main sequence to the right.
45. A typical person is about 60 kilograms, or 60,000 grams. Humans are about $\frac{2}{3}$ water, so we all contain about 40,000 grams of water. So the question is, what volume of an interstellar cloud contains 40,000 grams of water molecules? We are told then about a millionth of the clouds' masses are made of water and that the clouds have a density of 10^{-21} g/cm^3. So the density of water molecules is the overall density divided by a million, or 10^{-27} g/cm^3.

    Now we have the density and the mass of water we need, but we want the volume of the cloud required to supply that much water. Density is defined as mass over volume, so we can solve to get:

$$\text{volume} = \frac{\text{mass}}{\text{density}}$$

    Putting in our numbers, then:

$$\text{volume} = \frac{40{,}000 \text{ g}}{10^{-27} \text{ g/cm}^{-3}}$$
$$= 4 \times 10^{30} \text{ cm}^3$$

We would need 4×10^{30} cm^3 of cloud to get enough water for one human.

We are asked to compare this to the volume of the Earth, so let us set up a ratio.

$$\frac{\text{volume of cloud for one human's worth of water}}{\text{volume of the Earth}}$$

We know that the Earth's volume must be roughly that of a sphere, of $\frac{4}{3}\pi r^3$, where r is Earth's radius. From Appendix E we find that the Earth's radius is 6,378 kilometers, or 6.378×10^8 cm. This leads to a volume of 1.09×10^{27} cm^3. The ratio of the volume of cloud needed for a single human to that of the Earth is therefore:

$$\frac{4 \times 10^{30}\ \text{cm}^3}{1.09 \times 10^{27}\ \text{cm}^3} = 3{,}670$$

We would need a cloud 3,670 times the volume of the Earth to get enough water for one human.

46. a. From Appendix E we learn that the mass of the Sun is 2×10^{30} km, or 2×10^{33} g. A 1,000 M_{Sun} cloud then has a mass of 2×10^{36} g. If 1% of this material is in the form of dust, then there are 2×10^{34} g of dust in the cloud. Each particle has a mass of 10^{-14} g, so we get the number of grams by dividing:

$$\frac{2 \times 10^{34}\ \text{g/cloud}}{10^{-14}\ \text{g/grain}} = 2 \times 10^{48}\ \text{grains/cloud}$$

So there would be around 2×10^{48} grains in such a cloud.

b. To get the area in square light-years, we will first convert the radius of the individual particles to light-years. Appendix E tells us that a light-year is 9.46×10^{12} km, or 9.46×10^{15} m. The grains have radii of 10^{-7} meter and this converts to 1.06×10^{-23} light-years. The area of a single grain is πr^2, so putting in the radius we get an area of 3.51×10^{-46} square light-year. From part (a) there are 2×10^{48} grains, so multiplying the number of grains by the area per grain, we get the total surface area of all of the grains in the cloud: 702 square light-years.

c. Since density is defined as mass over volume:

$$\text{volume} = \frac{\text{mass}}{\text{density}}$$

We know that the cloud's density is about 10^{-21} g/cm^3 and we know that the mass is about 2×10^{36} g. Therefore, the volume of the cloud must be 2×10^{57} cm^3. If the cloud is spherical, the radius can expressed in terms of the volume by turning around the formula for the volume of a sphere:

$$r = \sqrt[3]{\frac{3 \times \text{volume}}{4\pi}}$$

Putting in the numbers, we find that the radius of the spherical cloud is about 7.82×10^{18} cm or 7.82×10^{16} m. Converting to light-years gives us 8.27 light-years. The cross-sectional area is πr^2, like a circle, so the area of the cloud is 214 square light-years (200 light-years, rounded off).

d. Since the area of the dust grains is more than three times the cross-sectional area of the cloud, any photon entering the cloud will probably collide with at least one grain as it passes through.

47. From Mathematical Insight 16.1, the critical mass needed for a cloud to collapse is:

$$M_{\text{balance}} = 18M_{\text{Sun}}\sqrt{\frac{T^3}{n}}$$

We are told that the temperature, T, is 30 K and the density, n, is 300 molecules per cubic centimeter, so we can find the critical mass: $171M_{\text{Sun}}$. The mass of the Sun is 2×10^{33} g, so the cloud's mass in grams is 3.42×10^{35} g. To get the radius, we will first need to convert this mass to a volume and then from volume to radius. The density of the cloud in grams per cubic centimeter can be found by taking the density in molecules per cubic centimeter and multiplying by the mass per molecule. We are told to assume that all of the molecules are hydrogen with a mass of 3.3×10^{-24} g so the density is 9.9×10^{-22} g/cm^3. Since density is mass over volume, we can solve to get:

$$\text{volume} = \frac{\text{mass}}{\text{density}}$$

The volume of the cloud is therefore 3.45×10^{56} cm^3.

Assuming that the cloud is spherical, we can solve the formula for the volume of a sphere to get the radius:

$$r = \sqrt[3]{\frac{3 \times \text{volume}}{4\pi}}$$

Inserting the volume, this leads to a radius of 4.35×10^{18} cm. Converting to meters gives us 4.35×10^{16} m. Converting to light-years gives 460 light-years. So the cloud would need to be at least 460 light-years in radius to start star formation.

48. From Mathematical Insight 16.1, the critical mass needed for a cloud to collapse is:

$$M_{\text{balance}} = 18M_{\text{Sun}}\sqrt{\frac{T^3}{n}}$$

We are told that the temperature, T, is 10 K and the density, n, is 100,000 molecules per cubic centimeter, so we can find the critical mass: $1.8M_{\text{Sun}}$. The mass of the Sun is 2×10^{33} g, so the cloud's mass in grams is 3.6×10^{33} g. To get the radius, we will first need to convert this mass to a volume and then from volume to radius. The density of the cloud in grams per cubic centimeter can be found by taking the density in molecules per cubic centimeter and multiplying by the mass per molecule. We are told to assume that all of the molecules are hydrogen with a mass of 3.3×10^{-24} g so the density is 3.3×10^{-19} g/cm^3. Since density is mass over volume, we can solve to get:

$$\text{volume} = \frac{\text{mass}}{\text{density}}$$

The volume of the cloud is therefore 1.09×10^{52} cm^3.

Assuming that the cloud is spherical, we can solve the formula for the volume of a sphere to get the radius:

$$r = \sqrt[3]{\frac{3 \times \text{volume}}{4\pi}}$$

Using our volume, this leads to a radius of 1.38×10^{17} cm. Converting to meters gives us 1.38×10^{15} m. Converting to light-years gives 14.5 light-years. So the cloud would need to be at least 14.5 light-years in radius to start star formation.

49. From Mathematical Insight 16.1, the critical mass needed for a cloud to collapse is:

$$M_{\text{balance}} = 18M_{\text{Sun}}\sqrt{\frac{T^3}{n}}$$

We are told that the temperature, T, is 200 K and the density, n, is 300,000 molecules per cubic centimeter, so we can find the critical mass: $93M_{\text{Sun}}$. So stars from the first generation were probably around 93 times the mass of the Sun or more. Looking it up in Chapter 15, we see that such stars have lifetimes of less than 10 million years.

50. a. We will take the mass density of the Sun and then divide by mass per particle to get the average number density of particles in the Sun. First, we need the density of the Sun. For this, we need the mass (2×10^{33} g) and the volume. We compute the volume by assuming that the Sun is a sphere and using:

$$\text{volume} = \frac{4}{3}\pi r^3$$

The radius of the Sun is 6.96×10^{10} cm, so the volume of the Sun is 1.41×10^{33} cm^3. Density is mass over volume, so the average density of the Sun is 1.42 g/cm^3. Finally, we just need to convert this to number density by dividing the mass density by the mass per particle. The mass per particle is given as 10^{-24} g, so we get the average number density of the Sun to be 1.42×10^{24} particles/cm^3.

b. From Mathematical Insight 16.1, the critical mass needed for a cloud to collapse is:

$$M_{\text{balance}} = 18M_{\text{Sun}}\sqrt{\frac{T^3}{n}}$$

Solving for the temperature gives us:

$$T = \sqrt[3]{n\left(\frac{M_{\text{balance}}}{18M_{\text{Sun}}}\right)^2}$$

We know the Sun is in balance, so M_{balance} must be the mass of the Sun ($1M_{\text{Sun}}$). We found the number density inside the Sun in part (a). Calculating, we find that the Sun must be 16 million K.

c. The Sun's core is actually around 15 million K, but this estimate is quite close. *[Caveat to instructor: The close agreement of the simple calculations in Questions 50–51 is something of a coincidence, since the mean mass per particle assumed in our Jeans Mass equation is larger than that for the ionized gas in the Sun.]*

51. a. We will take the mass density of the Sun and then divide by mass per particle to get the average number density of particles in the Sun. First, we need the density of the Sun. For this, we need the mass and the volume. The mass is 0.05 times the mass of the Sun. The Sun's mass is 2×10^{33} g, so the mass of the brown dwarf is 1×10^{32} g. We compute the volume by assuming that the Sun is a sphere and using:

$$\text{volume} = \frac{4}{3}\pi r^3$$

The radius of the brown dwarf is 0.1 times the radius of the Sun. The radius of the Sun is 6.96×10^{10} cm, so the radius of the brown dwarf is 6.96×10^{9} cm and the volume is 1.41×10^{30} cm^3. Density is mass over volume, so the average density of the Sun is 70.9 g/cm^3. Finally, we just need to convert this to number density by dividing the mass density by the mass per particle. The mass per particle is given as 10^{-24} g, so we get the average number density of the Sun to be 7.09×10^{25} particles/cm^3.

b. From Mathematical Insight 16.1, the critical mass needed for a cloud to collapse is:

$$M_{\text{balance}} = 18 M_{\text{Sun}} \sqrt{\frac{T^3}{n}}$$

Solving for the temperature gives us:

$$T = \sqrt[3]{n \left(\frac{M_{\text{balance}}}{18 M_{\text{Sun}}} \right)^2}$$

If we assume that the brown dwarf is in balance, M_{balance} must be the mass of the brown dwarf ($0.05 M_{\text{Sun}}$). We found the number density inside the Sun in part (a). Calculating, we find that the brown dwarf must be about 8.2 million K. *[Instructors, please read caveat in solution for Question 50.]*

c. The temperature calculated in part (b) is less than the 1×10^7 K needed to start fusion, which explains why the brown dwarf never becomes a star.

d. If we raised the mass by a factor of two, the internal temperature would go up. A slight increase in temperature would cause the core to pass the critical temperature needed to start nuclear fusion and we would have a star.

52. a. Newton's version of Kepler's third law states that the period, p, the orbital separation, a, and the total mass, M, are related with the formula:

$$a^3 = \frac{GM}{4\pi^2} p^2$$

We are told that the total mass is twice the mass of the Sun. The mass of the Sun is 2×10^{30} kg, so the total mass in this system is 4×10^{30} kg. The period is 10 days, which we need to convert to seconds: 8.64×10^5 seconds. We can therefore find the orbital separation by plugging in the numbers. We get 1.72×10^{10} m.

b. The average speed will be the distance that the planets travel around their orbits over the period. We know the period (in both days and in seconds), so we just need the average distance that the stars each travel. If we assume that the orbits are circular, we can take the distance traveled to be the circumference of a circle with a diameter equal to the separation found in part (a). The circumference is πa. Using the separation in part (a), this tells us that each star travels 5.40×10^{10} m in each orbit. Dividing by the period gives us the average speed: 62,500 meters per second.

c. Angular momentum is mass × speed × radius. We have all three quantities now: The mass of each star is 2×10^{30} kg, the speed is 62,500 meters per second, and the radius is half the separation, 8.6×10^9 m. Calculating the angular momentum, we get 1.08×10^{45} kg/m^2/s^{-1}.

d. Let us assume that all of the Sun's mass was located a solar radius from the Sun's rotation axis. In this case, the angular momentum will be mass × speed × radius, where the mass is the mass of the Sun and the radius is the radius of the Sun (6.96×10^8 m). We have the angular momentum from part (c), so we can solve for the speed that the Sun's material would have to move at:

$$\begin{aligned} \text{speed} &= \frac{\text{angular momentum}}{\text{mass} \times \text{radius}} \\ &= \frac{2.16 \times 10^{45}\ \text{kg}^2/\text{m}^2/\text{s}^{-1}}{(2 \times 10^{30}\ \text{kg}) \times (6.96 \times 10^8\ \text{m})} \\ &= 1{,}550{,}000\ \text{m/s} \end{aligned}$$

We estimate that the Sun's material would have to move at 77,600 meters per second to have the same angular momentum as each of the two stars.

e. The escape velocity from the Sun's surface is:

$$\text{escape velocity} = \sqrt{\frac{2\,GM}{r}}$$

Putting in values for the Sun, we get an escape velocity of 619,000 m/s. Our estimate of the speed that the material would need in order to have the same angular momentum as the two orbiting stars is more than twice this value. So if the Sun were spinning this fast, the material would escape off the surface and the Sun would fall apart.

53. We could solve for the total mass of the system using Newton's version of Kepler's third law. However, we do not need to. We know the total mass in a system where the period is 1 year and the separation is 1 AU because that is the period and orbital radius of the Earth! The total mass must be the same as the mass of the Sun and Earth combined. Since the Sun is much, much more massive than the Earth, this is approximately equal to the mass of the Sun. So the system has a total mass of 2×10^{30} kg. All that remains is to find out how much of the mass is in the brown dwarf. Since the brown dwarf moves 20 times faster than the main-sequence star, we conclude that it must be 1/20 of the mass in order for them to conserve momentum. The total mass of the system is 21 times the mass of the brown dwarf. To get the mass of the brown dwarf, we multiply the total mass of the system by the ratio of the mass of the brown dwarf to the total mass of the system (1/21):

$$\begin{aligned} \text{mass} &= \frac{1}{21}(2 \times 10^{30}\ \text{kg}) \\ &= 9.5 \times 10^{28}\ \text{kg} \end{aligned}$$

So the brown dwarf has a mass of 9.5×10^{28} kg.

54. To answer this, let us make a table. We will list the star type, the number of stars of that type in the cluster, the luminosity of a star of that type (we can get this off of the H-R diagram in Chapter 15), and the total luminosity of all the stars of that type (the product of the luminosity of one star and the number of stars).

Star Type	Number	Luminosity per Star (solar units)	Total Luminosity (solar units)
O	1	1.00E+05	1.00E+05
A	10	1.00E+02	1.00E+03
G	100	1	1.00E+02
M	1000	1.00E–03	1.00E+00

So we see that in spite of the smaller stars having larger numbers, the O star wins with the largest total luminosity, thanks to its being so bright. If we looked at this cluster from a great distance, we would mainly see the O star's light, so the cluster would look blue like the O star.

Chapter 17. Star Stuff

This chapter covers stellar evolution, highlighting the role that stars play in creating the elements necessary for life. As in Chapter 16, the battle between pressure and gravity provides the drama, and the saga of the lives of stars continues. In this chapter we begin by discussing the importance of star masses, and we provide the narrative for the two main categories of stars, the high-mass and the low-mass stars.

As always, when you prepare to teach this chapter, be sure you are familiar with the relevant media resources (see the complete, section-by-section resource grid in Appendix 3 of this Instructor's Guide) and the online quizzes and other study resources available on the Mastering Astronomy Web site.

What's New in the Fourth Edition That Will Affect My Lecture Notes?

Those who have taught from previous editions of *The Cosmic Perspective* should be aware of the following organizational or pedagogical changes to this chapter (i.e., changes that will influence the way you teach) from the Third edition:

- The new chapter structure has four sections rather than five, because material on star formation in old Section 17.2 was moved to its own chapter, Chapter 16 on Star Birth.
- There are eight brand-new calculation questions at the end of the chapter.

Teaching Notes (By Section)

Section 17.1 Lives in the Balance

This section introduces the subject of stellar evolution by returning to the idea of gravitational equilibrium and explaining that the tug-of-war between gravity and thermal pressure is what governs a star's life. If students understand that all the changes a star goes through are driven by the need to balance pressure and gravity, they will have a much easier time understanding how stars evolve.

- In this section, we divide stars into three categories of initial mass: low-mass stars ($<2M_{\text{Sun}}$), intermediate-mass stars ($2M_{\text{Sun}} - 8M_{\text{Sun}}$), and high-mass stars ($>8M_{\text{Sun}}$). Later in the chapter we discuss the lives of low-mass stars and high-mass stars explicitly in order to point out the sharp contrasts between them. Because intermediate-mass stars behave sometimes like high-mass stars and sometimes like low-mass stars, we do not discuss them separately but instead include them in our discussion of high-mass stars, pointing out their similarities and differences where appropriate.

Section 17.2 Life as a Low-Mass Star

This section describes how a low-mass star progresses from birth to death, covering the transition to the red giant stage, hydrogen shell burning, helium burning, the horizontal branch of the H-R diagram, and planetary nebulae.

- High-mass stars are often given all the credit for producing the elements necessary for life, but most of the carbon in the universe does not come from the stars that explode. Late in their lives, the more massive low-mass stars (those beginning at around $2M_{\text{Sun}}$) and intermediate-mass stars expel dredged-up carbon from the carbon-burning core via winds and planetary nebulae. These stars are the source of most carbon in the universe.
- The atmospheres of some red giant stars are more oxygen-rich than carbon-rich, and these stars tend to produce silicate dust grains rather than carbon grains. We avoid mentioning them here in order to keep the discussion simple and focused on carbon production, but the mass of interstellar dust in the form of silicates is roughly equal to the mass in the form of carbon particles.
- We have found that one of the best ways to engage students is to relate astronomical phenomena to life on Earth. To that end, this section concludes with speculation about the fate of the Earth once the Sun exhausts its core hydrogen. Vividly telling this story in class is a good way to help students remember the evolutionary stages of low-mass stars.

Section 17.3 Life as a High-Mass Star

This section traces the life of a high-mass star. It includes discussions of the CNO cycle, advanced nuclear burning, the difficulty of extracting energy from iron, and supernovae.

- The text explains that higher-mass stars fuse hydrogen via the CNO cycle because their higher core temperatures enable hydrogen to fuse to heavier nuclei, but it does not explain why lower-mass stars prefer the proton-proton chain to the CNO cycle. The reason has to do with the need for protons to decay to neutrons via weak interactions. In the proton-proton chain, fusion of two protons into deuterium requires this decay to happen on a very short time scale, making successful fusion highly improbable in any one proton-proton collision. The CNO cycle circumvents this bottleneck because it creates unstable isotopes that can take their time decaying. Thus, it is strongly preferred when core temperatures are high enough.

- Intermediate-mass stars burn via the CNO cycle during their main-sequence lives but never burn all the way to iron and therefore do not explode as supernovae. The initial mass cutoff separating these stars from high-mass stars that do explode is still somewhat uncertain. Analyses of the main-sequence turnoff points in star clusters that contain white dwarfs indicate that the cutoff must be at least as high as $8M_{Sun}$.
- Supernovae remnants are discussed near the end of this section, but they are covered in more detail in Chapter 19.
- Figure 17.20 provides a two-page pictorial summary of stellar evolution that can serve as a study aid for visually oriented students.

Section 17.4 The Roles of Mass and Mass Exchange

This last section briefly summarizes why mass is the single most important factor in determining a star's fate, and then describes how mass transfer in close binary systems can alter the standard pathways of stellar evolution.

- This section helps prepare students for the following chapter, in which mass transfer onto white dwarfs, neutron stars, and black holes receives considerable attention.
- In the interests of focusing on the "big picture," we avoid going into the nomenclature of close binaries (e.g., contact binaries, common-envelope systems). Our primary goal is to provide an example of what can happen when stars transfer mass, not to enumerate all the possibilities.

Answers/Discussion Points for Think About It Questions

The Think About It questions are not numbered in the book, so we list them in the order in which they appear, keyed by section number.

Section 17.2

- (p. 552) A star grows larger and brighter after core hydrogen is exhausted because hydrogen then begins to burn in a shell around the core. Shell burning can proceed at a much higher temperature than core burning, resulting in a much larger luminosity that causes the star to expand. The red giant stage halts either when helium starts to burn in the core or when the overlying hydrogen envelope is gone. If the temperature required for core helium to ignite were larger, helium core fusion would ignite later, after the red giant star had grown even larger and more luminous.
- (p. 554) If the universe contained only low-mass stars, elements heavier than carbon would be very rare, because the core temperatures of most low-mass stars are insufficient for fusing other nuclei to carbon. (However, stars on the upper end of the low-mass range can fuse helium to carbon, making some oxygen.)

Section 17.3

- (p. 558) The very first high-mass stars in the universe did not have any carbon, nitrogen, or oxygen. They were pure hydrogen and helium, so they could not have used the CNO cycle during their main-sequence lifetime.

- (p. 560) If hydrogen had the lowest mass per nuclear particle, nuclear fusion would be impossible, so stars would not give off any energy other than that released by gravitational contraction. All stars would be like brown dwarfs—bad news for life!
- (p. 562) This is a good topic for class discussion about how important the night sky was to various cultures.

Solutions to End-of-Chapter Problems (Chapter 17)

1. Star mass is important because larger stars have higher temperatures in their cores, making fusion proceed more rapidly and allowing fusion of heavier elements in their cores. We categorize stars by mass into three groups: low-mass stars (less than 2 solar masses), intermediate-mass stars (between 2 and 8 solar masses), and high-mass stars (above 8 solar masses).
2. Low-mass stars all have long lifetimes on the main sequence and go through the same basic life stages: main sequence, red giant with hydrogen shell burning, helium flash, then white dwarf. They differ in details like the depth of their convective zone and their rotation rates. These two factors in turn affect how active the stars are. Flare stars are very-low-mass stars (M stars) with fast rotation rates and deep convection zones. Such stars have spectacular outbursts in X rays.
3. The core of a red giant contracts because there is no more hydrogen fusion to heat the core and raise thermal pressure to resist gravity. However, the shell of hydrogen outside the core heats up to very high temperatures (hotter than the core during the main-sequence phase), and hydrogen fusion is occurring quickly. The star as a whole expands because the energy transport cannot keep up with the shell's increasing energy generation rate, so the thermal energy is trapped in the star and builds up, pushing the surface outward.
4. The overall reaction involved in helium burning is to combine three helium nuclei into one carbon nucleus. Because helium nuclei have two protons, and therefore twice the charge of hydrogen nuclei, they repel one another more strongly. Therefore, the nuclei must slam into one another at much higher speeds than is needed for fusion, requiring much higher temperatures. Before helium fusion begins, the core is supported by degeneracy pressure. This means that it does not expand as the core heats up, so that when the helium fusion begins, the core is very dense and very hot. This causes the helium fusion rate to rocket upward rapidly, resulting in the helium flash.
5. H-R diagrams for globular clusters show a horizontal branch because helium-burning stars all have about the same luminosity, but they vary in their surface temperature. Stars on the horizontal branch are all burning helium in their cores.
6. After a low-mass star exhausts it core helium, fusion ceases and the star core contracts. Because degeneracy pressure halts the collapse before the carbon becomes hot enough to sustain carbon fusion, the core becomes inert. The layers about the core, where shells of hydrogen and helium both continue fusing, will keep generating heat, causing the star to expand further than in the red giant phase. But this fusion cannot last long, perhaps a few million years, and then the star is dead, leaving behind a white dwarf.

7. A carbon star is a low-mass star in its final stages of life, when double shell burning makes the surface expand and causes strong convection. The convection dredges up carbon from the core and brings it to the surface. Because the carbon can then be lost via the stellar winds, these stars seed the interstellar medium with carbon, including the carbon that is used for life on Earth.
8. A planetary nebula is a glowing shell of gas that was once the outer layers of a star. The gas glows because the hot core's ultraviolet light ionizes the gas. The core, now exposed, will cool with time and become a white dwarf.
9. As the Sun evolves and becomes more luminous, the Earth's climate will eventually no longer be able to regulate itself, and Earth will experience a runaway greenhouse like what occurred on Venus. As the Sun becomes a red giant, it will expand to nearly Earth's orbit. The Earth may be destroyed in this stage.
10. The star begins on the main sequence in Figure 17.13 and spends most of its life there. Eventually, hydrogen fusion in the core must end and the star expands to a subgiant as the core shrinks and the overall star expands, powered by shell burning. Over a period of about a billion years, the star will grow in radius into a red giant. Eventually, the core temperature reaches about 100 million K and the helium there can begin fusion. When this happens, the fusion starts suddenly and strongly, producing a great deal of energy, which then heats and expands the shell outside the core, causing fusion there to stop. When this happens, the star becomes less luminous and actually shrinks until helium fusion ceases. When the helium fusion ends, the core again contracts and shell burning begins again. This time, there will be two shells of fusion: one of hydrogen and one of helium. The star again swells into a red giant. The outer layers of the star are no longer attached to the star very strongly and can blow off into space, creating a planetary nebula. Eventually, the shell burning also ends and the core of the star, now exposed, becomes a white dwarf.
11. High-mass stars go through their lives more quickly than low-mass stars. Part of this is because during their main-sequence lifetimes, they burn hydrogen via the CNO cycle, which produces more energy. After their main-sequence lifetimes, high-mass stars begin burning a series of heavier and heavier elements in their cores. But eventually, high-mass stars reach a stage when they have iron cores and cannot fuse any elements together to produce more energy. When this happens, the star explodes as a supernova.

    Intermediate-mass stars have similar lives up through the phase where the cores create carbon and oxygen. At this point, the intermediate-mass stars can no longer fuse elements to produce energy and die as white dwarfs.
12. The simplest sequences of fusion are helium capture reactions, where helium nuclei fuse with other nuclei. This builds carbon into oxygen, oxygen into neon, neon into magnesium, and so on.

    Also, at the high temperatures in the high-mass stars' cores, heavy nuclei can be fused together. So carbon can be combined with oxygen to form silicon, two oxygen nuclei can create sulfur, and so forth.

    These reactions all require high temperatures, which low-mass stars cannot produce. So low-mass stars are never able to use these reactions to power themselves like the high-mass stars do.

13. Iron cannot be fused to release energy because for elements heavier than iron, the mass per nuclear particle increases, so fusing two iron nuclei requires more energy than it produces.
14. One piece of evidence that supports our theories about how heavy elements form in high-mass stars is the chemical composition of older stars. Our theory predicts that the older stars should have fewer heavy elements in their compositions. Observations indicate that this is so. Another piece of evidence supporting our theories is the relative abundances of the various elements. For example, since the helium-capture reactions are an important series of reactions in high-mass stars, we expect to see more elements with even numbers of protons than odd numbers of protons. This predicted pattern agrees with the observations quite well.
15. When high-mass stars reach the stage of iron cores, degeneracy pressure will briefly support the core against collapse. However, this situation cannot last as gravity pushes the electrons past the limits and degeneracy pressure fails. In a fraction of a second, the iron core shrinks from the size of the Earth to a ball a few kilometers across. The contraction is halted by neutron degeneracy pressure. The contraction releases an enormous amount of energy that blows the outer layers away from the star in a supernova explosion.

    After the supernova, the core is left exposed. If the neutron degeneracy pressure is strong enough to resist gravity, a neutron star is left over. However, if even the neutron degeneracy pressure is insufficient to resist gravity, the core collapses into a black hole.

    Theoretical models reproduce the energy outputs of real supernovae, indicating that our understanding of supernovae is pretty good. Also, when Supernova 1987A occurred, we were able to look at the pre-supernova star and see that most of our predictions about the evolution of the star were pretty accurate.
16. The Algol paradox is that in the binary system Algol, the stars should be the same age, yet the bigger star is still on the main sequence and the smaller star is in the subgiant phase. Stellar evolution models say that the more massive star should live its life faster and die more quickly, yet the reverse appears to have occurred. The resolution to this paradox is that the subgiant star was once more massive and lived its life faster. As it expanded, it spilled its mass onto its companion. Mass transfer causes the companion to grow and the subgiant to shrink. This mass exchange allows stars with companions to change their masses throughout their lives, altering the life tracks in ways that single isolated stars cannot.
17. *The iron in my blood came from a star that blew up over 4 billion years ago.* This statement is sensible. The iron in the solar system was created before our Sun was formed about 4.6 billion years ago. Because iron is created in high-mass stars and delivered into interstellar space by supernova explosions, the supernova (or supernovae) responsible for creating the solar system's iron must have occurred before the Sun formed.
18. *Humanity will eventually have to find another planet to live on, because one day the Sun will blow up as a supernova.* This statement is not sensible. The Sun will eject a planetary nebula and fade away as a white dwarf. It is not massive enough to explode as a supernova. However, if humanity survives that long, we will probably have to find another place to live.

19. *I sure am glad hydrogen has a higher mass per nuclear particle than many other elements. If it had the lowest mass per nuclear particle, none of us would be here.* This statement is sensible. Iron has the lowest mass per nuclear particle, making it the end of the road for stellar energy production through fusion. If hydrogen had the lowest mass, it would be the end of the road, and none of the other elements would form via fusion in stars. Stars would be powered only by gravitational contraction, an energy source that does not last very long compared to the time needed for humans to evolve.
20. *I just discovered a 3.5M_{Sun} main-sequence star orbiting a 2.5M_{Sun} red giant. I'll bet that red giant was more massive than the 3M_{Sun} when it was a main-sequence star.* This statement is sensible. The 2.5M_{Sun} red giant had to be more massive than its companion at some point in the past in order for it to be more advanced in its evolutionary state than its companion.
21. *If the Sun had been born as a high-mass star some 4.6 billion years ago, rather than as a low-mass star, the planet Jupiter would probably have Earth-like conditions today, while Earth would be hot like Venus.* This statement is not sensible. If the Sun had been born as a high-mass star 4.6 billion years ago, it would have exploded as a supernova a long time ago.
22. *If you could look inside the Sun today, you'd find that its core contains a much higher proportion of helium and a much lower proportion of hydrogen than it did when the Sun was born.* This statement is sensible. Because the Sun is about halfway through its hydrogen-burning life, it has turned about half the core hydrogen into helium.
23. *Virtually all the supernova explosions that occur in a star cluster happen during its first 40 million years.* This statement is sensible only if you mean all the massive star supernovae explosions, since most of the stars with 8 solar masses will have evolved away from the main sequence by this time.
24. *Globular clusters generally contain lots of white dwarfs.* This statement is sensible, since globular clusters contain lots of old stars, so since the end-state of low-mass stars are white dwarfs, one would expect globular clusters to have lots of them.
25. *After hydrogen fusion stops in a low-mass star, its core cools off until the star becomes a red giant.* This statement is not sensible, because hydrogen fusion stops because the core has run out of hydrogen to fuse but is not yet hot enough to fuse helium. So the core of a red giant shrinks and heats up (it does not cool off).
26. *The gold in my new ring came from a supernova explosion.* This statement makes sense because elements heavier than iron are formed in supernova explosions. A gold atom is heavier than an iron atom, so gold falls into this category.
27. a. a white main-sequence star
28. b. A red supergiant is undergoing advanced nuclear burning, which requires higher temperatures, even if it is not currently burning anything in its core.
29. a. a red giant
30. b. Its luminosity goes down (its size drops).
31. c. Nuclear fusion would not occur in stars of any mass.
32. a. Supernovae would be more common.
33. b. a white dwarf

34. c. a 1-solar-mass star in a close binary with a 2-solar-mass star, since the 2-solar-mass star will evolve more quickly; then the 1-solar-mass star will gain material from its companion and evolve more quickly than the other systems.
35. a. It shrinks and heats up.
36. b. Uranium—the other elements can be made inside stars.

The answers to Problems 37–43 revolve around several features of stars: their lifetimes compared to the time presumably required to spawn an advanced civilization, their effects on complex life forms that rely on a protective environment such as an atmosphere or an ocean and a steady source of energy to survive. All of these assumptions are, of course, debatable. Informed answers to these problems will address what stage of life each star is in, what stages of life it has passed through, what may have happened to a planet in that time, and how long it has lived so far. Here are sample answers to each question:

37. A 10-solar-mass star has a very short lifetime. It also produces copious amounts of ultraviolet radiation that may discourage living organisms.
38. A 1.5-solar-mass star has a lifetime of a few billion years and produces light of similar character to that of our Sun. It seems reasonable to imagine it being orbited by a planet with a civilization.
39. A 1.5-solar-mass red giant is a temporary stage of life for a low-mass star. If an advanced civilization had already developed around this star, which is possible, then it may have had the resources to respond to its expanding, reddened sun.
40. A 1-solar-mass horizontal branch star is a late-stage low-mass-star, burning helium. Life had time to develop, but it would have had to be very clever and have sufficient natural resources and probably a lot of cooperation to persist.
41. A red supergiant is a late-stage high-mass star in the advanced state of nuclear burning—that is, burning elements heavier than helium in its core. Its envelope is gigantic. Its age at this point is rather young because massive stars live short lives. With our stated assumptions, an advanced civilization probably does not have enough time to develop.
42. A flare star is a star with less mass than the Sun, and very high coronal activity. The star would be very long-lived, but life on this planet may be plagued by frequent solar storms, and the planet's atmosphere would be more vulnerable to its star's weather. However, life could be protected from the star's moodiness by magnetic fields and water, and so while I wouldn't look there first, I wouldn't rule it out.
43. A carbon star is an evolved star, a red giant. Our own Sun could become something like this. While the Earth itself will not be a happy place when the Sun becomes a red giant, such a star could still support a civilization that developed the capability to move to the outer planets or moons of its solar system.
44. Helium fuses into carbon by combining three helium nuclei (atomic number 2) into one carbon nucleus (atomic number 6) and therefore bypassing the elements lithium, beryllium, and boron with atomic numbers 3 through 5. Therefore, fusion processes in the cores of stars do not form these three elements. (Beyond the scope of this book: Trace amounts of lithium and perhaps beryllium and boron formed in the Big Bang. Most of the beryllium and boron may have formed via cosmic ray collisions with heavier elements. The exact origin of these elements is still a topic of astronomical research. These three elements are also rather fragile and tend to be destroyed in the cores of stars rather than being created there.)

45. The Sun will have an angular size of 30°. The setting position moves through the sky—360°—once every 24 hours. So at 30°, or 1/12 of 360°, the Sun will take 1/12 of 24 hours = 2 hours to set, from the moment the limb touches the horizon to the time it vanishes below the horizon. (Compare that to the approximately 2 minutes it takes the $\frac{1}{2}$ degree Sun to set now.)

46. This question involves independent research. Answers will vary, depending on the choice of historical supernova chosen.

47. If the Sun expanded so that its radius was the same as Earth's orbital distance (1 AU), then we can use the formula for the volume of a sphere to find the volume of the Sun:

$$\text{volume} = \frac{4}{3}\pi r^3$$

We know, from Appendix E, that 1 AU is 1.496×10^8 km or 1.496×10^{13} cm. (We will use centimeters to make it easier to compare the densities later.) So we can calculate the volume to get 1.40×10^{20} cm^3.

What will the Sun's density be if this is the volume? To find density, we take the mass over the volume. We have the volume and we know, from Appendix E again, that the mass of the Sun is 2×10^{13} g. So the density is 1.43×10^{-7} g/cm^3. This density is much, much smaller than the density of water and quite a bit smaller than the density of air at sea level.

48. We recall that the formula for escape velocity is:

$$\text{escape velocity} = \sqrt{\frac{2\ GM}{r}}$$

We know that the mass of the star is 1 solar mass, 2×10^{30} g. The radius of the red giant is 100 solar radii, or 6.96×10^{10} m. So we calculate the escape velocity to be 6.19×10^4 m/s. We can also use the Sun's radius (and the same mass, of course) to find the escape velocity from the surface of the Sun: 6.19×10^5 m/s. So the escape velocity from the Sun is 10 times more than the escape velocity from the red giant. This fact tells us that material will have a much easier time escaping the surface of the red giant than material trying to escape the surface of the Sun. It makes sense that red giants have strong winds, then, because smaller velocities are required to escape red giants.

49. In order for the Earth to maintain a balance, it will have to emit 3,000 times as much power from each square meter as it does today. (We can go from total power emitted to power per unit area because the Earth's size is not changing.) We know that the formula for power per unit area emitted by a black body is σT^4, where σ is a constant. The ratio of the power emitted when the Sun is a red giant to the power emitted today is therefore:

$$\frac{P_{\text{red giant}}}{P_{\text{today}}} = \frac{\sigma T^4_{\text{red giant}}}{\sigma T^4_{\text{today}}}$$

We know that the ratio of the powers is 3,000. We can also simplify the right-hand side by canceling and collecting, leaving us with:

$$3{,}000 = \left(\frac{T_{\text{red giant}}}{T_{\text{today}}} \right)^4$$

We can now rearrange this to solve for the temperature when the Sun is a red giant:

$$T_{\text{red giant}} = T_{\text{today}} \sqrt[4]{3{,}000}$$

Earth's temperature today is around 300 K, so we plug in the numbers and find that the Earth will be around 2,200 K when the Sun is a red giant. (Actually, today's temperature is as high as 300 K because of our atmosphere, which is unlikely to survive this phase of the Sun's life. So the proper temperature to scale up is Earth's "no atmosphere" temperature of around 255 K rather than 300 K, for a red giant phase Earth temperature of 1,900 K—as far as we, the oceans, and the atmosphere are concerned, this subtlety is of little concern.)

50. We know from Mathematical Insight 15.1 that the apparent brightness of a star is given by the formula:

$$\text{Apparent brightness} = \frac{L}{4\,\pi d^2}$$

We can set up a ratio of the brightness of Betelgeuse to the brightness of Sirius:

$$\frac{\text{Apparent brightness of Betelgeuse}}{\text{Apparent brightness of Sirius}} = \frac{\dfrac{L_{\text{Betelgeuse}}}{4\,\pi d^2_{\text{Betelgeuse}}}}{\dfrac{L_{\text{Sirius}}}{4\,\pi d^2_{\text{Sirius}}}}$$

We can simplify this with a bit of algebra to:

$$\frac{\text{Apparent brightness of Betelgeuse}}{\text{Apparent brightness of Sirius}} = \frac{L_{\text{Betelgeuse}}}{L_{\text{Sirius}}} \left(\frac{d_{\text{Sirius}}}{d_{\text{Betelgeuse}}} \right)^2$$

Putting in the numbers given:

$$\frac{\text{Apparent brightness of Betelgeuse}}{\text{Apparent brightness of Sirius}} = \frac{10^{10} L_{\text{Sun}}}{26 L_{\text{Sun}}} \left(\frac{26 \text{ light-years}}{427 \text{ light-years}} \right)^2$$

$$= 8.44 \times 10^5$$

If Betelgeuse exploded as a supernova, it would appear 844,000 times brighter in our sky than Sirius.

51. a. Since each carbon atom has six protons, the total number of protons in two carbon atoms is 12, making a magnesium nucleus.
    b. A carbon nucleus has six protons and a neon nucleus has 10, so the combination of these nuclei makes a sulfur nucleus, with 16 protons.
    c. An iron nucleus has 26 protons and a helium nucleus has just two. When we combine these nuclei, we get 28 protons—a nickel nucleus.

52. We can look at this as a rate problem: The material travels at 0.11 arcsecond per year and it has traveled a total of 100 arcseconds. So we can calculate the time it has taken to get there by dividing the distance traveled by the rate:

$$\text{time} = \frac{100''}{0.11''/\text{yr}}$$
$$= 909 \text{ yr}$$

If "now" is the year 2006, subtracting, we find that the supernova should have been observed in around 1097 A.D. This date is close to the actual year it was observed (1054 A.D.) [The difference is attributed to the questionable assumption that the material has been moving at a constant rate over the last 1,000 years—measurement error is not the issue here since such rates and positions can be measured fairly accurately.]

53. Newton's version of Kepler's third law tells us that:

$$p^2 = \frac{4\pi^2}{GM_{\text{total}}} a^3$$

We can solve for the orbital separation:

$$a = \sqrt[3]{\frac{GM_{\text{total}}}{4\pi^2} p^2}$$

The total mass of the two stars is $4.6M_{\text{Sun}}$. The mass of the Sun is 2×10^{30} kg, so the total mass of the system is 9.2×10^{30} kg. The period is given as 2.87 days, which we can convert to 248,000 seconds. Putting this numbers in, we find that the orbital separation is 9.85×10^9 m. The text tells us that the Sun will eventually swell to 100 times its radius, making it about 7×10^{10} m when it becomes a red giant. Thus, the distance between these two stars is less than the typical radius of a red giant.

54. The material has traveled 0.7 light-year in about 18 years (2005 AD – 1987 AD = 18 years). The speed is easy to determine, just divide distance by time to get 0.047 light-year/year. This is 4.7% the speed of light, a pretty good speed but not unreasonable.

Chapter 18. The Bizarre Stellar Graveyard

This chapter covers the end points of stellar evolution: white dwarfs, neutron stars, and black holes. Students often enter an astronomy course interested in and enthusiastic about these objects, and may find this to be one of their favorite chapters in the book.

As always, when you prepare to teach this chapter, be sure you are familiar with the relevant media resources (see the complete, section-by-section resource grid in Appendix 3 of this Instructor's Guide) and the online quizzes and other study resources available on the Mastering Astronomy Web site.

What's New in the Fourth Edition That Will Affect My Lecture Notes?

Those who have taught from previous editions of *The Cosmic Perspective* should be aware of the following organizational or pedagogical changes to this chapter (i.e., changes that will influence the way you teach) from the Third edition:

- The discussion of the sources of gamma-ray bursts has been updated to emphasize the growing evidence that supernovae are responsible for at least some of these bursts.
- The short overview in Section 18.1 has been assimilated into the rest of the chapter, so the chapter now has four sections rather than five.

Teaching Notes (By Section)

Section 18.1 White Dwarfs

This section discusses white dwarfs and the consequences of mass transfer in a close binary that contains a white dwarf. Here is where we cover white dwarf supernovae (a.k.a., Type Ia) and the differences between supernova light curves.

- In order to keep the terminology as descriptive as possible, we call $1.4M_{Sun}$ the *white dwarf limit* rather than the *Chandrasekhar limit.*
- In the same vein, we use the term *white dwarf supernova* when referring to the supernovae that come from exploding white dwarfs, and *massive star supernova* when referring to the supernovae that come from exploding massive stars. Even professional astronomers sometimes have trouble keeping the Type Ia, Type Ib, Type II nomenclature straight! A footnote explains these terms, should you want to use them in class.
- A simple in-class way to reinforce the idea that friction is what heats accretion disks is to have the students rub their hands together until they heat up.
- This section glosses over the fact that astronomers are uncertain about the progenitors of white dwarf supernovae. Most astronomers think that gradual accretion drives a single white dwarf past the $1.4M_{Sun}$ limit; however, a few argue that these explosions could arise from the merger of two white dwarfs in the same binary system.

Section 18.2 Neutron Stars

This section covers neutron stars and their manifestations as pulsars, X-ray binaries, and X-ray bursters. The story of Jocelyn Bell and her advisor's Nobel Prize engages students, and addresses the societal aspect to science: Who gets credit for a discovery and why they get credit may surprise your students, many of whom might assume that ascertaining credit should be obvious and objective.

Section 18.3 Black Holes: Gravity's Ultimate Victory

This section introduces students to the weird world of black holes, the highlight of an astronomy course for many students.

- The masses we quote for the Cygnus X-1 system are somewhat smaller than those you might find elsewhere. We have taken them from the work of Herrero et al. (1995, *A&A* 297, 556), who measured the mass of the O star in this system by analyzing its spectrum to determine its surface gravity.

Section 18.4 The Mystery of Gamma-Ray Bursts

This section introduces the mystery of gamma-ray bursts. Their energy output suggests that they are produced during the formation of a black hole, and at least some gamma-ray bursts have been associated with supernova explosions in distant galaxies. However, the mechanism that produces the gamma-ray burst remains mysterious. Also, gamma-ray bursts come in two types, based on their duration and their spectrum, and only one of these types has been tied to supernovae. The other type may come from merging neutron stars in a binary system.

- Our knowledge about gamma-ray bursts has continued to advance very rapidly since the first discoveries of their optical counterparts in 1997. This field is an extremely active one, and therefore it is worth looking for the most-up-to date material available when you present this subject. In particular, the identity of the short-term bursts may be known by the time you read this, since *SWIFT* has enabled for the first time the follow-up of the short-time-scale bursts.

Answers/Discussion Points for Think About It Questions

The Think About It questions are not numbered in the book, so we list them in the order in which they appear, keyed by section number.

Section 18.1

- (p. 575) No, both novae and white dwarf supernovae arise from mass transfer from a companion star, and thus cannot occur outside of binary systems.

Section 18.2

- (p. 577) If a neutron star is a pulsar but its rotating beams of radiation never touch the Earth, then we will not see it as a pulsar. However, if these rotating beams periodically point at some other civilization, then that civilization will see the neutron star as a pulsar.

Section 18.3

- (p. 585) X-ray bursts are bursts of fusion on the surface of an accreting neutron star. A black hole does not have a surface, only an event horizon that accreting material passes right through. Thus, black holes can never be X-ray bursters.

Solutions to End-of-Chapter Problems (Chapter 18)

1. Degeneracy pressure is a kind of pressure that arises when subatomic particles are packed as closely as the laws of quantum mechanics allow. Degeneracy pressure is important to neutron stars and white dwarfs because it is what allows them to resist the pull of gravity. In the case of white dwarfs, the degeneracy pressure is provided by electrons, so that version is called "electron degeneracy pressure." For neutron stars, it is the neutrons that provide the pressure and this version of degeneracy pressure is therefore called "neutron degeneracy pressure."

2. A typical white dwarf has about the mass of the Sun packed into a ball the radius of the Earth. This compact object has a very high density: A teaspoon of a white dwarf would weight as much as a small truck. Because the white dwarf is supported by degeneracy pressure, adding mass to the white dwarf causes it to shrink in radius.
3. As the mass of a white dwarf increases, the pressure must increase to resist gravity. To do this, the electrons must move faster. However, there is a limit on how fast the electrons are allowed to move: the speed of light. If mass of the white dwarf becomes so great that the electrons would have to move faster than light to resist the gravity, the white dwarf must collapse into a neutron star. This limit is about $1.4M_{Sun}$.
4. An accretion disk is a disk of orbiting material that is falling toward a central body, like a white dwarf. We see these only in close binary systems (not isolated stars) because they require material to be transferred from one star to another. As the material falls onto the white dwarf, gravitational energy is turned into heat. The heat provides the white dwarf with a new energy source, allowing it to glow in the ultraviolet.
5. A nova is the glow from the thermonuclear flash from the onset of fusion in a hydrogen shell on the surface of a white dwarf. The hydrogen shell comes from accretion when the white dwarf steals material from its companion. As the hydrogen builds up on the surface, the pressure and temperature rise, and eventually hydrogen fusion becomes possible.
6. A white dwarf supernova occurs when the white dwarf gains enough mass for the carbon interior of the star to begin carbon fusion. The fusion begins almost instantly throughout the star, so the entire star ignites and the white dwarf explodes completely. These supernovae, unlike the massive star supernovae, lack hydrogen lines in their spectra, allowing astronomers to tell the two types of supernovae apart.
7. A neutron star packs a greater mass than the Sun into a ball about 10 kilometers in radius. Something so massive and compact is extremely dense: A paper clip made of neutron star material would weigh more than a mountain. If a neutron star came to my hometown, it would destroy my town and the entire Earth, crushing the planet into a shell no thicker than my thumb.
8. We know that pulsars are neutron stars because we have found pulsars at the centers of supernova remnants, right where we expect to see neutron stars. We are also confident that pulsars are neutron stars because we know of no other objects that could spin as fast as pulsars must.

   Not all neutron stars are pulsars. Older neutron stars have lost so much of their magnetic fields and their spins have slowed so much that we would not see pulses. Even a young neutron star would not necessarily be a pulsar: If the beam of radio waves was not oriented to sweep over the Earth, we would not see a pulsar even though some other planet might.
9. In a close binary, a neutron star can accrete matter from its companion. As this material falls down in the neutron star's intense gravity, its potential energy is converted to heat. This process makes the inner region of the disk so hot that it glows in the X rays. We call such a system an "X-ray "binary." In some systems, we see bursts of X rays. These bursts are caused by a similar process to white dwarf novae, except that in this case it is helium fusion, not hydrogen fusion, that powers the burst. Because a steady stream of hydrogen pours onto the neutron star, the pressure and temperature at the bottom of the hydrogen shell are high enough for hydrogen fusion. Accretion builds up a layer of helium, which can eventually ignite, releasing a burst of X rays.

10. A black hole is like a hole in spacetime because it represents a part of the universe we can never observe and from which we could never return if we went in. The event horizon is the boundary between the inside of the black hole and the outside universe. The radius of the event horizon is called the "Schwarzschild radius." Black holes have only three measurable properties: mass, spin, and charge.
11. A singularity is a mathematical concept where the math becomes undefined (like division by zero, or infinity). In this chapter, a singularity is the point where all of the mass of a black hole is crushed into a point. Unfortunately, general relativity (which describes gravity) disagrees with quantum mechanics (which describes the physics of the very small) at this extreme limit of tiny sizes but the $F_g = \infty$ at the center of a black hole, so our current theories are inadequate to describe what happens precisely at the singularity.
12. If we were to fall into a black hole, we would perceive our own time as passing normally. We would see the rest of the universe run faster and faster around us as we fell farther into the black hole. Unfortunately, as we got very close to the black hole, the tidal forces would stretch us out and eventually tear us apart, if the extreme radiation environment did not kill us first.
13. We think that black holes should sometimes be formed by supernovae because models indicate that in some supernovae the outer layers of the star are not completely blown away. The extra mass can push the neutron star core over the mass limit for neutron stars and make it a black hole. We have strong evidence for black holes in X-ray binary systems where the mass of the unseen companions are larger than the mass limit for neutron stars. The only kind of object we currently know of that these companions could be are black holes. We also have evidence for supermassive black holes in the centers of many galaxies.
14. We know that gamma-ray bursts do not come from the same source as X-ray bursts because their distribution in our sky is uniform rather than concentrated in the disk of the Milky Way as X-rays bursts are. We hypothesize that gamma-ray bursts may be caused by a supernova that forms a black hole, releasing many times more gravitational potential energy than the kind of supernova that forms a neutron star. It is also possible that some gamma-ray bursts are caused by neutron stars colliding with each other. But at the moment, we still do not fully understand the source of gamma-rays bursts.
15. *Most white dwarfs have masses close to that of our Sun, but a few white dwarf stars are up to three times as massive as the Sun.* This statement is not sensible. A $3M_{\text{Sun}}$ star would exceed the white dwarf mass limit of 1.4 solar masses.
16. *The radii of white dwarf stars in close binary systems gradually increase as they accrete matter.* This statement is not sensible. The higher gravity of the more massive white dwarfs compresses the white dwarf material to a higher density and a smaller, not larger, radius.
17. *If you want to find a pulsar, you might want to look near the remnant of a supernova described by ancient Chinese astronomers.* This statement is sensible. Other pulsars have been discovered in historical supernova remnants such as the Crab Nebula. Pulsars are the product of supernova explosions; therefore, it makes sense to look for them in supernova remnants or in locations where supernovae were noted historically.
18. *If a black hole 10 times more massive than our Sun were lurking just beyond Pluto's orbit, we'd have no way of knowing it was there.* This statement is not sensible. A black hole of 10 solar masses would exert a profound gravitational influence on the orbits of the planets, even if the black hole lurked beyond the orbit of Pluto.

19. *If the Sun suddenly became a $1M_{Sun}$ black hole, the orbits of the nine planets would not change at all.* This statement is sensible. The orbits of the planets depend only on the mass of the object they are orbiting, regardless of whether it is a black hole, a neutron star, a main-sequence star, or anything else.
20. *We can detect black holes with X-ray telescopes because matter falling into a black hole emits X rays after it smashes into the event horizon.* This statement is not sensible. The black hole has no surface for material to smash into. X-ray telescopes can detect black holes because the gas falling into a black hole can be very hot despite the black hole's lack of a surface.
21. *If two black holes merge together, the resulting black hole is even smaller than the original ones because of its stronger gravity.* This statement is not sensible, since the size of a black hole is proportional to its mass. So the more massive a black hole is, the larger its event horizon.
22. *If gamma-ray bursts really channel their energy into narrow beams, then the total number of gamma-ray bursts that occur is probably far greater than the number we detect.* This statement is sensible. If we detect a gamma-ray burst only when its "beam" is pointed at us, then there must be a lot of other gamma-ray burst events where their beams are not pointed at us.
23. *If the Sun suddenly became a $1M_{Sun}$ black hole, Earth's tides would become much greater.* This statement is not sensible because the tidal force arises from the difference in gravitational pull between the Sun and the Earth across the radius of the Earth, not the Sun. (The tidal forces on the Sun itself would be smaller, since the Sun would be smaller.)
24. *The pulsation period of a pulsar appears to speed up if the pulsar is moving toward us.* This statement is sensible. The speed at which the light pulses travel will stay the same, but the time separation between the pulses will decrease if the pulsar is moving toward us.
25. a. A 1.2-solar-mass white dwarf is smallest. (It is about Earth-size, and the 0.6-solar-mass white dwarf has a larger radius than a 1.2-solar-mass white dwarf.)
26. a. A 1.2-solar-mass white dwarf is largest (the other two are about the same size, 10 kilometers).
27. b. The event horizon of a 3-solar-mass black hole.
28. c. The Earth's orbit would not change.
29. b. The isolated pulsar that pulses 600 times per second.
30. b. Its spin would slow down.
31. a. An X-ray binary containing an O star and another object of equal mass
32. c. Its flashes would shift to the infrared part of the spectrum (and slow down).
33. a. A 10-solar-mass black hole
34. c. Extremely distant galaxies

35–38. These are extended essays; answers will vary. (*Note:* Many students really enjoy these problems, but you will need adequate resources to grade these essays if you assign them.) The student's story should identify key points in stellar evolution that lead to these various end-states.

39. White dwarfs are more common than neutron stars and black holes because they are the result of the evolution of low-mass stars, which are far more numerous than high-mass stars.

40. If X-ray bursts are not powerful enough to accelerate material beyond the escape velocity of the surface of a neutron star, any accreted material will eventually settle back on the neutron star, even if it experiences violent nuclear fusion explosions. The neutron star increases in mass until it exceeds the mass density that can be supported by neutron degeneracy pressure and becomes a black hole.

41. In order to fall into a black hole, an object must lose enough of its initial angular momentum to get close enough to cross the event horizon. The event horizon of a stellar black hole is tiny compared to typical orbits, so the orbit would have to become extremely eccentric (almost falling directly at the black hole) in order for it to have a chance of crossing the horizon.

42. The life preserver would have to counteract the stretching force of the tide. Let's say the person is falling in feetfirst. The hoop around the person's waist would have to pull on the head of the person (toward his waist) and pull the feet of the person up toward his waist. The stretching force of the tide is pulling harder on the person's head than on his feet, so the life preserver's gravitational forces subtract from the tidal forces, and make it possible for the person to live a tiny bit longer.

43. a. To calculate the sum of the masses, we simply solve Kepler's law algebraically and convertthe period to seconds (4 days = 345,600 seconds) and the separation to meters (2×10^7 km = 2×10^{10} m):

$$(m_1 + m_2) = \frac{4\,\pi^2\,a^3}{G\;p^2} = \frac{4\,\pi^2}{\left(6.67 \times 10^{-11}\,\dfrac{\text{m}^3}{\text{kg} \times \text{s}^2}\right)}\,\frac{(2 \times 10^{10}\ \text{m})^3}{(345{,}600\ \text{s})^2} = 4 \times 10^{31}\ \text{kg}$$

The sum of the masses of the two stars is 4×10^{31} kg. Dividing the mass sum of 4×10^{31} kg by the solar mass of 2×10^{30} kg, we find that the mass sum is equivalent to $20M_{\text{Sun}}$.

b. Because the combined mass of the two stars is about $20M_{\text{Sun}}$ and the B2 star has a mass of $10M_{\text{Sun}}$, the unseen companion also has a mass of about $10M_{\text{Sun}}$—far too large for a neutron star. It must be a black hole.

44. a. To calculate the sum of the masses, we simply solve Kepler's law algebraically and convert the period to seconds (5 days = 432,000 seconds) and the separation to meters (1.2×10^7 km = 1.2×10^{10} m):

$$(m_1 + m_2) = \frac{4\,\pi^2 a^3}{Gp^2} = \frac{4\,\pi^2}{\left(6.67 \times 10^{-11}\,\dfrac{\text{m}^3}{\text{kg} \times \text{s}^2}\right)}\,\frac{(1.2 \times 10^{10}\ \text{m})^3}{(432{,}000\ \text{s})^2} = 5 \times 10^{30}\ \text{kg}$$

The sum of the masses of the two stars is 5×10^{30} kg. Dividing the mass sum of 5×10^{30} kg by the solar mass of 2×10^{30} kg, we find that the mass sum is equivalent to $2.5M_{\text{Sun}}$.

b. Because the combined mass of the two stars is about $2.5M_{\text{Sun}}$ and the G2 star has a mass of $1M_{\text{Sun}}$, the unseen companion also has a mass of about $1.5M_{\text{Sun}}$—within the range of a neutron star. So these data are consistent with the companion being a neutron star, and it is probably not a black hole.

45. a. To find the density of the neutron star, we will need the total volume. We can use the formula for the volume of a sphere:

$$\text{volume} = \frac{4}{3}\pi r^3$$

All we need is the radius in centimeters. We are told that the radius is 10 kilometers and converting tells us that this is the same as 1×10^6 cm. Plugging in the numbers, we learn that the volume of a typical neutron star is 4.19×10^{18} cm³. The mass of a typical neutron star is $1.5M_{\text{Sun}}$, or 3×10^{30} kg (since we know that the mass of the Sun is 2×10^{30} kg, given in Appendix A). The density is therefore 7.16×10^{11} kg/cm³.

b. Since the density of neutron star material is 7.16×10^{11} kg/cm³, the mass of 1 cubic centimeter of neutron star material is 7.16×10^{11} kg. This is more than 10 times the mass of Mount Everest's 5×10^{10} kg.

46. a. For the $10^8 M_{\text{Sun}}$ black hole, the Schwarzschild radius is:

$$R_S = 3 \times \frac{10^8 M_{\text{Sun}}}{M_{\text{Sun}}} \text{ km} = 3 \times 10^8 \text{ km}$$

The Schwarzschild radius of a $10^8 M_{\text{Sun}}$ black hole is about 300 million kilometers, or about 2 AU (twice the distance from the Earth to the Sun). Because of the relatively large size of such a black hole, tidal forces across a small object—such as a person or a spaceship—will be less significant than those caused by a smaller black hole. It *might* be possible to survive a trip across the event horizon of a massive black hole, but what would you find when you got there?

b. The Schwarzschild radius of a $5M_{\text{Sun}}$ black hole is:

$$R_S = 3 \times \frac{5M_{\text{Sun}}}{M_{\text{Sun}}} \text{ km} = 15 \text{ km}$$

The Schwarzschild radius of a $5M_{\text{Sun}}$ black hole is about 15 kilometers.

c. The first formula in Mathematical Insight 18.1 is more useful in this case. The mass of the Moon is about 7.4×10^{22} kg, so its Schwarzschild radius is:

$$R_S = \frac{2\,GM}{c^2} = \frac{2 \times \left(6.67 \times 10^{-11} \frac{\text{m}^3}{\text{kg} \times \text{s}^2}\right)(7.4 \times 10^{22} \text{ kg})}{\left(3 \times 10^8 \frac{\text{m}}{\text{s}}\right)^2} \approx 1.1 \times 10^{-4} \text{ m} = 0.11 \text{ mm}$$

The Schwarzschild radius of the Moon is barely a tenth of a millimeter. The Moon would have to be crushed to smaller than a pinhead to become a black hole.

d. Your Schwarzschild radius will depend slightly on your mass. Let's take 50 kilograms as a typical mass for a person. Then your Schwarzschild radius would be about:

$$R_S = \frac{2\,GM}{c^2} = \frac{2 \times \left(6.67 \times 10^{-11} \frac{\text{m}^3}{\text{kg} \times \text{s}^2}\right)(50 \text{ kg})}{\left(3 \times 10^8 \frac{\text{m}}{\text{s}}\right)^2} = 7 \times 10^{-26} \text{ m}$$

Your Schwarzschild radius is about 7×10^{-26} m. Recall that the typical size of an atom is about 10^{-10} meter and the typical size of an atomic *nucleus* is about 10^{-15} meter. You would have to be crushed to some *10 billion times smaller* than an atomic nucleus to become a black hole.

47. The estimated age of the Crab Pulsar is:

$$\begin{aligned} \text{age} &= \frac{p}{2\,r} \\ &= \frac{0.0333 \text{ s}}{2 \times (4.2 \times 10^{-13} \text{s/s})} \\ &= 3.96 \times 10^{10} \text{ s} \end{aligned}$$

We can convert to years in the usual way to get 1,250 years for the age of the Crab Pulsar. This implies that the Crab Pulsar formed 750 A.D., which is about three centuries earlier than when the supernova was actually observed.

48. Mathematical Insight 18.1 tells us that the Schwarzschild radius of a black hole is given by the expression:

$$R_S = (3 \text{ km}) \frac{M}{M_{\text{Sun}}}$$

For this problem, we will convert the 3 kilometers into 3×10^5 cm since we want a volume in g/cm^3. The volume of a sphere of this radius is $\frac{4}{3} \pi r^3$. Knowing that the density is the mass over the volume, we can write an expression for the density:

$$\begin{aligned} \text{density} &= \frac{\text{mass}}{\text{volume}} \\ &= \frac{M}{\frac{4}{3} \pi R_S^3} \end{aligned}$$

Using the formula for R_S:

$$\begin{aligned} \text{density} &= \frac{3M}{4\,\pi \left((3 \times 10^5 \text{ cm}) \frac{M}{M_{\text{Sun}}} \right)^3} \\ &= \frac{3M_{\text{Sun}}^3}{4\,\pi M^2 (2.7 \times 10^{16} \text{ cm}^3)} \end{aligned}$$

For this problem, we are given the density and asked to find the mass of the black hole. We therefore solve for the mass:

$$\frac{M}{M_{\text{Sun}}} = \sqrt{\frac{3M_{\text{Sun}}}{4\,\pi \times (\text{density}) \times (2.7 \times 10^{16} \text{ cm}^3)}}$$

The mass of the Sun is 2×10^{30} kg, or 2×10^{33} g. We are told that the density we seek is 1 g/cm^3. The mass of the black hole is therefore:

$$\frac{M}{M_{\text{Sun}}} = \sqrt{\frac{3(2\times10^{33}\text{ g})}{4\ \pi\times(1\text{ g/cm}^3)\times(2.7\times10^{16}\text{ cm}^3)}}$$

$$= 1.33\times10^{8}$$

The black hole would need a mass of 133 million times the mass of the Sun to have an average density of water. (There are supermassive black holes that are this massive. So this is not unreasonable.)

49. To answer this question, we will use $1.5M_{\text{Sun}}$ as the typical mass of a neutron star. This comes out to be about 3×10^{30} kg. The radius of the neutron star is given as 10 kilometers, but we will convert it to 1×10^4 m. Using the formula for gravitational potential energy released:

$$\text{energy released} = \frac{GM^2}{r}$$

we find that about 6.00×10^{46} joules are released.

In order to compare this figure to the total energy output of the Sun over its lifetime, we will need the Sun's expected lifetime. The Sun is expected to live about 10 billion years, or 3.16×10^{17} seconds. The Sun puts out 3.8×10^{26} watts of energy, so the total energy output of the Sun over its entire lifetime is the product of the energy generation rate and the lifetime. Putting in the numbers, we find that the Sun will produce 1.20×10^{44} joules of energy in its entire life. The energy produced in a supernova is around 500 times the Sun's total lifetime energy output.

50. If a neutron star suddenly appeared in your hometown, its mass (and therefore its gravity) would be far larger than that of the Earth. The entire Earth would wrap itself around the neutron star in a thin layer. The hint describes the method for determining the thickness of the layer formed by the Earth: The total mass of the Earth ($M_{\text{Earth}} \approx 6\times10^{24}$ kg) would be compressed to neutron-star density and thus have a volume of:

$$V_{\text{shell}} = \frac{M_{\text{Earth}}}{\text{density}}$$

$$= \frac{6\times10^{24}\text{ kg}}{7\times10^{11}\ \frac{\text{kg}}{\text{cm}^3}}$$

$$= 9\times10^{12}\text{ cm}^3$$

You can then calculate the thickness of the shell on the neutron star from the formula given, where r is the 10-kilometer radius of the neutron star:

$$V_{\text{shell}} \approx 4\ \pi r^2_{\text{shell}} \times \text{thickness} \Rightarrow \text{thickness} = \frac{V_{\text{shell}}}{4\ \pi r^2_{\text{shell}}}$$

Plugging in the volume found above and 10 kilometers (= 10^6 cm) for the shell radius, we find that the thickness of the layer formed by the Earth would be about 0.7 centimeter, or 7 millimeters.

Part VI: Galaxies and Beyond

Chapter 19. Our Galaxy

Chapter 19 introduces students to galaxies by detailing how our Milky Way works. The agenda of this chapter is twofold: to inform students about our home galaxy and to set the stage for the coverage of galaxy evolution in the next chapter. In our view, galaxies are no less important than stars in preparing the conditions for life, because a large star system is needed to retain and recycle the elements created in the cores of stars. We find that students respond positively to this idea because it connects the behavior of galaxies, and the Milky Way in particular, to their own lives.

As always, when you prepare to teach this chapter, be sure you are familiar with the relevant media resources (see the complete, section-by-section resource grid in Appendix 3 of this Instructor's Guide) and the online quizzes and other study resources available on the Mastering Astronomy Web site.

What's New in the Fourth Edition That Will Affect My Lecture Notes?

Those who have taught from previous editions of *The Cosmic Perspective* should be aware of the following organizational or pedagogical changes to this chapter (i.e., changes that will influence the way you teach) from the Third edition:

- Material in the chapter was reordered, and some of the material was shuttled to other chapters. Material introducing the ISM was moved to the Star Birth chapter, and deeper discussion of the phases of the ISM are now completely contained in Section 19.2 on galactic recycling.
- Section 19.1 now contains both the appearance of the Milky Way and the discussion of stellar orbits that was formerly in Section 19.4. The coverage of rotation curves and dark matter in the Milky Way was moved to Chapter 22 (Dark Matter, Dark Energy and the Fate of the Universe) to provide a more compact discussion of the evidence for dark matter.
- The galactic center material moved from Section 19.5 to Section 19.4.
- There is no longer a Section 19.5.

Teaching Notes (By Section)

Section 19.1 The Milky Way Revealed

This section introduces the Milky Way by summarizing its structure and its contents. For students who have covered Part I of the text, this will be largely a review lesson.

- The historical material on humanity's discovery of the Milky Way's true size and scope appears in a Special Topic box titled "Discovering the Milky Way."

Instructors who are pressed for time can skip this material, which is not essential to understanding the Milky Way itself, even though it's an interesting tale about how science progresses.

- The appearance of the galaxy (halo and disk) now directly precedes the discussion of stellar orbits (halo and disk).
- While we mention dark matter here (including a Mathematical Insight), since stellar orbits provide evidence for its existence, we postpone a detailed discussion of what dark matter is and galactic rotation curves to Chapter 22.

Section 19.2 Galactic Recycling

This section describes the workings of our galaxy's interstellar medium. We have found that tracing the star-gas-star cycle from stellar mass ejection, through the subsequent cooling of the gas and its collapse into clouds, to the process of star formation is an excellent way to tie together the various states of the interstellar medium and to show the dynamism of our galaxy. Students who understand how our galaxy cycles gas into stars are well prepared to understand how galaxies evolve from pristine gas clouds to vast collections of stars enriched with heavy elements.

- We have chosen to avoid the astronomical term *metals* in this book, preferring the term *heavy elements* instead. The term *metals* can confuse students who already know what real metals are, but we do provide the astronomical definition in a footnote.
- The text states that hot-gas bubbles fill roughly 20–50% of the Milky Way's disk, but the topology of the hot-gas distribution is still debated among experts. Certainly some of the hot gas is in well-defined bubbles, but in other parts of the galaxy, hot bubbles may be so interconnected that it makes more sense to speak of cooler clouds embedded in a hot substrate.
- Figure 19.12, which is drawn from a poster produced by NASA, is one of the best tools we have run across for demonstrating the power of multiwavelength astronomy. Consider spending some time going over this set of pictures in class. They summarize the various states of the interstellar medium while showing students how much richer our view of the cosmos becomes when we broaden our view to other parts of the electromagnetic spectrum.
- Spiral density waves are one of the most difficult topics for students. We have improved our discussion and figures.

Section 19.3 The History of the Milky Way

This section discusses the two major stellar populations and how they help lead us to a model for how our galaxy formed.

- We avoid using the terminology of *population I* and *II,* instead calling them by what they are: the *disk population* and the *spheroidal population*, respectively. Since some instructors may insist students learn these terms, we have left them as italicized terms (as opposed to boldface terms).

Section 19.4 The Mysterious Galactic Center

This section focuses on the galactic center. Recent observations of stellar proper motions in the vicinity of Sgr A^* have greatly strengthened the case for a black hole of 3 to 4 million solar masses within it.

Answers/Discussion Points for Think About It Questions

The Think About It questions are not numbered in the book, so we list them in the order in which they appear, keyed by section number.

Section 19.1

- (p. 595) Stars are so far apart, relative to their sizes, that there is little chance that a halo star will collide with the Sun or the Earth.

Section 19.2

- (p. 597) Stars in open clusters are generally much younger than those in globular clusters, having formed after many heavy elements accumulated in the interstellar medium. Thus, open clusters have higher proportions of heavy elements than globular clusters.
- (p. 604) Many different patterns are evident in Figure 19.12. The most prominent is the band of molecular clouds across the midplane, which appears bright in molecular emission, atomic hydrogen 21-centimeters emission, far-infrared dust emission, and gamma-ray emission, but dark in optical and X-ray light because of its obscuring effects.
- (p. 605) The red color is from hydrogen emission, the blue is starlight scattered by dust, and the black arises because dust obscures light from background stars.

Section 19.3

- (p. 608) In the far future, after all the gas in the disk has been transformed into stars, the disk environment will look more like today's halo environment because it will no longer harbor young stars.
- (p. 609) The universe was much denser in the past, so its galaxies were much closer together, making collisions much more frequent.

Solutions to End-of-Chapter Problems (Chapter 19)

1. Student sketches should look like Figure 19.1.
2. The Large and Small Magellanic clouds and the Sagittarius and Canis Major dwarfs are small galaxies orbiting the Milky Way.
3. Stars in the disk of the galaxy have nearly circular orbits that are mostly in the plane of the galactic disk. The disk stars have vertical motions out of the plane, making them appear to bob up and down, but they never get "too far" from the disk. Orbits of stars in the bulge and the halo of the galaxy are much less orderly, traveling around the galactic center on elliptical orbits with more or less random orientations.
4. If we know a star's orbital period and orbital radius, we can use Newton's version of Kepler's third law to determine the mass of the galaxy, with a minor warning: We get only the mass of the galaxy within the orbit of the star we examine. What we have learned from our studies to find the distribution of mass in the galaxy is that the stars in the galactic disk orbit at about the same speeds. This observation tells us that most of the galaxy's mass resides far from the center and is distributed throughout the halo.

5. Stars are born from collapsing molecular clouds. They shine for millions or billions of years through their nuclear fusion, turning lighter elements into heavier elements. When the stars die, much of their mass is returned to space, enriching the gas clouds with heavier elements to form the next generation of stars.
6. Bubbles of hot, ionized gas are created by supernova explosions and powerful stellar winds. As the bubbles expand, they create shock waves in the interstellar medium that drain away their energy. Eventually, the bubbles' expansions slow and the bubbles merge with interstellar gas.
7. Superbubbles form when the hottest, most massive stars in a cluster explode as supernova within a few hundred thousand years of each other. Their bubbles merge into a giant bubble, called a "superbubble." If the superbubbles grow bigger than the thickness of the galaxy, there is no longer anything (other than gravity) to keep them from expanding in the vertical direction, resulting in blowouts.
8. The galactic fountain causes gas to spew from the disk up into the halo. The material eventually falls back down into the disk, mixing with the gas that's already there. This process mixes the halo and disk gases.
9. Cosmic rays are electrons, protons, and neutrons that move at close to the speed of light and permeate interstellar space. We think that they come from supernova explosions.
10. Atomic hydrogen gas is gas that is cool enough for atoms of hydrogen to hold on to their electrons rather than be ionized. Atomic hydrogen gas clouds are not composed entirely of hydrogen: There is quite a bit of helium and a small amount of other elements in the clouds. Atomic hydrogen gas is spread through the entire galactic disk and makes up a few percent of the mass of the galaxy.
11. There are several kinds of gas present in the disk of the galaxy. First, there are hot bubbles, which are composed of low-density hot gas. This gas is best viewed in the X-ray portion of the spectrum. At lower temperatures we find atomic hydrogen gas (warm and cool). This gas is cool enough to be un-ionized and is best viewed in the radio part of the spectrum, thanks to the 21-centimeter line of atomic hydrogen. Molecular clouds are even cooler than atomic hydrogen, allowing molecules to form. This gas is also best viewed in the radio wavelengths. The coolest parts of the molecular clouds are the molecular cloud cores that go on to form stars.
12. Ionization nebulae are colorful, wispy blobs of gas found near hot stars. They require the hot stars' ultraviolet light to raise the electrons in the atoms of the gas to high energy levels or to ionize the atoms completely. When the electrons fall back down to the lower energy levels, they emit photons, causing the nebulae to glow.
13. We know that the spiral arms cannot rotate like pinwheels around the center of the galaxy because the inner stars would finish several orbits while the outer stars complete one. This would wind the spiral pattern up, which we do not see. The arms appear bright because of the enhanced star formation in them. The bright, bluer stars in particular make the arms bright since these stars do not live long and so die before spreading far from the arm in which they formed.
14. Stars are created more readily in spiral arms because collisions between gas clouds compress the gas in the clouds, increasing the strength of gravity and triggering star formation.

15. The disk population of stars contains both young and old stars, all of which are made up of about 2% heavy elements. The spherical population stars are always old and have low masses. The stars in the halo are also lower in heavy elements than the disk population of stars.
16. According to the protogalactic cloud model, the spherical population of stars formed first, when the cloud that went on to form our galaxy was still blob-shaped and had little rotation. This explains why the stars on random orbits are older than the disk stars. It also explains why heavy elements make up so little of their compositions: The gas that formed them had not been through many repetitions of the star-gas-star cycle yet.
17. We think that the protogalactic cloud that formed the Milky Way was formed from several smaller protogalactic clouds because stars in different globular clusters have different heavy-element contents. The variations do not seem to depend on the distance from the center of the galaxy, as we would expect if the galaxy had formed from a single protogalactic cloud.
18. Sgr A^* is the name of the radio source we find at the center of our galaxy. The motions of the stars in this region indicate that it contains a few million solar masses of matter within a small space. Observations show that there are not nearly enough stars to account for all of the mass, so we suspect that it contains a supermassive black hole.
19. *We did not understand the true size and shape of our galaxy until NASA satellites were launched into the galactic halo, enabling us to see what the Milky Way looks like from outside.* This statement is not sensible, because NASA has not yet been able to get satellites much beyond our own solar system, let alone into the Milky Way's halo.
20. *Planets like the Earth probably didn't form around the very first stars because there were so few heavy elements back then.* This statement is sensible, because the Earth formed through the accretion of smaller, rocky objects made from heavy elements.
21. *If I could see infrared light, the galactic center would look much more impressive.* This statement is sensible, because infrared light from the galactic center can penetrate the Milky Way's disk much more easily than visible light can.
22. *Many spectacular ionization nebulae can be seen throughout the Milky Way's halo.* This statement is not sensible. Virtually no star formation is happening in the galactic halo, so there are no hot, short-lived stars there to generate ionization nebulae.
23. *The carbon in my diamond ring was once part of an interstellar dust grain.* This statement is sensible. Much of the carbon in the interstellar medium is in the form of dust grains, so the carbon in the interstellar cloud out of which the Earth formed must also have been largely in the form of dust grains.
24. *The Sun's velocity around the Milky Way tells us that most of our galaxy's dark matter lies within the solar circle.* This statement is not sensible. The Milky Way's rotation curve remains flat well beyond the orbit of the Sun, indicating that the majority of the Milky Way's mass lies beyond the Sun's orbit.
25. *We know that a black hole lies at our galaxy's center because numerous stars near it have vanished over the past several years, telling us that they've been sucked in.* This statement is nonsense. The orbital velocities of stars at the galactic center are what indicate a black hole. None of these stars has vanished from sight.
26. *If we could watch a time-lapse movie of a spiral galaxy over millions of years, we'd see many stars being born and dying within the spiral arms.* This statement is sensible. Spiral arms are bright because they contain many short-lived blue stars that shine for only a few million years.

27. *The star-gas-star cycle will keep the Milky Way looking just as bright in 100 million years as it looks now.* This statement is sensible. Star formation in the Milky Way is not proceeding so fast as to deplete the gas in the Milky Way in a time scale of only 100 million years—it will take several tens of billions of years for the Milky Way to fade.
28. *Halo stars orbit around the center of our galaxy much faster than the disk stars.* This statement is not sensible. At a given distance from the center of the Milky Way, the stars feel the same gravity and move at rougly the same speed.
29. c. the halo
30. a. because the gravitational pull of other disk stars always pulls them toward the disk
31. c. from the orbits of stars and gas clouds orbiting the galactic center at greater distances than the Sun
32. b. the halo
33. c. 70% hydrogen
34. c. infrared light
35. a. in the halo
36. c. in a spiral arm
37. c. an M star
38. a. the orbits of stars in the galactic center
39. A star made of only helium and hydrogen would have to be among the first generation of stars ever born, arising out of the primordial mix of elements that came from the Big Bang. The oldest stars we know of are about 13 billion years old—a star made of only helium and hydrogen would have to be at least this old. (No such star has ever been discovered.)
40. If one supernova can blow out all of the interstellar gas from a globular cluster, no gas remains from which subsequent generations of stars can form. Therefore, a globular cluster may consist primarily of the original gas cloud's first (and only) generation of stars. The fact that one supernova can do so much damage to a cluster's interstellar gas may explain why stars ceased forming in clusters long ago. It also explains why globular clusters are rather deficient in heavy elements. Heavy elements collect as gas is processed by multiple generations of stars. If only one generation has passed, very few heavy elements build up.
41. Because stars that are traveling along with us in the disk of the Milky Way move at a velocity relative to us of only about 20 km/s, we would reason that a star observed to be moving relative to us at a velocity of 200 km/s was *not* traveling with us in the disk but was part of some other component of the Milky Way, likely the halo. The orbits of halo stars are not concentrated in the flattened disk but are distributed more like a sphere. So a halo star flying through the disk would appear to us to have the rotational speed of the disk, 200 km/s.
42. This question involves independent research regarding the opinions of humans about the nature of the Milky Way.
43. In 100 billion years, the Milky Way might resemble an elliptical galaxy. The gas supply for star formation in the Milky Way will have been exhausted. No new stars will have formed for many billions of years. The only stars burning hydrogen in their core will be the lowest mass stars of M spectral type. More massive stars will have either obliterated themselves in supernova explosions or become white dwarfs, cooling embers in outer space. The oldest M stars will

also now have become helium white dwarfs. Brown dwarfs, planets, comets, asteroids, and the like are simply cooling off, radiating any residual heat they may have to the universe. So in 100 billion years, the Milky Way will be much dimmer (at all wavelengths). Its optical light will be much redder. The bluish color of the spiral arms and disk will be gone; gas and dust may not be completely gone, but there probably won't be dust lanes. The question of whether the Galaxy will still have spiral arms and a disk is interesting—the pattern and the disk would have to survive multiple galaxy collisions between the Galaxy and its Local Group neighbors—in 100 billion years, the Milky Way Galaxy and the Andromeda Galaxy alone will have collided, perhaps 10 to 20 times by then!

44. From Figure 19.22, the two stars that reach the highest orbital speeds must be SO-16 and SO-4. Since angular momentum is conserved, the quantity ($r^x \times v$) is the same at the nearest and farthest separations for each orbit. So the stars that eventually achieve the fastest speeds are the ones with the orbits with the largest ellipticities . . . or it's the stars that get the closest, with the smallest average velocities. This question is harder than it looks.

45. The angular momentum is what drove the formation of a disk, so the Milky Way without ANY angular momentum doesn't get around to forming a disk, and perhaps it would look more spheroidal—all bulge and halo, no disk.

46. The student is asked to sketch the locations of interstellar clouds (molecular and atomic), and locations of hot bubbles in the Milky Way based on Figure 19.12, and explain why. The molecular gas is coldest and therefore doesn't puff out as much as the warmer atomic gas. The hot bubbles are created by coalitions of hot stars blowing up, so the hot gas is not smoothly distributed.

47. To find the mass within 160,000 light-years of the center of the galaxy, we will use the orbital velocity law:

$$M_r = \frac{rv^2}{G}$$

To use this law, we need to convert the radius and the speed of the Large Magellanic Cloud into meters and meters per second. We are told that the orbital radius is 160,000 light-years. Since Appendix A tells us that 1 light-year is 9.46×10^{15} m, this is the same as 1.51×10^{21} m. The speed is given as 300 kilometers per second, which we can easily convert to 3.0×10^5 m/s. Plugging into the orbital velocity law, we learn that the approximate mass of the galaxy within 160,00 light-years of the center is 2.0×10^{42} kg, or about 1 trillion times the mass of the Sun.

48. To find the mass of the central black hole, we will use the orbital velocity law:

$$M_r = \frac{rv^2}{G}$$

To use this law, we need to convert the radius and the speed of the star into meters and meters per second. The speed we find to be 1.0×10^6 m/s. The radius is a bit more work, since it is given as 20 light-days. There should be about 365 light-days in a light-year, so we can convert and find that the orbital radius is also 2.5×10^{-2} light-years. A light-year is 9.46×10^{15} m, so the orbital radius is 5.2×10^{14} m. We therefore find that the mass of the central black hole is 7.8×10^{36} kg, or about 3,900,000 times the mass of the Sun.

49. To find the mass of the globular cluster, we will use the orbital velocity law:

$$M_r = \frac{rv^2}{G}$$

To use this law, we need to convert the radius and the speed of the star into meters and meters per second. The speed is about 10,000 kilometers per second. Recalling that 1 light-year is 9.46×10^{15} m the 50-light-year radius is equivalent to 4.7×10^{17} m. Plugging in the numbers, we find that the mass of the globular cluster is 7.1×10^{35} kg or 350,000 times the mass of the Sun.

50. To find the mass of Saturn, we will use the orbital velocity law:

$$M_r = \frac{rv^2}{G}$$

To use this law, we need to convert the radius and the speed of the particle into meters and meters per second. The speed is 2.38×10^4 m/s and the radius is 6.70×10^7 m. Putting in those values, we find that the mass of Saturn is about 5.69×10^{26} kg. This mass agrees with the value in Appendix E.

51. The formula that tells us the mass of a cloud in which pressure balances gravity was given in Mathematical Insight 16.1 as:

$$M_{\text{balance}} = (18M_{\text{Sun}})\sqrt{\frac{T^3}{n}}$$

where T is the temperature and n is the number density. For a hot ionized gas cloud, the temperature is 10^6 Kelvins and the density is 0.01 particle per cubic centimeter. Putting in these values, we find that the mass of the cloud in which gravity and pressure would exactly balance is 180 billion times the mass of the Sun. This cloud would be roughly the same size as the galaxy itself, so we would be unlikely to see a cloud of hot ionized gas that was large enough to collapse. In order to collapse, such hot gas must cool.

52. The formula that tells us the mass of a cloud in which pressure balances gravity was given in Mathematical Insight 16.1 as:

$$M_{\text{balance}} = (18M_{\text{Sun}})\sqrt{\frac{T^3}{n}}$$

where T is the temperature and n is the number density. For a warm atomic gas cloud, the temperature is 10^4 Kelvins and the density is 1 particle per cubic centimeter. Putting in these values, we find that the mass of the cloud in which gravity and pressure would exactly balance is 18 million times the mass of the Sun. This mass is very large for an interstellar cloud, so in order for a less massive cloud to collapse, it would first have to cool down further.

53. The formula that tells us the mass of a cloud in which pressure balances gravity was given in Mathematical Insight 16.1 as:

$$M_{\text{balance}} = (18M_{\text{Sun}})\sqrt{\frac{T^3}{n}}$$

where T is the temperature and n is the number density. For a cool atomic cloud, the temperature is 10^2 Kelvins and the density is 100 particles per cubic centimeter. Putting in these values, we find that the mass of the cloud in which gravity and pressure would exactly balance is 1,800 times the mass of the Sun. This is a very reasonable size for an interstellar cloud, so we would not be surprised to see larger clouds that would be unstable and collapse under their own gravity. Since most stars are much less massive than the Sun, around a few tenths of the Sun's mass, this means that a cloud just a bit more massive than this critical size might form several thousand stars when it collapsed.

54. To find the velocity of the supernova gas, we first solve the formula for kinetic energy for the velocity:

$$v = \sqrt{\frac{2E_{\text{kinetic}}}{M}}$$

A typical supernova ejects about $10M_{\text{Sun}}$ of gas. Converting, we see that this is about 2×10^{31} kg. The debris has about 10^{44} joules of energy, so plugging in these numbers we find that the debris should move at 3.2×10^6 m/s. This is 3,200 kilometers per second. Orbital speeds of stars in the galaxy are typically about a tenth of this speed, so it seems unlikely that the galaxy's gravity would be able to hold on to the debris if the debris were not slowed by the interstellar medium.

Chapter 20. Galaxies and the Foundation of Modern Cosmology

This chapter describes the observed properties of galaxies and their implications. We first examine the morphology and color of galaxies and then explain how their distances are measured. Rather than postpone the implications of these distance measurements to a later chapter, we continue on and explain Hubble's law, how it tells us the age of the universe, and how the age of the universe limits how far we can see. By the end of that section, students have learned about the size, scope, age, and expansion history of the observable universe, preparing them for the sections that follow on galaxy evolution and quasars.

As always, when you prepare to teach this chapter, be sure you are familiar with the relevant media resources (see the complete, section-by-section resource grid in Appendix 3 of this Instructor's Guide) and the online quizzes and other study resources available on the Mastering Astronomy Web site.

What's New in the Fourth Edition That Will Affect My Lecture Notes?

Those who have taught from previous editions of *The Cosmic Perspective* should be aware of the following organizational or pedagogical changes to this chapter (i.e., changes that will influence the way you teach) from the Third edition:

- We have consolidated the first two sections of the old Chapter 20 into the new Section 20.1.
- We have reorganized the material on galactic distances and cosmic ages in old Sections 20.3 and 20.4 into two sections: 20.2, "Measuring Galactic Distances," and 20.3, "Hubble's Law." The content and level, however, are the same as in the previous edition.
- We now discuss Hubble's system of galaxy classification as part of the main text flow, including a new figure showing a pictorial tuning fork diagram.
- We now quote Hubble's constant in units of *kilometers per second per million light-years*, rather than per megaparsec. This maintains consistency with our preference for light-years over parsecs as distance units throughout the book, and is now permissible (in our opinion) because many news sources (including the *New York Times*) have converted to these units in the past couple of years. Thus, it can no longer be argued that km/s/Mpc are the units that students are more likely to encounter in the news.

Teaching Notes (By Section)

Section 20.1 Islands of Stars

This section introduces students to the subject of galaxies by way of the Hubble Deep Field.

- In our experience, the Hubble Deep Field is an excellent pedagogical tool for students of all ages. People love the image and often start classifying the galaxies in it into systems of their own. Consider having students do some sort of classification exercise with this image.
- The even newer Hubble Ultra Deep Field goes about four times as deep, but we chose not to use it in this chapter for a simple pedagogical reason: To understand how *tiny* a piece of the sky they are seeing, students need to be able to relate the photo to a familiar pattern of stars in the night sky, and the Big Dipper (from which the HDF is drawn) is far better known to students than Fornax (constellation of the HUDF). But we don't ignore the HUDF: It is shown on page 1 of our book as the chapter opener for Chapter 1.
- In this book we do not delve very deeply into galaxy classification because, unlike stellar classification, which prepares students to understand stellar evolution, galaxy classification does not prepare students to understand galaxy evolution.
- Lenticular (SO) galaxies defy easy pigeonholing into the categories "spiral" and "elliptical" because they share some characteristics with each type. We include them with spirals in order to keep the discussion compact, but since discussion of their formation is beyond the scope of our course, the term is not boldface.
- Elliptical galaxies have an undeserved reputation for having no interstellar medium. In fact, elliptical galaxies tend to be filled with hot gas that radiates profusely in the X-ray band but is invisible in the optical band. (Furthermore, elliptical galaxies can have some tiny amounts of dust, molecular, and neutral gas, so saying they have "no" ISM is simply incorrect.)

Section 20.2 Measuring Galactic Distances

In order to know more about galaxies than just shape and color, we need to know their distances. This section describes the chain of distance measurement techniques that gives us the distances of galaxies, and how we then use distances to infer ages.

- We have had success in class likening the Hubble constant to the scale of a map. Redshift measurements accurately give us the relative distances of galaxies, enabling us to make a three-dimensional map of the universe. But until we know the scale of the map, we don't know the absolute distances of all the galaxies. However, just a handful of accurate measurements of absolute distances tells us the scale of the map, which provides absolute distance measurements for all the rest of the galaxies.
- Regarding white dwarf supernovae as distance indicators: Fusion in the incredibly hot, dense environment inside an exploding white dwarf creates radioactive nickel-56. After the explosion, the luminosity of the supernova's expanding cloud of gas comes from radiation emitted as this nickel-56 decays into iron. First, the nickel-56 (28 protons, 28 neutrons) decays into cobalt-56 (27 protons, 29 neutrons) with a half-life of 6.1 days. The cobalt-56 then decays into iron-56 (26 protons, 30 neutrons) with a half-life of 77 days. Because every white dwarf explosion produces the same amount of nickel-56—about 1 solar mass of it—the light curves of all white dwarf supernovae are nearly identical.
- Emphasizing the idea of *lookback time* usually clears up a lot of questions that students have about what is meant when they hear distances quoted in news articles about extragalactic astronomy.
- FYI: The Tully-Fisher relation is not unique. An analogous relation (Faber-Jackson, or arguably, the fundamental plane for ellipticals) exists between the luminosity of an elliptical galaxy and the characteristic velocities of its stars.

Section 20.3 Hubble's Law

Hubble's law has deep implications for the ages of galaxies. In this section we introduce the concept of lookback time and its relationship to a galaxy's redshift. The stage is then set for the discussion of galaxy evolution in Chapter 21.

Answers/Discussion Points for Think About It Questions

The Think About It questions are not numbered in the book, so we list them in the order in which they appear, keyed by section number.

Section 20.1

- (p. 618) The purpose of this classification question is just to get students to look at the Hubble Deep Field and consider what it means.

Section 20.3

- (p. 632) A dot 9 cm from Dot B would move at 9 cm/3 s = 3 cm/s according to scientists on Dot B.
- (p. 633) Distances in an expanding universe where the light travel times *are* significant, and the galaxy that we see *is* no longer at the same distance from us as it was when the light left it. Distances therefore must be defined precisely in order that everyone knows what you're talking about. What is the distance? Is it the distance the galaxy was when the light was emitted? The distance the galaxy is right now, today? The distance the light traveled from there to here?

Solutions to End-of-Chapter Problems (Chapter 20)

1. We must understand the evolution of the universe, or the history of the expansion of the universe, in order to understand the relationship between the age of the universe and the redshifts of the distant galaxies under study. As we look farther away in distance, we are looking further back in time, but to quantify just how far back in time we are seeing, we have to understand the overall evolution of the universe—its expansion history. Furthermore, understanding the expansion history is critical to knowing the density of matter and of galaxies as that changes with cosmic time.
2. The three major types of galaxies are spirals, ellipticals, and irregulars. Spiral galaxies, like our own, are flat white disks with yellowish bulges in the centers and contain cool gas interspersed with hotter ionized gas. Elliptical galaxies are redder, more rounded, and often football-shaped. They contain very little cool gas and often contain hot, ionized gas. Irregulars do not fit either of the other categories.
3. Normal spiral galaxies have disk and spheroidal components. Barred spirals also have a straight bar of stars cutting across their centers, with spiral arms curling away from the ends of the bar. Galaxies that have a disk and a spheroidal component but lack spiral arms are called "lenticular galaxies."
4. The disk component of spiral galaxies is a flat disk in which stars follow orderly, nearly circular orbits around the galactic center. The disk contains cool gas and active star formation. The spheroidal component of spiral galaxies has a round shape with stars orbiting at many different inclinations.
5. The major difference between spiral and elliptical galaxies is that elliptical galaxies lack a significant disk component, although both types have the spheroidal component. Like the halos of spiral galaxies, elliptical galaxies lack cool gas and so do not have much star formation. As a result, we see few hot, young stars in these galaxies since such stars are born rarely and die quickly.
6. Elliptical galaxies are much more common in large galaxy clusters than they are among isolated galaxies. Half of the large galaxies in large clusters are ellipticals, while among isolated galaxies they make up only about 15%.
7. A standard candle is a light source of known, "standard" luminosity. Since we know the luminosity, we can use the apparent brightness and the inverse square law for light to determine the object's distance.
8. For planets, we can use radar ranging to determine distances. For the closest stars, we can use parallax to determine the distances directly. For more distant stars within our galaxy, we use main-sequence fitting, a technique that uses the known brightness of the various classes of stars to turn main-sequence stars into standard candles. This does not work to find the distance to other galaxies since most main-sequence stars are too faint to be seen in other galaxies. In this case, we use Cepheid variables, variable stars whose luminosities are related to the period of their brightness variations. These stars have been very important since they have allowed us to measure the distances to nearby galaxies and played a key role in Edwin Hubble's discoveries.

   Even Cepheid variables cannot tell us the distance to most galaxies, however, because we cannot see them from so far away. White dwarf supernovae are much brighter and all have nearly the same peak luminosity. Although such supernovae are rare, they are bright enough to see to the far reaches of the universe, making them an extremely powerful tool in determining distances.

9. Hubble was able to see Cepheid variables in the Andromeda Galaxy. When he determined their luminosities from their periods, he was able to find the distance to the Andromeda Galaxy. Since the distance put it far beyond the reaches of our own galaxy, he showed that it must be a separate galaxy.
10. Hubble's law states that the speed at which more distant galaxies move away from us is proportional to their distance. The proportionality factor is called "Hubble's constant." So when we say that Hubble's constant is between 20 and 24 kilometers per second per million light-years, we mean that for every million-light-years-distance away from us, a galaxy's speed away from us increases by 20 to 24 kilometers per second.
11. The Cosmological Principle states that the matter in the universe is uniformly distributed without a center or an edge. This idea is important to our understanding of the universe because it presumes that we do not live in a special place in the universe, and thus allows us to make models of how the universe as a whole behaves.
12. The expansion of the universe is like the expansion of the surface of a balloon because both expand without having edges or centers. In this analogy, we can see how Hubble's constant can tell us the age of the universe. If a scientist on the balloon saw another scientist 6 centimeters away moving away from her at 2 centimeters per second, she would conclude that the scientist had been in contact with her 3 seconds ago. She would also find a Hubble constant of 2 cm/s/6 cm or 1/3 (1/s), So the inverse of her Hubble constant would give her the time when all of the scientists were in one spot. This time is also the expansion time or the age of the surface of the balloon, her "universe."
13. The lookback time to a distant galaxy is the difference between the present age of the universe and the age of the universe when the light left the galaxy. The lookback time is less ambiguous than the distance to a galaxy because in the time that the light has been traveling to us, the universe has expanded, changing its distance, but the time that the photons have been traveling is definite.
14. The cosmological horizon is the limit of the observable universe. Light from anything from beyond this limit cannot have reached us over the age of the universe because the light can only travel 300,000 km/second, and the universe has a limited age. The limit is easier to think of as a boundary in time.
15. The cosmological redshift is the stretching of photons due to the expansion of the universe. A Doppler shift is caused by motion toward or away from us. The galaxies are not projectiles in a static universe as implied by the interpretation of redshift as a Doppler shift; they are being carried along while spacetime itself is expanding. So while redshift and the Doppler shift are sometimes used interchangeably when cosmic distances are small, when they get large, it is important to remember that the redshifts are cosmological in origin, not due to galaxies physically flying about in space.
16. *If you want to find a lot of elliptical galaxies, you'll have better luck looking in clusters of galaxies than elsewhere in the universe.* The statement makes sense because while there are plenty of ellipticals outside of clusters, the fraction of galaxies that are ellipticals inside of clusters is much higher.
17. *Cepheid variables make good standard candles because they all have exactly the same luminosity.* This statement does not make sense. Cepheids are good standard candles because their light curve pattern tells you what their luminosity is.

18. *If the standard candles you are using are less luminous than you think they are, then the distances you determine from them will be too small.* This statement does not make sense because if the object is less luminous than you think it is, then you will think it is farther away than it really is. A more luminous object must be farther away to get the same observed brightness.
19. *Galaxy A is moving away from me twice as fast as galaxy B. That probably means it's twice as far away.* This statement makes sense, since Hubble's Law says that a galaxy's distance is proportional to its recession velocity.
20. *After measuring the galaxy's redshift, I used Hubble's law to estimate its distance.* This statement makes sense since Hubble's Law gives the relationship between redshift and distance, as calibrated by various standard candles.
21. *The center of the universe is more crowded with galaxies than any other place in the universe.* This statemtent makes no sense. There is no "center of the universe," since all parts of the universe are expanding away from all other parts.
22. *The lookback time to the Andromeda Galaxy is about 2.5 million light-years.* This statement does not make sense. A lookback time is a time, not a distance, so the appropriate unit would be 2.5 million years. (But the light travel distance to Andromeda is about 2.5 million light-years.)
23. *I'd love to live in one of the galaxies near our cosmological horizon, because I want to see the black void into which the universe is expanding.* This statement does not make sense because if I were living in one of those galaxies, my own cosmological horizon would center on me. It also does not make sense because the cosmological horizon is a boundary defined by light travel time, not in space. The universe "beyond the horizon" is the universe before the Big Bang. Finally, the universe is expanding, but it could be an infinite universe that is expanding, in which case it is easier to understand that it is not expanding into a dark void. (If it is not infinite, it's a little harder to visualize.)
24. *If someone in a galaxy with a lookback time of 4.6 billion years had a superpowerful telescope, they could see our solar system in the process of its formation.* This statement makes sense because the light travel times between two galaxies should be nearly symmetric, that is, the same whether light travels from galaxy 1 to galaxy 2 or from galaxy 2 to galaxy 1. So a galaxy 4.6 billion light-years away would see "us" as we were 4.6 billion years ago.
25. *We can't see galaxies beyond the cosmological horizon because they are moving away from us faster than the speed of light.* This statement does not make sense. We can't see beyond the cosmological horizon because that boundary is defined by the time light has had since the Big Bang to travel to us. [Note to the instructor: The expansion "speeds" of galaxies well inside the cosmological horizon can exceed the speed of light, but as you might guess, those "speeds" aren't proper relativistic velocities with respect to an inertial rest frame.]
26. a. a galaxy in the Local Group
27. b. a large elliptical galaxy
28. b. a large elliptical galaxy
29. c. a few hundred
30. c. 10 times as far as the Hyades's distance
31. c. a white dwarf supernova
32. c. visible light

33. c. Observers in all galaxies observe a similar phenomenon.
34. a. twice as large as its current value
35. c. 28 billion years
36. The reason is similar to the reason a pollster might want to query a sample of typical voters, or typical consumers, or typical children: One wants statistics that tells one something about people or objects that are not in the statistical sample.
37. A large spiral galaxy is more likely to host a massive star supernova, since an elliptical galaxy has very few if any massive stars remaining.
38. Classification of galaxies in the Hubble scheme is not exact! Even the professional astronomers disagree about the classification of a given galaxy, often by only one class or so. The grader here should look for a sensible attention to features and the student's discussion and less on a "right" answer. The classification system is defined only qualitatively in the text.
    a. NGC4594 is an Sab because it's an edge-on spiral with a large bulge. It would be difficult to say whether it has a bar or not because it is edge-on.
    b. NGC6744 looks like a face-on SBab or an Sab. It has a bar (might be faint), and its arms seem tightly wrapped.
    c. NGC4414 looks like a face-on Sa or Sb , but the arms are not as loosely wrapped as an Sc. There doesn't seem to be an obvious bar.
    d. NGC1300 is identified in the caption as a barred spiral. It's an SBa or SBb, given how tightly its spiral arms are wrapped.
    e. M87 is an elliptical, pretty round so it's an EO probably.
39. For an oncoming car at night, one might use several means to judge its distance. (1) The brightness of its headlights, (2) the separation of its headlights, and (3) the loudness of the sound of the car. The two techniques that are most like standard candles are the brightness and the loudness. The brighter and louder the car, the closer it is.
40. From Figure 15.12, one can read off the intrinsic luminosity of a Cepheid, given a certain period. An approximate reading shows that a Cepheid with a period of 8 days has a luminosity of 2,000 solar luminosities and that a Cepheid with a period of 35 days has a luminosity of a little over 10,000 solar luminosities.
41. Galaxy light is dominated by the light from stars, which is the most prominent in the visible. But the light from the most distant galaxies has been shifted to longer wavelengths, so the most distant galaxies we know about are brightest in the infrared, not in the visible.
42. The surface of a balloon is a good analogy for the expanding universe because the surface itself has no center (ignoring the entry position), and as the balloon expands, every piece of the ideal balloon surface stretches, not just one piece. The analogy is limited because a real balloon is not perfectly symmetric, and can stretch in goofy ways—the area around the input valve stretches less than the rest of the balloon. It's also limited because the surface is two-dimensional, while the universe has three space dimensions. It's also limited because the actual universe might be infinite, while the surface of the balloon is, of course, finite.
43. The thing that is moving the fastest (the most distance in the least amount of time) will have the shallowest slope on a diagram where distance is on the *x*-axis and time is on the *y*-axis. Since nothing can travel faster than a photon, the path of the photons will always have less steep paths than a galaxy on a spacetime diagram.

44. Again, the exercise of counting galaxies in this picture will vary with individual students. It will be interesting to the instructor to determine *how* the students counted the galaxies. Did they estimate from counting a few galaxies and estimating a mean density? Did they try to count them all?

45. Distances to star clusters. The distance to the Pleiades is 2.75 farther away than the Hyades, because $2.75^2 = 7.5$. 7.5 is the factor brighter the Hyades main sequence is over the Pleiades main sequence, from Figure 20.11. If the Hyades is 151 light-years away, then the Pleiades is 2.75×151 light-years or 415 light-years away.

46. We can solve the luminosity-distance formula for the distance d:

$$d = \sqrt{\frac{\text{luminosity}}{4\ \pi \times \text{apparent brightness}}}$$

Substituting the given values, we find:

- Cepheid 1: $d = 5.8 \times 10^{23} = 6.1 \times 10^7$ light-years
- Cepheid 2: $d = 5.0 \times 10^{23} = 5.3 \times 10^7$ light-years
- Cepheid 3: $d = 4.8 \times 10^{23} = 5.1 \times 10^7$ light-years

The results do not perfectly agree because of observational uncertainties. Taking the average, the distance is 5.6×10^7 light-years with a spread of ±4.5 million light-years, or a spread of a little less than 10% of the total distance to the galaxy M 100.

47. From Mathematical Insight 20.1, we know that distance, luminosity, and apparent brightness are related via:

$$\text{distance} = \sqrt{\frac{\text{luminosity}}{4\ \pi \times (\text{apparent brightness})}}$$

or, equivalently:

$$\text{apparent brightness} = \frac{\text{luminosity}}{4\ \pi \times (\text{distance})^2}$$

Knowing this, there are two ways to approach this problem. The first is to find the minimum apparent brightness that the Hubble Space Telescope can see. We would do this by using the luminosity of and distance to the Cepheids and the second equation. This apparent brightness can be plugged into the first equation to get the maximum distance to a fading white dwarf supernova. However, we can cut down on the calculation with a little bit of algebra by plugging the second equation into the first, remembering to keep track of which luminosities and distances are for the white dwarf supernova and which are for the Cepheid. Following the second approach (which cuts down on the errors in calculation, not to mention all of the conversions), we find:

$$\text{distance}_{\text{white dwarf supernova}} = \sqrt{\frac{\text{luminosity}_{\text{white dwarf supernova}}}{4\ \pi \times \left(\dfrac{\text{luminosity}_{\text{Cepheid}}}{4\ \pi \times (\text{distance}_{\text{Cepheid}})^2}\right)}}$$

$$= \text{distance}_{\text{Cepheid}} \sqrt{\frac{\text{luminosity}_{\text{white dwarf supernova}}}{\text{luminosity}_{\text{Cepheid}}}}$$

So using the values given, the maximum distance at which Hubble can see a white dwarf supernova is 10 billion light-years. This is about $\frac{2}{3}$ the size of the observable universe.

48. From the data we have provided for each galaxy, we can make a table of derived values. The lab (rest) wavelength is 656.3 nanometers for the hydrogen emission line of interest here. Note that the redshift z we compute is equivalent to the derived recession velocity divided by the speed of light. The speed of light is 300,000 km/sec. The Hubble constant we assume here is 24 km/sec/Mega light-year, so the distance we quote is in Mega-light-years (Mlt-yr).

	Wavelength (nm)	**a.** $z = \frac{\lambda_{obs} - \lambda_{lab}}{\lambda_{lab}}$	**b. v = cz**	**c.** $d = \frac{y}{H_0}$
Galaxy 1	659.6	0.0050	1,500 km/s	62.5 Mlt-yr
Galaxy 2	664.7	0.0128	3,840 km/s	160 Mlt-yr
Galaxy 3	679.2	0.0349	10,500 km/s	437.5 Mlt-yr

49. We estimate the age of the universe to be $1/H_0$. Before we can use the value given, we need to convert it:

$$H_0 = 33\frac{\text{km/s}}{10^6 \text{ ly}} = 33\frac{\text{km/s}}{10^6(3\times10^5 \text{ km/s})\times 1 \text{ yr}} = 1.1\times10^{-10} \frac{\text{km/s}}{\text{km/s}\times 1 \text{ yr}}$$

Inverting this result leads to an estimated age of the universe of 9 billion years.

50. The solid line in Figure 20.20 has a slope of about 500 kilometers per second per million parsecs. (The original y-axis for VELOCITY is labeled "km," but it should be "km/s.") Converting this to units we are more used to:

$$500 \frac{\text{km/s}}{10^6 \text{ pc}} \times \frac{10^6 \text{ pc}}{3.26\times10^6 \text{ ly}} = 150 \frac{\text{km/s}}{10^6 \text{ ly}}$$

We can use this to find the age of the universe he would have estimated. First, we convert:

$$H_0 = 150\frac{\text{km/s}}{10^6 \text{ ly}} = 150\frac{\text{km/s}}{10^6(3\times10^5 \text{ km/s})\times 1 \text{ yr}} = 5\times10^{-10} \frac{\text{km/s}}{\text{km/s}\times 1 \text{ yr}}$$

The age of the universe is $1/H_0$, so Hubble's original measurement predicts that the universe is 2 billion years old.

51. From Mathematical Insight 20.5, we know that the redshift of a galaxy, z, is related to the observed and rest wavelengths by the equation:

$$1 + z = \frac{\lambda_{\text{observed}}}{\lambda_{\text{rest}}}$$

We can solve for the observed wavelength:

$$\lambda_{\text{observed}} = (1 + z)\lambda_{\text{rest}}$$

If $z = 7$, then we find that the observed wavelength of the light we see should be eight times the original wavelength.

52. Mathematical Insight 20.5 tells us that the change in the spacing between galaxies that has occurred during the time it took for light to reach us is $1 + z$. If the redshift, z, is 1.7, then the galaxies were 2.7 times closer together when the light left. Density goes like

the inverse cube of this spacing: Making the average spacing twice as large makes the density one-eighth as much. So if the spacing was 2.7 times less in the past, then the density was 20 times larger in the past.

53. Lookback time and ages of galaxies. If the lookback time of the galaxy is 10 billion years, then the maximum age it could have is the age of the universe—10 billion years, or 14 billion years—10 billion years = 4 billion years, using our best estimate for the age of the universe.
54. Einstein's education and culture affected the way he thought the universe should be. This question challenges the student to think about the universe like a scientist-philosopher of the nineteenth century might. Does a universe with a finite age have philosophical implications? It might to your students.
55. The change of the Milky Way from a special galaxy to just one of billions was similar to the Copernican revolution in the sense that it moved the Earth and the Sun, and now the Galaxy, ever further from a special central place in the universe. *[Note to instructor: Remember that just because the Earth occupied a special place in the Ptolemaic universe, it wasn't necessarily the "best" place in the universe—the Earth was considered to occupy the lowliest position in the universe, the place where mere matter, the garbage substance of the universe, found its home. So the shift of the Milky Way to a less central position was much less of a theological or philosophical challenge than moving the Earth.]*

Chapter 21. Galaxy Evolution

This chapter surveys what we currently know about how galaxies evolve. It brings together material on distant galaxies, starburst galaxies, and active galaxies and explains how each of these manifestations might represent a different chapter in the life of a "normal" galaxy.

As always, when you prepare to teach this chapter, be sure you are familiar with the relevant media resources (see the complete, section-by-section resource grid in Appendix 3 of this Instructor's Guide) and the online quizzes and other study resources available on the Mastering Astronomy Web site.

What's New in the Fourth Edition That Will Affect My Lecture Notes?

As everywhere in the book, we have edited to improve the text flow, improved art pieces, and added new illustrations. In addition, those who have taught from previous editions of *The Cosmic Perspective* should be aware of the following organizational or pedagogical changes to this chapter (i.e., changes that will influence the way you teach) from the Third edition:

- Our discussion of what makes some galaxies spiral and others elliptical contains some new artwork. Figures 21.3a and 21.3b illustrate the roles of angular momentum and primordial density in determining the differences between spiral and elliptical galaxies.

- In order to connect the topic of active galactic nuclei more closely to galaxy evolution, we highlight the new evidence relating the mass of a central black hole to the mass of its spheroidal component, and we include a new figure developed for this purpose.
- The chapter has been refocused to address the questions about why some galaxies are spiral and some are not and what the roles of starbursts and active galactic nuclei are in galaxy evolution. Material about the Milky Way's fossil record was moved to Section 19.3 on the Milky Way. All of the other material in the previous Chapter 21 is still here, but repartitioned.
- Specifically, the former Chapter 21 had six sections; the new one has three. Section 21.1, "Looking Back Through Time," now includes the Learning Goal "How did galaxies form?" The material in former Sections 21.3, "Why do Galaxies Differ?," and 21.4, "Starbursts," is discussed in the new Section 21.2 called "The Lives of Galaxies." Finally, Section 21.3, "Quasars and Other Active Galactic Nuclei," now includes the Learning Goal "How do quasars let us study the gas between the galaxies?" (formerly covered in Section 21.6).

Teaching Notes (By Section)

Section 21.1 Looking Back Through Time

This section introduces the idea of galaxy evolution by presenting "family albums" of various types of galaxies at different lookback times. We emphasize that astronomers do not yet know many of the details of galaxy evolution, so the story presented in this chapter is incomplete.

- Students are generally captivated by the idea that we can look back in time by looking deeply into space. When explaining this concept, we find it helpful to speak about observing "old light from young galaxies." This approach counteracts the tendency for students to think of these galaxies as "very old" because we are seeing them as they were "billions of years ago." They are not ancestors of the Milky Way but rather its cousins, and by looking at their "baby pictures" we are seeing what our middle-aged Milky Way was like during its childhood.
- We have found that the term *active galaxy* can confuse students because they mistake it for a galaxy type different from the ellipticals and spirals they already know about. Because active galaxies, aside from their luminous nuclei, are just like other elliptical and spiral galaxies, we speak only about *active galactic nuclei* in order to keep students focused on the part of an active galaxy that makes it active.

This section also discusses what is currently known about the process of galaxy formation, highlighting the inferences we draw from theoretical modeling, the characteristics of galaxy disks and spheroids, and detailed observations of our own Milky Way.

Section 21.2 The Lives of Galaxies

This section presents various ideas about how galaxies evolved into their present form, including a discussion of starburst galaxies. We motivate the material by posing the question, "Why do spiral galaxies have gas-rich disks while elliptical galaxies do not?" In addressing this question, we explain how both the characteristics of the original protogalactic cloud and environmental effects such as collisions with other galaxies can influence the evolution of a galaxy.

Stars are currently forming in starburst galaxies at an unsustainably high rate, so we know that the starburst phenomenon is a transient episode in galaxy evolution. The effects of a starburst on the subsequent evolution of a galaxy can be profound because the many supernovae buffeting the interstellar medium can drive a galactic wind that carries gas away from the galaxy.

Section 21.3 Quasars and Other Active Galactic Nuclei

This section covers active galactic nuclei and the evidence for supermassive black holes at their cores. Many introductory texts treat active galactic nuclei as a sort of astronomical sideshow. However, current astronomical research is showing that the nuclei of virtually all galaxies might pass through an active stage at least once. Because nuclear activity would then be a common stage in the evolution of galaxies, we believe that quasars and other such beasts fall naturally into a discussion of galaxy evolution. Furthermore, the great prevalence of quasars in the early universe is telling us something very important about galaxy evolution, even if we don't yet know what it's saying.

- In order to keep students focused on the physics of active galactic nuclei, we have kept the classification and nomenclature of active galaxies to a minimum. We call all the most luminous active galactic nuclei *quasars*, even though some astronomers separate these into *QSOs*, which are radio quiet, and quasars, which are radio loud. We also avoid the terms *Seyfert galaxy*, which describes lower-luminosity active galactic nuclei, and *BL Lac object*, which describes active galactic nuclei with very weak emission lines. In all these objects, the underlying engines are likely to be quite similar, and a proliferation of terminology tends to distract students from that fact.
- We elected not to allocate much space to discussing whether quasars really are at cosmological distances because this issue has been settled to the satisfaction of all but a few individuals in the astronomical community. In our opinion, discussing whether black holes really lie at the cores of quasars is a much better use of the limited time available in an astronomy course.
- We link active galactic nuclei to galaxy evolution through the bulge mass–black hole mass relation.

We complete our discussion of galaxy evolution by describing how we can probe what's going on in protogalactic clouds by analyzing the absorption lines in the spectra of quasars.

Answers/Discussion Points for Think About It Questions

The Think About It questions are not numbered in the book, so we list them in the order in which they appear, keyed by section number.

Section 21.1

- (p. 641) Compared to the 14-billion-year lifetime of the universe, 20 million years is very short. Galaxies that close haven't had much time to change since their light left, so we are seeing them essentially as they are today.

Section 21.2

- (p. 648) Because star bursts in dwarf galaxies can eject heavy elements, they have fewer of these elements to incorporate into new stars.

Solutions to End-of-Chapter Problems (Chapter 21)

1. Galaxy evolution is the study of how galaxies form and develop in our expanding universe. Telescopic observations allow us to observe the history of galaxies since the farther we look into the universe, the further we can see back in time. Theoretical modeling helps us study galaxy formation because we cannot see the galaxies before the first stars formed, and the time scales over which galaxies change and interact with each other are much longer than an astronomer's lifetime. Theoretical modeling allows us to test our ideas about how galaxies might have formed and changed over time as they grew more massive (or not) and interacted with other galaxies.
2. The basic assumptions used for galaxy formation are that the universe was initially filled more or less uniformly with hydrogen and helium gas and that some regions of this gas were slightly denser than others. The extra density in some areas increased the gravitational pull, slowing and then reversing the expansion that occurred due to the cosmological expansion, forming protogalactic gas clouds. The gas cloud begins to form stars, but the star formation does not use up all the gas. The angular momentum of the original gas cloud causes the gas cloud to spin up and flatten out as it collapses, gas particles collide with themselves, and the gas cools, thereby forming the disk of a spiral galaxy out of the leftover gas. (A system of stars and no gas would not form a disk.)
3. There are two possible differences that could lead to an elliptical galaxy forming rather than a spiral:
   - Protogalactic spin—Fast spinning clouds tend to form a disk as they collapse, leading to a spiral galaxy. Clouds with little or no spin may not form a disk, causing them to form elliptical galaxies.
   - Protogalactic density—Denser clouds collapse faster, causing stars to form more quickly. If the star formation is sufficiently rapid, the gas will never have a chance to settle into a disk, making the galaxy an elliptical rather than a spiral.
4. When two spiral galaxies collide, tremendous tidal forces tear apart the two disks, randomizing orbits of their stars. Some of the gas sinks to the center of the collision and rapidly forms new stars. Eventually, supernovae and stellar winds blow away the rest of the gas. When things finally settle down, a new elliptical galaxy is formed with little gas left over and stars with randomized orbits. Evidence that suggests that this scenario has occurred includes the fact that elliptical galaxies dominate the population of galaxies in the cores of clusters where collisions are most likely to occur. Additionally, some elliptical galaxies include stars and gas clouds with orbits that indicate that they are leftover pieces of galaxies that merged in a past collision.
5. We expect collisions between galaxies to be relatively common (while star-star collisions are rare) because the typical distance between galaxies is comparable in scale to the size of the galaxies themselves. However, the typical distance between stars is millions of times larger than the size of the stars, so collisions betweens stars are relatively rare. Galaxy collisions should have been even more common in the past then they are today because the density of galaxies was larger. Since, approximately, a similar number of galaxies existed in closer proximity to each other, they were more likely to encounter each other.

6. A starburst galaxy is a galaxy in the present-day universe that is forming stars at a prodigious rate. Starbursts can explain the ages of the stars in some of the small elliptical galaxies in the local group: The large number of supernovae after the first starburst stage created a galactic wind that blew most of the gas out of the galaxy, ending star formation until the gas could cool and reaccumulate within the galaxy for the next starburst. If ellipticals in general experienced a massive starburst episode early in their lives, that could explain the lack of ongoing star formation in ellipticals and why their stars are on average fairly old.
7. Starburst galaxies are difficult to recognize in visible light because dust grains in their molecular clouds absorb most of the visible light coming from their young stars. However, the grains reemit the visible and ultraviolet light that they absorb as infrared light, so in that part of the spectrum they are many times brighter than our galaxy.
8. A galactic wind is hot gas that erupts into intergalactic space. The gas is driven by the kinetic and thermal energy that it received from supernovae, similar to the formation of a superbubble in the Milky Way. The difference between a galactic wind and a superbubble is scale: A superbubble is large, but much smaller than the galaxy as a whole. A galactic wind covers the entire galaxy and may push much of the gas out of the galaxy as a result.
9. In the 1960s, Maarten Schmidt discovered a radio source that appeared to be a blue star in an optical image. The spectrum of the object looked nothing like any element he knew of until he realized that it was the spectrum of hydrogen at enormous redshift. When he found the distance from Hubble's law, he was able to work out the luminosity. He found that the object was hundreds of times more luminous than our galaxy. Discovery of similar objects, now called quasars, followed. Improved images show that quasars really are the centers of extremely distant galaxies and are often members of distant galactic clusters, telling us that their redshifts really do imply their great distances. Quasars are difficult to study because of their distances, but we can study nearby active galactic nuclei, which are like quasars. Because nearby active galactic nuclei are not as bright as quasars, we can study the surrounding galaxies as well, allowing us to learn more about the nature of active galactic nuclei and quasars.
10. If we see an object brighten in 1 hour, we know that the object cannot be larger than 1 light-hour in size. If it were, light from the far side of the object would start arriving at our telescopes over an hour after the light from the front side and we would see the brightening take longer than an hour to occur.
11. Matter falling toward a supermassive black hole is converted in kinetic energy. Collisions between infalling particles convert the kinetic energy into thermal energy. The resulting heat causes the matter to emit the intense radiation we observe. To produce the luminosities we observe, an amount of matter greater than that of the Sun must fall through the accretion disk each year.

    Evidence for the existence of supermassive black holes can be found in the orbits of gas clouds and accretion disks in galaxies like M87 and NGC4258. Their orbital speeds show that they are orbiting extremely massive objects. The central objects must also be quite small for their masses in these cases. We do not know of any objects that could be so massive and so small apart from a supermassive black hole.
12. Radio galaxies are galaxies that emit unusually strong radio waves. Many radio galaxies have jets of plasma shooting out in opposite directions. The ends of the jets are radio lobes where the strong radio emission comes from. We think that the ultimate energy source is the galactic nucleus because that is where the jets originate.

13. *Galaxies that are more than 10 billion years old are too far away to see even with our most powerful telescopes.* This statement is not sensible. Hubble and other telescopes have detected many galaxies more than 10 billion light-years away, and some of those pictures appear in the textbook.
14. *Heavy elements ought to be much more common near the Milky Way's center than at its outskirts.* This statement is sensible. Stars create heavy elements, and there has been much more star formation near the galaxy's center than at its outskirts. Observations of stars in the Milky Way's halo show that they contain a much smaller percentage of heavy elements than stars near the galactic center.
15. *Elliptical galaxies are more likely to form in denser regions of space.* This statement is sensible, because unlike spiral galaxies, elliptical galaxies appear to prefer the cores of clusters, which are denser regions of space.
16. *If the Andromeda Galaxy someday collides and merges with the Milky Way, the resulting galaxy would probably be elliptical.* This statement is sensible, because computer simulations of collisions between two large spiral galaxies show that these collisions create elliptical galaxies.
17. *NGC9645 is a starburst galaxy that has been forming stars at the same furious pace for about 10 billion years.* This statement is not sensible. The bursts of star formation in starburst galaxies must be temporary because they can turn all of a galaxy's gas into stars in a time much shorter than 10 billion years.
18. *The energy from supernova explosions can drive a large proportion of the interstellar gas out of a small galaxy.* This statement makes sense because the escape speed from a small galaxy is less than the speeds generated by a supernova explosion. [Note: Another issue is how much the ejecta slow down because of encounters with other gas clouds.]
19. *Astronomers proved that the quasar 3C473 contains a supermassive black hole because its center is completely dark.* This statement is not sensible. On the contrary, many galaxies with supermassive black holes have centers that are quite bright. The evidence for supermassive black holes rests on the orbital speed of objects that orbit close to the centers of galaxies. [Note: there is no 3C473; the name was intended to be realistic but not something one could find on Google.]
20. *The black hole at the center of our galaxy may once have powered an active galactic nucleus.* This statement is sensible. The active nuclei in other galaxies appear to be powered by accretion of matter onto supermassive black holes, so it is quite possible that our own galaxy underwent a similar episode of nuclear activity.
21. *Radio galaxies emit only radio waves and no visible light.* This statement makes no sense because even radio galaxies have stars. The particles emitting radio emission emit very little optical light in comparison to their radio light, however.
22. *Analyses of quasar light can tell us about intergalactic clouds that might otherwise remain invisible.* This statement is sensible. Quasar light typically passes through many interstellar clouds on its journey to Earth, and each of these clouds absorbs a little bit of light. Analysis of the resulting absorption lines in the quasar's spectrum can tell us about the characteristics of these clouds.
23. c. 4 billion years old
24. b. Some regions of the universe were slightly denser than others.
25. a. a large elliptical galaxy
26. a. a large elliptical galaxy

27. c. infrared light
28. b. about 10 times higher than in the Milky Way
29. c. the solar system
30. c. gravitational potential energy
31. a. the mass of the galaxy's bulge
32. c. a quasar with a redshift of 6

33–34. Life stories of a spiral and an elliptical. These questions all ask students to briefly restate and explain ideas taken directly from the reading. The key in grading these questions is to make sure that students demonstrate that they *understand* the concepts about which they are writing.

35. The color of an elliptical has changed as the population of stars, most of which formed at some early time, ages together. So similarly to a globular cluster, an elliptical's color gets redder as the high-mass stars evolve off the main sequence to become red giants.

36. Very early collisions of protogalactic gas clouds could end up as a spiral galaxy if the collisions did not induce huge amounts of star formation right away. Collisions could cause the clouds to condense and cool, but depending on the timing might form a disk before they formed stars. [A student could argue that the collisions would induce star formation that would happen before the clouds settled into a disk configuration—this answer, suitably argued, also would get credit in my class, since it exhibits understanding of the main idea.]

37. Figure 20.7a of the LMC shows reddish blotches, which are emission-line nebulae: places of active star formation and hot, blue main-sequence stars. (The red is coming from a specific transition of the hydrogen atom.) The blotches appear scattered throughout the galaxy.

38. In order for an object to be physically close AND have a large redshift, the galaxy would have to be physically moving in space away from us, at speeds approaching the speed of light. One way to test the hypothesis that the redshifts are caused by physical speeds rather than cosmological recession is to ask why most of the redshifts are in the direction AWAY from the Earth. Another way to test this hypothesis is to predict and to look for the consequences of extremely high-speed collisions. But the reason this hypothesis is not viable is because the Hubble law has been measured with standard candles over a very large range of redshifts, and such high "peculiar" velocities have not been seen in these systems.

39. Since the mass inside that radius is, to first approximation, completely dominated by the mass of the supermassive black hole, we expect the orbits to be Keplerian—like those of the planets around the Sun, which completely dominates the mass of the solar system. In a Keplerian system, the quantity $v^2 r$ is constant, so at $\frac{1}{2}$ the radius (30 light-years), the velocity must be $2^2 = 4$ times higher; and at 2 times the radius (120 light-years), the orbital velocity will be $(1/2)^2 = \frac{1}{4}$ times slower.

40. The universe was denser at higher redshift, and therefore galaxies were closer together at higher redshift. If galaxies were about the same physical size, one would expect that the absorption lines in the spectrum of quasars would be higher at longer redshifts (closer to the quasar, earlier in time.) Even if the size of galaxies grew at the same rate as the rest of the universe (like $(1 + z)$), there would still be more systems along the line of sight at high redshift. Hydrogen absorption systems with cosmological redshifts greater than that of the quasar are going to be mostly

BEHIND the quasar, not in front of it (there may be a few hydrogen systems at higher redshifts, including both cosmological expansion and peculiar velocities, because they are physically moving in space toward the quasar).

41. If we traveled back in time to an era when the distance between galaxies was a quarter of what it is today, then we would find that the density of galaxies would be $4^3 = 64$ times higher since the galaxies would be four times closer together in each of the three dimensions. In a cube that today contains one galaxy the size of the Milky Way, we would instead find 64 galaxies this size. With this higher density, we expect that the number of collisions between galaxies would be much higher as well.

42. This is a rate problem. We are asked to find how long it would take to turn $10^9 M_{\text{Sun}}$ of gas into stars at a rate of $10 M_{\text{Sun}}$ per year:

$$\text{time} = \frac{10^9 M_{\text{Sun}}}{10 M_{\text{Sun}}/\text{yr}} = 10^8 \text{ yr}$$

The galaxy will run out of gas in about 100 million years. Not long after this happens, the biggest, hottest stars will start dying. With no more gas to make stars, they will not be replaced. Over time, as the massive stars die out, the galaxy's composition will shift toward the redder, smaller stars. This process will make the galaxy dimmer and redder over time.

43. a. One author's mass in kilograms is 120 lb ÷ 2.2 lb/kg = 54.5 kg. The radiative energy released by mass falling into a black hole is about 10% of the rest mass energy. Thus, the energy released is:

$$0.10 \times mc^2 = 0.10 \times 54.5 \text{ kg} \times \left(3 \times 10^8 \frac{\text{m}}{\text{s}}\right)^2 = 4.9 \times 10^{17} \frac{\text{kg} \times \text{m}^2}{\text{s}^2} = 4.9 \times 10^{17} \text{ joules}$$

b. A 100-watt light bulb uses 100 joules per second, so we divide the energy released by the rate of energy release to see how long a bulb would have to burn to release the same amount of energy:

$$\frac{4.9 \times 10^{17} \text{ J}}{100 \text{ J/s}} = 4.9 \times 10^{15} \text{s} = 1.6 \times 10^8 \text{ yr}$$

A bulb would have to burn for 160 million years to release the same amount of energy.

44. Using the orbital velocity law, we calculate the mass of this black hole to be:

$$M = \frac{(4.8 \times 10^{15} \text{m}) \times \left(1.0 \times 10^6 \frac{\text{m}}{\text{s}}\right)^2}{\left(6.67 \times 10^{-11} \frac{\text{m}^3}{\text{kg} \times \text{s}^2}\right)} = 7.2 \times 10^{37} \text{kg}$$

Converting to solar masses gives:

$$M = (7.2 \times 10^{37} \cancel{\text{kg}}) \times \frac{1 M_{\text{Sun}}}{2.0 \times 10^{30} \cancel{\text{kg}}} = 3.6 \times 10^7 M_{\text{Sun}}$$

45. Mathematical Insight 21.2 tells us that we can find the mass of the central black hole from the orbital speed and radius of the stars around it with the formula:

$$M = \frac{r \times v^2}{G}$$

We will need our values of the radius and speed in meters and meters per second to use this formula. The speed is given as 400 kilometers per second, so it is easy to convert this into 4×10^5 m/s. The radius requires us to recall that 1 light-year is 9.46×10^{15} m. Using this conversion factor, we see the 3-light-year orbital radius is equal to 2.84×10^{16} m. So plugging these values into the formula above, we discover that the mass of the central black hole in M31 is 6.8×10^{37} kg or 34 million times the mass of the Sun.

46. To find out how much energy was radiated by this black hole as it grew, we can use Einstein's equation:

$$E = mc^2$$

The black hole is 3 billion times the mass of the Sun, or around 6×10^{39} kg. If 10% of this mass was radiated away, that is about 6×10^{38} kg of mass to turn into energy. Plugging into Einstein's equation, we see that the energy released is 5.4×10^{56} joules. If this were released over 10 billion years (3.16×10^{77} seconds), the accretion disk would have shown with a luminosity of 1.7×10^{39} watts or 4.5 trillion times the luminosity of the Sun. The luminosity of the Milky Way is 15 billion times the luminosity of the Sun, so this accretion disk is about 300 times more luminous than our entire galaxy.

47. a. The probability that our Sun will collide with a star in the Andromeda Galaxy is the ratio of the cross-sectional area of all of the stars in the Andromeda Galaxy over the area of the entire disk of the galaxy:

$$\text{probability of collision} = \frac{\text{area of stars}}{\text{total area}}$$

The disk of the Andromeda Galaxy has a radius of 100,000 light-years, or 9.46×10^{20} m. The area, given by the area of a disk (πr^2), is 2.8×10^{42} m^2. The area of a single star is also given by the area formula for a disk, only this time the radius is the radius of the Sun: 6.96×10^8 m. The area of a single star is therefore 1.5×10^{18} m^2. With 100 billion such stars, the total area covered by the stars is 1.5×10^{29} m^2. Now we are ready to compute the probability of the Sun hitting a star in the Andromeda Galaxy. Using the above formula, we find that the probability is 5.4×10^{-25}.

b. We will adopt the same approach as above, only this time we will use a little algebra to save ourselves some number crunching:

$$\begin{aligned}\text{probability} &= \frac{\text{area of clouds}}{\text{total area}} \\ &= \frac{100{,}000 \times (\pi r_{\text{cloud}}^2)}{\pi r_{\text{galaxy}}^2} \\ &= 100{,}000 \times \left(\frac{r_{\text{cloud}}}{r_{\text{galaxy}}}\right)^2\end{aligned}$$

Having gone through this, we do not need to convert from light-years to meters since the units on the radii cancel anyway. Putting in the radius of a gas cloud (300 light-years) and the radius of the galaxy (100,000 light-years), we find that the probability of a gas cloud from the Milky Way hitting a gas cloud in the Andromeda Galaxy is 0.9 or 90%.

c. The extremely small probability of one star hitting another star in a galactic collision means that we do not expect any such collisions to occur. However, the probability of a gas cloud hitting another gas cloud is very high and there will most likely be many such collisions in a galactic interaction. Collisions between gas clouds compress the clouds and trigger star formation in the clouds, leading to many new stars just after the galactic collision. This explains the bursts of star formation in galaxies that have collided, such as those in Figure 21.5.

48. The case for supermassive black holes, as for many aspects of our astronomical knowledge, rests on inferential evidence, as we will not be able to visit a black hole in our lifetime. The key to believing that supermassive black holes exist after all the evidence collected so far rests in being secure that no other phenomenon can explain what we observe. It may be a bit much to ask a college student to trust that you are not hiding some other viable explanation from them; it may help (if it comes up) to discuss Sir Martin Rees's concept that "all roads lead to a supermassive black hole." In that concept, one can certainly imagine setting up a system of stars, an extreme and massive star cluster that has the same mass in the same volume as a supermassive black hole. But simulations show that such a star cluster is gravitationally unstable, and very quickly such a system relentlessly evolves to eventually become a supermassive black hole.

49. Life in Colliding Galaxies. In this question, we are asking students to imagine the consequences to life on Earth if the Milky Way were undergoing a major collision. Could this be yet another way for life as we know it to end? A valid answer would point out that direct collisions between stars are extremely rare, even in galaxy-galaxy collisions, because the distances between stars are so vast compared to their size. But such a collision could increase the star formation rate, and some of the side effects of increased star formation include the increased incidence of supernovae and gamma-ray bursts, either of which, if they occurred near the Earth, might be a bummer for life on the planet. The rate of stars passing close to the solar system would also increase, thus slightly increasing the chance of a "nemesis" star disrupting the solar system. One should reward thinking quantitatively and soundly, even if the actual calculation might show the increase of risk to be slight. The night sky might have an "X" where the two Milky Way-size disks intersected! Cool.

Chapter 22. Dark Matter, Dark Energy, and the Fate of the Universe

This chapter focuses on dark matter and dark energy and their presumed significance in the universe. Students are likely to find this chapter particularly relevant because these topics are so often in the astronomical news.

As always, when you prepare to teach this chapter, be sure you are familiar with the relevant media resources (see the complete, section-by-section resource grid in Appendix 3 of this Instructor's Guide) and the online quizzes and other study resources available on the Mastering Astronomy Web site.

What's New in the Fourth Edition That Will Affect My Lecture Notes?

Those who have taught from previous editions of *The Cosmic Perspective* should be aware of the following organizational or pedagogical changes to this chapter (i.e., changes that will influence the way you teach) from the Third edition:

- The biggest difference in this chapter is the inclusion of dark energy: We discuss the evidence for the accelerating universe and the ways astronomers and physicists are seeking out the culprits for the acceleration, almost all of which fall into the category of "dark energy." We have even changed the title, adding "Dark Energy" to it, to emphasize our greater coverage of dark energy in the new edition.
- An entirely new introduction, Section 22.1, gives students an overview of what astronomers mean by the terms *dark matter* and *dark energy*. We emphasize that they are each hypothesized to exist as a way of explaining particular sets of observations, but that we really have little clear idea of what they are.
- We have updated the sections on dark matter to be clearer about the nature and strength of the evidence for it.
- The final section on the fate of the universe is significantly updated to discuss the role of dark energy in more detail.
- The old Chapter 22 had six sections. The new material is covered in four. The evidence for dark matter and what it might be is now covered in Section 22.2; that discussion spanned sections 22.2–22.4 in the third edition. Section 22.3 is an updated version of old Section 22.5 on structure formation. New Section 22.4 is a significantly updated version of old Section 22.6 on the fate of the universe.

Teaching Notes (By Section)

Section 22.1 Unseen Influences in the Cosmos

Students have probably heard the terms *dark matter* and *dark energy*, and will naturally assume that the fact we have names for these things means that we understand them. Of course, the reality is that these are simply names given to whatever might be causing particular patterns of observation: high orbital velocities in the case of dark matter, and an accelerating expansion in the case of dark energy. Thus, we feel it is important to begin the discussion by making clear that dark matter and dark energy are nothing more than names for unseen influences, and that we know little more about them.

- Notice our emphasis on the scientific process, so that students can understand why dark matter and dark energy are taken seriously by scientists even though we have not identified the source of either one.

Section 22.2 Evidence for Dark Matter

This section presents the evidence for dark matter in galaxies and galaxy clusters, then discusses the possible nature of dark matter. We also discuss the possibility that dark matter does not exist at all and that we are instead misunderstanding the nature of gravity, but point out that the sheer weight of the evidence at this point means that any such alternative interpretation of the data will have to meet many observational constraints (i.e., alternative gravity theories must do more than simply explain flat rotation curves of spiral galaxies—the evidence for dark matter is far deeper and more diverse than that today).

- We have eliminated the jargon of *mass-to-light ratio* in this edition, although we still use the basic idea.
- Again, this is a great place to discuss the nature of science, and how we can admit the possibility of alternate interpretations of the same data; in this case, the data that suggest the evidence for dark matter might potentially be explained by an alternate theory of gravity, though no one has yet done so successfully.
- For simplicity, we define *ordinary matter* to be baryonic matter and *extraordinary matter* to be nonbaryonic matter. We prefer to speak of ordinary matter and extraordinary matter when possible because these terms convey the proper impression to that large group of students who has trouble remembering what a baryon is.

Section 22.3 Structure Formation

This section covers large-scale structure in the universe and the role of dark matter in creating it. The analogy for growth via gravity we use in class is that in certain economic systems, the rich get richer, the poor get poorer, and the dichotomy grows with time. Places in the universe that were a little denser than average get more dense and become the clusters and the galaxies; places that were a little less dense than average get more diffuse and become the voids.

Section 22.4 The Fate of the Universe

This section explains how we try to assess the fate of the universe by measuring the amount of dark matter it contains and by measuring how the universe's expansion rate has changed through time. We emphasize the potential role of dark energy, presuming it exists.

- In pre-2000 texts, we could make a one-to-one correspondence between three possible models of universal fate (recollapsing, critical, and coasting) and three possible geometries (closed, flat, open). The introduction of dark energy breaks this simple correspondence. We therefore find it easiest for students if we first discuss the three models as they would be in the absence of dark energy, then discuss how dark energy enters the picture and adds a fourth possible model of acceleration.
- Many popular articles and books have taken the evidence for an accelerating expansion to suggest that we now know that the universe will expand forever. The logic is simple: Evidence was already pointing to an open universe before the acceleration was discovered, so acceleration would seem only to further seal the case. We take a slightly different viewpoint, emphasizing that future discoveries could potentially change our interpretation of the current data. Furthermore, since we don't understand exactly what dark energy is, we can't predict how it will behave in the future. Certain theories for dark energy (quintessence) allow the dark energy effects to die away with time. There is a possibility the acceleration could increase or even reverse its sign! *Until we know the nature of dark energy, we cannot predict the fate of the universe.*

Answers/Discussion Points for Think About It Questions

The Think About It questions are not numbered in the book, so we list them in the order in which they appear, keyed by section number.

Section 22.2

- (p. 665) The rotation curve for the moons around Jupiter would decrease with increasing distance, just as the rotation curve for the planets around the Sun decreases with increasing distance.
- (p. 668) If you instantly removed the dark matter from a cluster of galaxies without changing the velocities of the galaxies, it would be as if you were swinging a bucket on a rope around your head, and you cut the cord. The bucket would go flying off, tangential to the point at which you cut the cord. The galaxies would go flying away from each other, continuing on in a straight line: The external force of the dark matter's gravity would no longer be there.
- (p. 671) This is a good discussion question. Most astronomers find the agreement between these different mass measurement techniques reassuring. (They do not agree for all clusters, but as these techniques become more refined, they agree for more and more clusters.)
- (p. 673) This is another good discussion question. One pertinent example to mention is the discovery of Neptune, which was based solely on inferences from Newtonian gravity.

Section 22.3

- (p. 674) The Earth is held together by its own gravity; a hurricane is not. It's not clear (at least as far as the book goes) whether or not the Orion Nebula is gravitationally bound or not. Certainly at one point it was, but it loses gas as the stars evolve and now may or may not be bound. A supernova is not gravitationally bound, since its shock wave propagates out and does not return. (A piece of a massive star supernova may be left behind, as a neutron star or a black hole is certainly itself gravitationally bound.)

Section 22.4

- (p. 678) Robert Frost's poem that opens this section is a good departure point for a discussion of the philosophical implications of the universe's fate. (Another question to ask is WHY is there a preferred fate?)

Solutions to End-of-Chapter Problems (Chapter 22)

1. Dark matter is matter that gives off little or no light. Dark energy is the name given to whatever it is that is causing the universe's expansion to accelerate. While they have similar names, they are not similar in nature. Dark matter is massive, and behaves like something that gravitates, and is really dark in the sense that we do not see it. Dark energy is not matter and it is dark in the sense that we can't see it.

   Dark matter has been detected by carefully observing the gravitational effects on matter we can see in clusters of galaxies and in galaxies. Dark energy has been detected by observing the rate of expansion of the universe, from studies of white dwarf supernovae. (There is also less direct evidence for dark energy that goes beyond the scope of the course.)

2. A rotation curve plots the rotational velocity of stars against their distances from the center of the galaxy. The Milky Way's rotation curve quickly rises as we move out from the center of the galaxy and then more or less levels off at a nearly constant speed. This tells us that most of the mass of the galaxy is *not* concentrated in the center of the galaxy the way that the Sun holds most of the mass in the solar system. In fact, most of the mass of the galaxy is located outside of the Sun's orbit. We can also use the rotational velocity to determine the mass of the galaxy. When we do, we discover that there is much more mass in the galaxy than we have been able to see with light. This tells us that there is a large amount of dark matter in the galaxy.
3. We construct rotation curves for other spiral galaxies by observing the 21-centimeters line of clouds of atomic hydrogen gas. The rotation curves of other galaxies also turn out to be remarkably flat like the Milky Way's rotation curve. Their flatness tells us that other spiral galaxies also have halos of dark matter that make up around 90% of their masses.
4. We measure the masses of elliptical galaxies by measuring how much the spectral absorption lines of stars in the galaxy are broadened in a spectrum that includes measurements of many stars at once. The broader the spectral line (composed of many lines from individual stars, blurred together), the faster the stars are moving relative to each other. Faster motion in the stars means that the galaxy is more massive. We can also use the orbital speeds and distance of globular star clusters to get the masses of elliptical galaxies. Like spiral galaxies, ellipticals also seem to contain far more matter than is accounted for by stars.
5. We can measure the masses of clusters of galaxies in three ways. The first method is to measure the speeds and positions of the galaxies in the cluster as they orbit. Applying Newton's law of gravitation to the speeds, we can deduce the mass of the cluster. The second method for finding the mass of a cluster is to measure the temperature of the gas between galaxies in the cluster and the approximate size of the cluster. The pressure of the gas must balance the pull of gravity from the mass in the cluster, so the temperature and size lead us to the mass. The final method of finding the mass of a galactic cluster is gravitational lensing. This technique measures the amount that a beam of light is bent as it passes near the cluster and uses Einstein's theory of general relativity to determine the mass of the cluster. The results from all three methods agree on the masses of the clusters, and these results tell us that the clusters hold substantial amounts of dark matter.
6. Gravitational lensing is the bending of beams of light by massive objects. This bending occurs because masses distort spacetime, according to general relativity. Since more massive objects distort spacetime more and therefore bend the light more, we can use the amount of bending to calculate the mass of the object doing the lensing.
7. Since we have measured the masses of galaxies and clusters using our current theories of gravity, it is possible that the masses of the galaxies and clusters are wrong if our theories of gravity are wrong. However, the theories we use have been repeatedly tested and we are very confident in them as a result of their successes. Also, alternative theories have been sought, but none has yet managed to unseat our current theory. So we conclude that the masses we have determined are accurate and that there is more mass in the galaxies and in the clusters than we can see.
8. Dark matter is dark in the sense that it does not emit enough light for us to see it. Since people and planets do not emit enough light to be seen across the vast

distances of space, they qualify as dark matter. Dim stars can also qualify, since they do not emit enough light for us to see them.

9. MACHOs are a possible form of dark matter that is made up of dim stars, brown dwarfs, and planets. Although these objects do not emit enough light for us to see them directly, they do occasionally cause stars to appear to brighten due to gravitational lensing. This possibility allows us to search for MACHOs. Current studies that have looked for dark matter using this technique have turned up objects, but not enough objects have been found to account for most of the dark matter in the galaxy.
10. We say that particles like neutrinos are weakly interacting because they interact only with other objects through gravity or the weak nuclear force. They do not interact with radiation or charged particles via the electromagnetic force. As a result, they tend to pass through most matter in the universe. Neutrinos cannot make up most of the dark matter in galaxies because they move too fast and easily escape a galaxy.
11. WIMPs, or weakly interacting massive particles, are a form of dark matter that is made up of particles that do not generally interact with baryonic matter much, except through their gravitational influence. These particles need to be more massive than neutrinos and therefore move more slowly so that mutual gravity could hold clumps of them together. Such matter seems the best candidate for the dark matter in the universe since it meets the requirements for dark matter well and because there are other reasons to think that such matter (nonbaryonic) should exist in the universe. (Stay tuned for Chapter 23.)
12. Dark matter is thought to have played a role in the formation of galaxies because its mass would have provided most of the gravitational attraction needed to make the protogalactic clouds and larger structures. Since the dark matter cannot radiate away its energy, it never collapses like the rest of matter, leaving it to form a cocoon around the galaxies and clusters, just as we observe.

    We think that large-scale structures may still be forming today because a careful look at the relative speeds of clusters, including our own, shows that they are not moving apart as fast as Hubble's law predicts. In fact, it appears that the rate of separation between our Local Group and the Virgo Cluster is slowing and in time might reverse so that we will head toward the Virgo Cluster some time in the future.
13. The large-scale structure of the universe shows galaxies arranged in huge chains and sheets that span millions of light-years. Between the chains and sheets are giant voids. This structure probably mirrors the original distribution of dark matter in the early universe. The denser regions in the early universe would have gone on to collapse and form galaxies, clusters, and superclusters while the less dense regions would have gone on to form the voids.
14. The critical density of the universe is the density that divides the lower densities where the universe will expand forever and the higher densities where it will eventually recollapse. Current studies indicate that the actual density of the universe is about 25% of the critical density, so we think that the universe will keep expanding forever.
15. There are four possible patterns for the expansion of the universe:
    - Recollapsing—We get a recollapsing universe if there is enough mass in the universe. If this were the case, eventually gravity will halt the expansion of the

universe and reverse it. All of the matter in the universe will eventually be crushed back together again, re-creating the conditions of the Big Bang.

- Critical—We get a critical universe if the density of the universe is exactly the critical density. In this case, the expansion will slow with time but never reverse. (It will halt after an infinite time.)
- Coasting—A costing universe occurs if the density of the universe is less than the critical density and there is no dark energy to accelerate the rate of expansion. The universe will continue expanding forever with little change in the rate of expansion.
- Accelerating—An accelerating universe occurs if there is dark energy in the universe to exert a repulsive force. This will increase the rate of expansion of the universe over time.

We can tell which of these models is the right one with white dwarf supernovae measurements. These supernovae are bright enough to be seen across vast distances and function as standard candles, letting us find their distances. Their redshifts tell us how much the universe has expanded since their light was emitted, so we can work out the rate of expansion of the universe in the past.

16. Whether the universe's expansion is accelerating or not, current evidence indicates that the universe will continue to expand forever. If the acceleration of the universe is real, it is even possible that the dark energy will cause galaxies, stars, and planets to break apart and disperse.
17. *Strange as it may sound, most of both the mass and energy in the universe may take forms that we are unable to detect directly.* This statement is true, because the amount of dark matter seems to far outweigh the ordinary matter and the inferred mass-energy of dark energy is even greater.
18. *A cluster of galaxies is held together by the mutual gravitational attraction of all the stars in the cluster's galaxies.* This statement is false. The amount of mass in a cluster's stars is much lower than the amount needed to hold the cluster together. That is why we believe that most of a cluster's matter is dark.
19. *We can estimate the total mass of a cluster of galaxies by studying the distorted images of galaxies whose light passes through the cluster.* This statement is true. Gravitational lensing is now commonly used to measure the masses of clusters.
20. *Clusters of galaxies are the largest structures that we have so far detected in the universe.* This statement is false. Superclusters and voids are much larger than clusters of galaxies.
21. *The primary evidence for an accelerating universe comes from observations of young stars in the Milky Way.* This statement is false. In order to measure accelerating expansion, we need to measure distances to objects billions of light-years away.
22. *There is no doubt remaining among astronomers that the fate of the universe is to expand forever.* This statement is false. Current evidence favors eternal expansion, but we have much more to learn about the universe before we can be certain of its fate.
23. *Dark matter is called "dark" because it blocks light from traveling between the stars.* This statement is false. Dark matter is "dark" because it does not emit much if any light. Weakly interacting particles don't interact with light at all.
24. *If the universe has more dark matter than we think, then it is also younger than we think.* This statement makes sense. If the universe actually has more dark matter,

then its initial expansion slowed down more than we estimate, and the expansion began more recently than we calculate today.

25. *The distance to a white dwarf supernova with a particular redshift is larger in an accelerating universe than in a universe with no acceleration.* This statement makes sense. The accelerating universe is a larger, older universe than a universe with no acceleration.
26. *If dark matter consists of WIMPs, then we should be able to observe photons produced by collisions between these particles.* This statement does not make sense. Since WIMPs are "weakly intereacting," they do not interact via the electromagnetic force, and therefore do not collide in the same way that charged particles can collide.
27. c. We can observe its gravitational influence.
28. a. observations suggesting that the expansion of the universe is accelerating
29. a. orbit the galactic center just as fast as stars closer to the center
30. c. clusters of galaxies
31. c. a gravitational lens
32. b. Yes, but only if there is something wrong with our current understanding of how gravity should work on large scales.
33. a. subatomic particles that we have not yet detected in particle physics experiments
34. b. a region whose matter density was higher than average
35. a. white dwarf supernovae
36. c. a recollapsing universe
37. The case for dark matter: The student should describe the evidence and at least two possible conclusions from the evidence (dark matter exists, or the theory of gravity must be revised).
38. The case for dark energy: The case for dark energy as presented at this level is the case made by the supernovae for the accelerating universe. (This case is much strengthened when evidence from the Cosmic Microwave Background and clustering is also brought to bear, but this convergence is beyond the scope of the usual course.)
39. The future universe: If the acceleration continues, the formation of large-scale structures in the universe ceases almost completely over the next 10 billion years. The universe continues to expand—clusters of galaxies get farther and farther apart, ever more rapidly. The universe is rapidly becoming a very empty and boring place! Because there are fewer mergers, the gas shortage in galaxies becomes increasingly problematic—star formation slows down to a crawl and will eventually cease altogether as galaxies will see no new sources of gas.
40. Dark matter and life. Two reasons I can think of for why dark matter is essential for life: Dark matter allows gravity to make galaxies in a reasonably sprightly time, and gives life a chance to develop (before the nasty boring phase of the universe happens). Dark matter also allows galaxies to remain bound and stable, and therefore keep gas circulating in the stellar life cycle of an ever-increasing heavy-element fraction. And life needs those heavy elements.
41. Drawn and labeled rotation curves. We can't draw them here, but we'll describe them.
    a. All mass concentrated in the center of the galaxy: Decreasing rotation curve, where v is proportional to $1/\sqrt{r}$: Keplerian, like the planets around the Sun.
    b. Constant mass density inside 20,000 light-years, zero outside: v increases like r^2 out to 20,000 light-years, falls off like $1/\sqrt{r}$ outside of that.

c. Constant mass density inside 20,000 light-years, and $M(<r)$ is proportional to r outside of that: v increases like r^2 out to 20,000 light-years, v is constant outside of that.

42. In this situation, the supernovae turned out to be fainter than they expected. (The universe was a bigger, older place than they thought it would be.) The position of the supernovae on the horizontal axis was farther away from "now" than the astronomers expected.
43. What is Dark Matter? One possible constituent is Weakly Interacting Massive Particles, which do not interact with light. We can detect them in similar ways as we detect neutrinos, or by their gravitational effects on objects that do interact with light. Massive Compact Halo Objects were another possible source of dark matter; such objects can be detected via gravitational lensing. Such objects interact with light much as a rock does (it blocks and may also faintly emit thermal radiation). Other forms that are technically "dark" but are unlikely candidates for the majority of dark matter include neutrinos (which are detected by neutrino detectors—except for the remnant cosmological neutrinos, which are so low energy that they are probably forever beyond our direct detection capability) and cold, diffuse baryonic gas that is perhaps clumped and very difficult to detect, even in absorption.
44. Alternative gravity: To explain flat rotation curves without dark matter, gravity would have to be stronger than expected at large distances because you need more gravity for less mass.
45. The mass-to-light ratio of a $1M_{\text{Sun}}$ white dwarf with a luminosity of $0.01L_{\text{Sun}}$ is the mass over the luminosity, so:

$$\frac{M}{L} = \frac{1M_{\text{Sun}}}{0.01L_{\text{Sun}}} = 100\,\frac{M_{\text{Sun}}}{L_{\text{Sun}}}$$

So the mass-to-light ratio of the white dwarf is $100M_{\text{Sun}}/L_{\text{Sun}}$.
46. The mass-to-light ratio of a $30M_{\text{Sun}}$ supergiant with a luminosity of $300{,}000L_{\text{Sun}}$ is the mass over the luminosity, so:

$$\frac{M}{L} = \frac{30M_{\text{Sun}}}{300{,}000L_{\text{Sun}}} = 10^{-4}\,\frac{M_{\text{Sun}}}{L_{\text{Sun}}}$$

So the mass-to-light ratio of the white dwarf is $10^{-4}\,M_{\text{Sun}}/L_{\text{Sun}}$.
47. To find the mass-to-light ratio of the solar system, we need the mass and the luminosity. The luminosity is essentially just that of the Sun since the Sun is the only significant light source in the solar system. (Recall that planets are considered dark matter because they emit so little light.) The mass of the solar system is also basically the mass of the Sun since the planets add up to less that 1% of the Sun's mass. So the ratio becomes:

$$\frac{M}{L} = \frac{1M_{\text{Sun}}}{1L_{\text{Sun}}} = 1\,\frac{M_{\text{Sun}}}{L_{\text{Sun}}}$$

So the mass-to-light ratio of the white dwarf is $1M_{\text{Sun}}/L_{\text{Sun}}$.

48. We can derive a general-purpose formula for this problem and for the next problem by using the mass/velocity/radius relation and plugging in 100 km/sec for the velocity and 10,000 light-years for the radius. Then, for this and the next problem, we need to do fewer computations:

$$M_r = \frac{rv^2}{G} = \frac{\left[r \times \left(\frac{9.46 \times 10^{19}\text{ m}}{10{,}000\text{ ly}}\right)\right] \times \left[v^2 \times \left(\frac{1000\text{ m}}{1\text{ km}}\right)^2\right]}{\left(6.67 \times 10^{-11}\frac{\text{m}^3}{\text{kg} \times \text{s}^2}\right) \times \left(2 \times 10^{30}\frac{\text{kg}}{M_{\text{Sun}}}\right)}$$

$$= 7.1 \times 10^5 M_{\text{Sun}} \times \underbrace{r}_{r\text{ in }10^4\text{ ly}} \times \underbrace{v^2}_{v\text{ in km/s}}$$

$$= 7.1 \times 10^9 M_{\text{Sun}} \times \left(\frac{r}{10{,}000\text{ ly}}\right) \times \left(\frac{v}{100\frac{\text{km}}{\text{s}}}\right)^2$$

For NGC7541 from Figure 21.4, the velocity at 30,000 light-years is 200 km/sec, and the velocity at 60,000 light-years is about 220 km/sec. Plugging into our handy formula, we can solve parts (a), (b), and (c).

a. Within 30,000 light-years:

$$M_r = 7.1 \times 10^9 M_{\text{Sun}} \times \left(\frac{30{,}000\text{ ly}}{10{,}000\text{ ly}}\right) \times \left(\frac{200\frac{\text{km}}{\text{s}}}{100\frac{\text{km}}{\text{s}}}\right)^2 = 8.5 \times 10^{10} M_{\text{Sun}}$$

b. Within 60,000 light-years:

$$M_r = 7.1 \times 10^9 M_{\text{Sun}} \times \left(\frac{60{,}000\text{ ly}}{10{,}000\text{ ly}}\right) \times \left(\frac{220\frac{\text{km}}{\text{s}}}{100\frac{\text{km}}{\text{s}}}\right)^2 = 2.1 \times 10^{11} M_{\text{Sun}}$$

c. The mass at 30,000 light-years is about half the mass at 60,000 light-years (since the velocity curve is flat, the velocity isn't much different, and thus the mass enclosed increases proportionally to the radius). The mass is not concentrated in the center of the galaxy.

49. For this problem, we assume that the ↑ cluster ← gas is all at one temperature (isothermal) and that we have X-ray data from the cluster out to at least 2 Mpc (= 6.2×10^{22} m). The temperature that we measure for the gas is 8×10^7 K. We can use the formula from Mathematical Insight 22.2 to compute an equivalent velocity for the particles in the gas; thatis:

$$v_{\text{thermal}} = 100\frac{\text{m}}{\text{s}} \times \sqrt{T} = 8.94 \times 10^5\frac{\text{m}}{\text{s}}$$

We plug this velocity into the orbital velocity law to find the mass:

$$M = \frac{rv^2}{G} = \frac{(6.2 \times 10^{22}\ \text{m}) \times \left(8.94 \times 10^5 \frac{\text{m}}{\text{s}}\right)^2}{\left(6.67 \times 10^{-11} \frac{\text{m}^3}{\text{kg} \times \text{s}^2}\right)} = 7.4 \times 10^{44}\ \text{kg}$$

Dividing by 2×10^{30} kg/solar mass, we have 3.7×10^{14} solar masses inside 2 Mpc. (Note that this answer depends on the Hubble constant and maybe even other cosmological constants that must be assumed to figure out how "big" the cluster of galaxies is.)

50. Mathematical Insight 22.3 tells us that we can find the average speed of hydrogen nuclei at a given temperature with the equation:

$$v_H = 140\ \frac{\text{m}}{\text{s}} \times \sqrt{T}$$

This tells us that hydrogen nuclei in a gas at 9×10^7 Kelvins have an average velocity of 1.3×10^6 m/s. We can use this speed along with Newton's law of universal gravitation to find the mass of the cluster:

$$M = \frac{r \times v^2}{G}$$

The radius is 15 million light-years or 1.4×10^{23} m. Plugging in the speed found above and this radius, we find that the mass of the cluster is 3.5×10^{45} kg or $1.8 \times 10^{15} M_{\text{Sun}}$.

51. If all the stars were identical, Sun-like stars, their mass-to-light ratios in solar units would be 1. If you measured a mass-to-light ratio of 30 solar masses per solar luminosity, you could find the amount of dark matter by assuming the dark matter contributes nothing to the luminosity. You can divide the total mass into stellar mass (M_*) and dark matter mass (M_{DM}). The galaxy has luminosity L, completely provided by stars. Therefore:

$$\frac{M_* + M_{DM}}{L} = \frac{M_*}{L} + \frac{M_{DM}}{L} = 30 \text{ (measured)}$$

We estimate that, for stars:

$$\frac{M_*}{L} = 1$$

So,

$$\frac{M_{DM}}{L} = 29 \quad \text{and thus} \quad \frac{M_{DM}}{M_*} = \frac{29}{1} = 29$$

a. Therefore, the ratio of dark matter to "bright" matter in this case is 29-to-1.

b. If all of the dark matter is in Jupiter-size objects with 0.001 solar mass each, the number (N) of MACHOs per star is:

$$\frac{M_{DM}}{M_*} = \frac{N \times (0.001 M_{\text{Sun}})}{1 M_{\text{Sun}}} = 29$$

so

$$N = \frac{29}{0.001} = 29{,}000$$

If MACHOs make up the dark matter in this hypothetical galaxy, there are 29,000 MACHOs per star.

52. We can derive the expression used in Mathematical Insight 22.3 (and elsewhere) that gives us the mass of a galaxy or cluster from Newton's version of Kepler's third law. Newton's version of Kepler's third law is:

$$a^3 = \frac{GM}{4\pi^2} p^2$$

If we assume that our orbiting body is on a nearly circular orbit, we know that the speed of the object is the distance it covers in one orbit divided by the orbital period. The distance that such an object covers is $2\pi a$, the circumference of a circle. The period is p, so the speed is:

$$v = \frac{2\pi a}{p}$$

With this in mind, we regroup some of the variables in Newton's version of Kepler's third law to get:

$$GM = a\left(\frac{4\pi^2 a^2}{p^2}\right)$$

$$= a\left(\frac{2\pi a}{p}\right)^2$$

$$= a \times v^2$$

In this case, we can rename a and call it r, since they mean the same thing: the distance between the center of mass and the orbiting body. With that, we rearrange to solve for M to get:

$$M = \frac{r \times v^2}{G}$$

This is the expression we sought to devise.

53. Why do astronomers find dark matter a more appealing explanation than modifications to gravity? (What do you think?)

54. The fate of the universe: Why do we have a preference? (This is also a time-out-to-think question.)

Chapter 23. The Beginning of Time

This chapter concentrates on the Big Bang model for the universe's origin and presents the evidence in favor of it.

As always, when you prepare to teach this chapter, be sure you are familiar with the relevant media resources (see the complete, section-by-section resource grid in Appendix 3 of this Instructor's Guide) and the online quizzes and other study resources available on the Mastering Astronomy Web site.

What's New in the Fourth Edition That Will Affect My Lecture Notes?

Those who have taught from previous editions of *The Cosmic Perspective* should be aware of the following organizational or pedagogical changes to this chapter (i.e., changes that will influence the way you teach) from the Third edition:

- The most significant change is the inclusion of new results from WMAP and greater discussion of the evidence these new results offer for dark energy.
- The structural changes include combining old Sections 23.1 and 23.2 into a single Section 23.1, "The Big Bang." The section numbering then maps general topic coverage from old Section 23.3 to new Section 23.2, "The Evidence for the Big Bang," from old Section 23.4 to new Section 23.3, "The Big Bang and Inflation"; and from old Section 23.5 to new Section 23.4, "Observing the Big Bang for Yourself."

Teaching Notes (By Section)

Section 23.1 The Big Bang

This section outlines the scientific history of the universe according to the Big Bang theory.

- If you are pressed for time or do not want to devote time to deeper cosmological issues, this section may be the only one you want to cover.
- Inflation of the universe does not necessarily coincide with the end of the GUT era, but we discuss it here because the idea of inflation arose out of GUT physics.
- The mass-energy density of the universe goes from radiation-dominated to matter-dominated at $z \approx 10{,}000$. While the distinction between a matter-dominated universe and a radiation-dominated universe is important to astronomers, causing a change in the relation between time and redshift (for one thing), it is a relatively subtle concept to explain to students. We chose not to mention it here because it does not bring about a qualitative change in the contents of the universe and because students already have enough eras to keep in mind.

Section 23.2 Evidence for the Big Bang

In this section, we present the two strongest pieces of evidence in favor of the Big Bang theory: the cosmic background radiation and the helium abundance of the universe.

- The relationship between the microwave background fluctuations seen by *COBE* and the underlying mass distribution is less direct than it might seem at first glance.

In perturbations larger than the cosmological horizon, matter and radiation are strongly coupled, and density perturbations increase in amplitude ($=\delta\rho/\rho$). As time passes, the horizon grows to encompass ever larger scale perturbations, and at $z > 1{,}000$, perturbations of ordinary matter that have just become smaller than the horizon start to oscillate because (a) sound waves can now cross the horizon and (b) the photons, trapped by free electrons, resist gravitational compression. However, if structure formation in the early universe involves WIMPs, as many astronomers suspect, the dark-matter components of these perturbations continue to grow while the baryons oscillate. The temperature fluctuations recorded in the microwave background therefore depend on the phase of the oscillating compression waves at the time ofrecombination as well as the gravitational redshift owing to the growing dark-matter potential wells.

- The moment at which microwave background photons begin to stream freely across the universe is often said to be simultaneous with the moment of recombination. However, while most electrons are captured into atoms at $z \approx 1{,}500$, the remaining free electrons in the universe continue to scatter background photons until $z \approx 1{,}000$. We do not draw attention to this discrepancy in the text, but it leaves us with two different redshifts to choose from when defining the end of the era of nuclei. We have chosen to define the end of this era to be at $z \approx 1{,}000$ because it makes the arithmetic and memorization easier for students (i.e., 3 K now implies 3,000 K at the time of last scattering).

Section 23.3 The Big Bang and Inflation

This section presents three well-known problems with the original Big Bang model—the origin of structure, the smoothness problem, and the flatness problem—and then discusses how an early episode of inflation might solve all three.

- We pay little attention to the steady state theory of the universe in this chapter because it is no longer a viable alternative to the Big Bang theory; we now have ample evidence that galaxies in the universe evolve in rough synchronization. A more current controversy to focus on is the question of whether inflation really happened. Should we prefer the classic Big Bang theory, with its attendant problems, or the inflation-modified Big Bang theory, which might solve these problems but depends on physics that we don't quite understand?
- Density perturbations in the early universe must all have had roughly the same amplitude as they are first encompassed by the cosmological horizon. This type of uniformity arises naturally in inflationary models—a point in their favor. However, naive models of inflation overpredict the amplitudes of these fluctuations by a few orders of magnitude—a strike against them.
- Inflation naturally explains the uniformity of the microwave background temperature and the geometric flatness of the universe. There is currently no other plausible explanation for this uniformity. All other explanations to date must rely on "fine-tuning" or similar ad hoc starting conditions.
- Increasingly sophisticated observations of variations in the microwave background temperature across the sky will test various predictions of inflation over the next several years. We encourage students to be on the lookout for news items about these experiments.

Section 23.4 Observing the Big Bang for Yourself

In this section, the last on cosmology and the universe, we want to leave the students looking up at the sky and thinking for themselves, so we pose the question, "Why is the night sky dark?" This leads to a discussion of Olbers' paradox and its implications about the evolution of the universe.

- The text states that the darkness of the night sky implies that "the universe has either a finite number of stars, or it changes in time in some way that prevents us from seeing an infinite number of stars." Note that while the gross characteristics of the universe in the steady state model do not change with time, the universe does change on smaller scales in a way that solves Olbers' paradox. Old galaxies are flying apart and are redshifted out of view with the expansion of the universe, while new galaxies appear in the growing gaps. For more on the subject of Olbers' paradox, consult Edward Harrison's excellent book *Darkness at Night.*

Answers/Discussion Points for Think About It Questions

The Think About It questions are not numbered in the book, so we list them in the order in which they appear, keyed by section number.

Section 23.2

- (p. 692) A variety of stars and galaxies spanning a wide range of redshifts would produce thermal radiation at many different temperatures, quite unlike the pure thermal spectrum we observe. However, the Big Bang theory successfully predicts that the universe should be permeated by thermal radiation at a uniform temperature.
- (p. 695) The Big Bang fused 25% of the normal matter in the universe into helium. Galaxies that have not cycled very much of their gas through stars should not have much more than 25% helium. No galaxies should have less than this amount, because stars can only increase the helium fraction—they cannot diminish it.
- (p. 696) The density of normal matter we infer from the abundances of deuterium and other light elements and from the CMB observations is a few times less than the density of dark matter we infer from mass-to-light ratios and other techniques. The difference between those two densities is why astronomers suspect that the majority of the mass in the universe is made up of weakly interacting particles. The mean density of baryons is now well established through two completely independent means, and it is still only about 10% of the total amount of matter density in the universe. So astronomers are fairly confident in the following claim: that either dark matter is largely nonbaryonic matter or our theory of gravity is incomplete.

Solutions to End-of-Chapter Problems (Chapter 23)

1. An antimatter particle has the same mass as its particle equivalent, but all of the rest of the properties are precisely opposite. A particle-antiparticle pair can be produced when high-energy photons collide. Such pairs can destroy each other by meeting back up again. When they meet, they annihilate each other totally, turning their mass back into energy.
2. The Big Bang theory is the scientific theory for the universe's earliest moments. The theory is based on the idea that the universe began as an incredibly hot and dense collection of matter and radiation. Everything we can observe today was

crammed into a tiny space. The expansion and cooling of this unimaginably intense mixture of particles and photons could have led to the present universe and it explains several aspects of the present universe with impressive accuracy.

3. The universe has gone through several major eras over its history. These eras are:
   - The Planck Era—The time before 10^{-43} second when we require a "theory of everything" to describe what happened. As yet, we lack such a theory.
   - The GUT Era—At the start of this era, gravity became distinct from the other forces so that there were two forces: gravity and the GUT force (a combination of the electromagnetic, the weak, and the strong forces). The era ended when the strong force froze out of GUT force, leaving three forces operating (gravity, strong, and the electroweak force). When this happened, the energy released may have driven inflation, causing the universe to go from the size of an atom to the size of the solar system in 10^{-36} second.
   - The Electroweak Era—This era was dominated by three forces: the strong, the electroweak, and gravity. At the end of this era, all four forces became distinct and have remained so for the rest of the history of the universe.
   - The Particle Era—This era started with the separation of the electromagnetic and the weak forces and ended when the universe became cool enough that particle production ceased. At the end of the particle era, all of the quarks in the universe combined into protons and neutrons.
   - The Era of Nucleosynthesis—During this era, the temperature of the universe remained so high that nuclei broke apart as quickly as they formed. This era ended when the universe became too cool for fusion to occur. At this time, about 75% of the baryonic matter of the universe was in hydrogen nuclei, with the rest mainly in helium, deuterium, and a small amount of lithium nuclei.
   - The Era of Nuclei—During this era, atomic nuclei remained totally ionized and roamed free from their electrons. Photons bounced rapidly from one electron to the next, never getting far. This era ended when the temperature fell low enough for the electrons to combine with the nuclei and photons were free to stream across the universe.
   - The Era of Atoms—During this era, baryonic matter in the universe formed atoms. The atoms formed into galaxies due to slight density enhancements.
   - The Era of Galaxies—This is our current era. Stars build nuclei into heavier elements and planets form around the stars.
4. Our current theories cannot describe the universe during the Planck era because we require a theory that links quantum mechanics and general relativity. We do not yet have such a theory, so we are currently unable to describe the universe in this era.
5. The four forces that operate in the universe today are gravity, electromagnetism, the strong nuclear force, and the weak nuclear force. We know that at high enough energies the electromagnetic and weak forces merge. We think that at higher temperatures the strong force may also merge with these two, leaving just two forces to operate.
6. Grand unified theories are theories that predict the merger of the strong and the electroweak forces. According to these theories, only two forces operated during the GUT era of the universe: the GUT force and gravity. The GUT force is the combination of electromagnetism and the strong and weak nuclear forces.
7. Inflation refers to the dramatic expansion of the universe that brought it from the size of an atom to the size of the solar system in a mere 10^{-36} second. We think that this occurred at the end of the GUT era when the strong force froze out of the GUT force, around 10^{-38} second into the universe's history.

8. We think that there must have been a slight imbalance of matter and antimatter in the early universe because some matter has survived annihilation to this day. It appears that there was about one part in a billion more protons than antiprotons. So when all of the antiprotons annihilated with protons, the excess protons remained.
9. The era of nucleosynthesis lasted about 3 minutes, until the universe cooled down enough to end fusion. This era was important to the present universe because it set the relative amounts of hydrogen, helium, deuterium, and lithium that were present when the first stars formed. In fact, these relative amounts have not changed much even today.
10. When we observe the cosmic microwave background, we are seeing the universe as it was when it was 380,000 years old. The photons we see have been traveling for about 14 billion years, since not long after the Big Bang.
11. The cosmic microwave background was discovered accidentally by two physicists working for Bell Labs. They were calibrating a sensitive microwave antenna and found that they could not get rid of some of the noise despite their best efforts. Ultimately, after a chance meeting with an astronomer, they discovered that the noise that they had been detecting was the cosmic microwave background.
12. The Big Bang theory, along with the observed current temperature of the cosmic microwave background, tells us how much helium should have been made by the end of the era of nucelosynthesis. The result is about 25%, which agrees with observations very well.
13. Calculations based on the abundance of deuterium, lithium, and helium-3 indicate that the density of baryonic matter in the universe should be about 4% of the critical density. We know from our observations of galactic masses and the masses of cluster that the actual density of the universe is about 25% of the critical density. Therefore we conclude that the remaining 21% is not baryonic and must be in the form of WIMPs.
14. The three questions left unanswered by the Big Bang theory before inflation were:
    - Where did structure come from? We need to explain the origin of the density enhancement that led to the formation of galaxies and larger structures. Quantum fluctuations in energy fields explain this conundrum. The distribution of energy in space is always fluctuating and therefore the distribution is always slightly irregular. The scale of these irregularities is small, but inflation would have amplified them many times, making them large enough to become the density fluctuations that formed galaxies and larger structures.
    - Why is the universe so uniform? We need to explain why the universe has the same density in regions separated by vast distances. Inflation explains this because those regions were in contact right before the inflation, allowing them to develop the same temperature and density.
    - Why is the density of the universe so close to the critical density? The density of the universe is relatively close to the critical density when it could easily be many thousands or millions of times more or less. It is difficult to consider this a coincidence. Inflation also explains this mystery. When spacetime expanded, it flattened out to the point where any curvature that would have existed before the inflation would be noticeable only on scales much larger than the observable universe.
15. We can test inflation by looking at the fluctuations in the cosmic microwave background. The tiny temperature differences tell us about the structure of the early universe before inflation. So far, we have learned that the universe should have a flat geometry, a total matter density of 27% (4.4% ordinary matter, 23% dark matter), a dark-energy density of 73%, and an age of about 13.7 billion years.

16. Olber's paradox notes that if the universe and the number of stars in the universe were infinite, then in every direction we looked, every line of sight would end up hitting a star eventually (like looking around in a dense forest: in every direction you see a tree trunk). However, the night sky is dark, so either the universe is finite in size or there are a finite number of stars. The resolution of this is the Big Bang: The universe has a finite age so we can see only out to the cosmological horizon. The observable universe is finite, resolving the paradox.
17. *Although the universe today appears to be made mostly of matter and not antimatter, the Big Bang theory suggests that the early universe had nearly equal amounts of matter and antimatter.* This statement is true. According to the Big Bang theory, the universe was hot enough to generate matter-antimatter pairs of particles very early in time, ensuring that the amounts of matter and antimatter were virtually equal.
18. *According to the Big Bang theory, the cosmic microwave background was created when energetic photons ionized the neutral hydrogen atoms that originally filled the universe.* This statement is false. On the contrary, the cosmic microwave background was created when there were too few energetic photons in the universe to keep the hydrogen ionized. After the transition from ionized hydrogen to neutral hydrogen, the universe became much more transparent, and the photons that eventually became the cosmic microwave background began to stream freely across the universe.
19. *While the existence of the cosmic microwave background is consistent with the Big Bang theory, it is also easily explained by assuming that it comes from individual stars and galaxies.* This statement is false. The cosmic microwave background is extremely smooth across the sky, and its blackbody spectrum corresponds to a very precise temperature. The characteristics of this background would not be so uniform if it came from many different objects.
20. *According to the Big Bang theory, most of the helium in the universe was created by nuclear fusion in the cores of stars.* This statement is false. According to the Big Bang theory, most of the helium in the universe was created by fusion in the uniform gas that filled the universe during the first few minutes after the Big Bang.
21. *The theory of inflation suggests that the structure in the universe today may have originated as tiny quantum fluctuations.* This statement is true. If inflation really happened, then tiny quantum fluctuations present in the universe before inflation were stretched to enormous size during the episode of inflation. These fluctuations would then be large enough to eventually grow into galaxies and clusters of galaxies.
22. *The fact that the night sky is dark tells us that the universe cannot be infinite, unchanging, and everywhere the same.* This statement is true. Olbers' paradox states that if the universe were infinite, unchanging, and everywhere the same, then the entire sky would blaze as brightly as the Sun.
23. *We'll never know whether inflation actually happened because our model of inflation does not make any predictions we can test.* This statement is false. The inflation model makes testable predictions. All of its testable predictions so far have turned out to be true, which is why it is still a viable model.
24. *In the distant past, the radiation that we call the cosmic microwave background actually consisted primarily of infrared light.* This statement makes sense. In the distant past, the cosmic background was not as redshifted as it today, so its radiation consisted of shorter wavelength light than it did today. Therefore the light at some point in the distant past was in the infrared.

25. *The main reason that the night sky is dark is that stars are generally so far away.* This statement does not make sense because even if stars were far away, if there were an infinite number of stars, eventually every clear line of sight would land on a star.
26. *The cosmic microwave background is our main source of information about what the universe was like at an age of about 3 minutes.* This statement does not make sense since the CMB is a primary source of information about what the universe was like at an age of about 380,000 years. The main source of information about the universe at 3 minutes comes from theories of nuclear physics (nucleosynthesis) and observations of helium, lithium, and deuterium.
27. b. a few degrees
28. b. negative
29. c. They convert into two photons.
30. b. observations of the amount of hydrogen in the universe
31. b. less deuterium
32. a. the uniformity of the cosmic microwave background
33. b. the origin of galaxies
34. c. observations of the CMB that indicate a flat geometry for the universe
35. b. a few hundred thousand years after the Big Bang
36. a. The universe is not infinite in space.
37. The ideal story will begin with the genesis of a proton as a partner in a proton-antiproton pair shortly after the Big Bang and will include incorporation into a star massive enough to create oxygen nuclei.
38. This is basically a creative writing project that should be constrained by science. It can be fun for students, and revealing as to what they do and do not understand.
39. Particle accelerators can create new particles from the energy of high-speed collisions. They are an example of $E = mc^2$ in action.
40. The student is asked to justify a wager of up to $100 on a bet that we understand the universe at 1 minute. A similar question to the instructor is quite useful: How would the instructor bet his or her $100 on the theory being correct at various epochs of the early universe? This chapter is often uncomfortable for students who view science as holding all of the "answers" and have not yet accepted that real science is always a work in progress: It is about seeking and asking the right questions about nature.
41. We cannot ever detect the early universe because it is hidden to our gaze, behind an impenetrable veil of ionized gas and radiation, similar to the way the core of a star cannot be directly seen by us.
42. According to Figure 17.14, the elements with atomic numbers between those of helium and carbon, lithium, boron, and beryllium, are at a local maximum in mass per nuclear particle. Extra energy is required for nuclear fusion reactions to fuse hydrogen and helium to lithium, boron, and beryllium, so such reactions were rare under the conditions present in the early universe.
43. Kepler might have imagined that there was a giant sphere hiding the rest of the infinite heavens from us. Scientists today would test this hypothesis under the principles of thermodynamics, which say that such a sphere would eventually come to equilibrium at the temperature of the infinite number of hidden stars, and the sphere would glow. Kepler also might have speculated that out to a certain distance, there aren't any more stars, just dark void. That hypothesis could be tested by looking for the edge of the void. Students don't have to imagine they are

Kepler—the key to this question is imagining what it might be like to ponder the night sky but not know anything about the expanding universe or the Big Bang (like Olber himself, or Newton, or a young Einstein for that matter).

44. Some of the features of the universe that are satisfactorily described by the Big Bang plus inflation (and the same features would have to be explained by alternative models): the cosmic microwave background, the helium-to-hydrogen ratio of at least 25% everywhere, the Hubble law (the expansion of the universe), the homogeneity and synchronicity of the universe in all directions, the geometric flatness of the universe (or, equivalently, the mass-energy density of the universe), the origin of the seeds for galaxy formation, the darkness of the night sky, the abundance of the light elements deuterium and lithium, and the average density of baryons. [The question asks for at least seven features—there are more than seven listed here.]

45. To find out how much matter and antimatter we would need to supply the energy needs of the United States for 1 year, we will use Einstein's equation, $E = mc^2$. We can solve for the mass by dividing both sides by c^2. The total energy use for the United States in 1 year is 2×10^{20} joules, leading to a required mass of 2,000 kilograms (only half of which would have to be antimatter). My car's gas tank holds around 13 gallons of gas. At 4 kilograms per gallon, that's around 52 kilograms of fuel. The total amount of matter and antimatter needed to supply the energy requirements for the United States in 1 year is only about 40 times this.

46. a. Using Newton's universal law of gravitation we find that the force of gravity due to the Earth on a 60-kilogram person is about 590 newtons.

    b. At 5×10^7 C/kg, a 60-kilogram person would have a total charge of 3×10^9 Coulombs while the Earth would have a charge of 3×10^{32} Coulombs. Plugging into the law for electromagnetic force, we find that the force of attraction between the person and the Earth would be 2×10^{38} newtons.

    c. From the calculations in parts (a) and (b), it is clear that the force of gravity would be much weaker than electromagnetism in this case.

47. To answer this question, we will first use the equation given in Mathematical Insight 23.1 to calculate the peak wavelength of the microwave background at $z = 7$:

$$\begin{aligned} \lambda_{\text{max}} &= \frac{1.1 \text{ mm}}{1+z} \\ &= \frac{1.1 \text{ mm}}{1+7} \\ &= 0.14 \text{ mm} \end{aligned}$$

The peak wavelength of the microwave background at the time when the light left the most distant galaxies we can see was 0.14 mm. We can also find the temperature at that time with another equation from Mathematical Insight 23.1:

$$\begin{aligned} T_{\text{universe}} &= 2.73 \text{ K} \times (1+z) \\ &= 2.73 \text{ K} \times (8) \\ &= 21.8 \text{ K} \end{aligned}$$

The temperature of the universe at this time was around 21.8 Kelvins.

48. a. The wavelength of maximum intensity for thermal radiation with a temperature of 3,000 K is:

$$\lambda_{\text{max}} = \frac{2.9\times10^6}{T(\text{K})}\text{nm} = \frac{2.9\times10^6}{3000\text{ K}}\text{nm} = 970\text{ nm}$$

b. If the current temperature of the microwave background is 2.73 K:

$$\lambda_{\text{max}} = \frac{2.9\times10^6}{T(\text{K})}\text{nm} = \frac{2.9\times10^6}{2.73\text{ K}}\text{nm} = 1.06\times10^6\text{ nm} = 1.06\text{ mm}$$

The peak of the microwave background is near 1 mm.

c. The ratio between the peak wavelength at the time of recombination, when the universe was at 3,000 K, and now is:

$$\frac{\lambda_{\text{max},T=2.73}}{\lambda_{\text{max},T=3000}} = \frac{1.1\times10^6\text{ nm}}{970\text{ nm}} = \frac{3000\text{ K}}{2.73\text{ K}} \approx 1000$$

This change is consistent with an expansion factor (of $1 + z$) of 1,000. (Hence you may hear astronomers talking about the recombination epoch at $z = 1{,}000$.) As the universe expands, so does the wavelength of the photons in the microwave background.

49. We know that a redshift of z, the average spacing between galaxies, was $1/(1 + z)$ what it is today. So for a spacing of two times what we see today, we can solve for z and find that $z = -0.5$. Plugging this into the formula for the temperature of the cosmic microwave background from Mathematical Insight 23.1:

$$\begin{aligned} T_{\text{universe}} &= 2.73\text{ K}\times(1+z) \\ &= 2.73\text{ K}\times(0.5) \\ &= 1.37\text{ K} \end{aligned}$$

So the temperature of the microwave background will be 1.37 K at that time.

50. If the largest bumps on a 1-meter table were at a scale of one part in 100,000, their sizes would be:

$$1\text{ m}\times\frac{1}{100{,}000} = 1\times10^{-5}\text{ m} = 0.01\text{ mm}$$

The scale would be 0.01 mm, a size much too small for us to even see.

51. a. This problem might help students only a very little in "visualizing" 10^{100} years, but it helps some students to practice with scientific notation and powers of 10. A trillion years is 10^{12} years, 100 times longer than 10^{10} years (10 billion years). A quadrillion years, 10^{15}years, is 100,000 times longer than 10 billion years, and 10^{20} years is 10 billion times longer than 10 billion years.

b. If protons have a half-life of 10^{32} years, the number of remaining protons will be one-half the current number in 10^{32} years; this is the definition of a half-life. The number of remaining protons will be one-fourth the current number in 2×10^{32} years. By 10^{34} years, 100 half-lives of 10^{32} years will have gone by. By that time, $(1/2)^{100} = 7.9\times10^{-31}$ of the original protons exist. By 10^{40} years, 10^8 half-lives will have gone by. The fraction $\frac{1}{2}$ raised to the power of 100 million is a very small number, so 100 million half-lives is more than sufficient for all of the protons in the universe to be gone.

c. 10^{100} zeros are a lot of zeros. If one zero could be written into 1 cubic micrometer, how many zeros could fit into the observable universe? One cubic micrometer is 10^{-6} m × 10^{-6} m × 10^{-6} m = 10^{-18} m^3, a very small thing. How many cubic micrometers are there in the observable universe? If we take the approximate size of the universe to be 15 billion light-years in radius, then the volume of the universe in cubic micrometers is:

$$\frac{4}{3}\pi r^3 = \frac{4}{3}\pi\left[1.5\times10^{10}\ \text{ly}\times\left(9.46\times10^{15}\ \frac{\text{m}}{\text{ly}}\right)\times\left(10^6\ \frac{\mu\text{m}}{\text{m}}\right)\right]^{-3} \approx 10^{97}\ \mu\text{m}^3$$

So only 1/1,000 of 10^{100} zeros of one cubic micrometer would fit into the observable universe. (*Note:* Carl Sagan performed a short "skit" illustrating this point in his *Cosmos* series.)

52. a. The apparent brightness formula says that:

$$\text{apparent brightness} = \frac{\text{luminosity}}{4\ \pi \times d^2}$$

Using the luminosity of the Sun (3.8 × 10^{26} watts) and the distance to the Sun (1.496 × 10^{11} m), this yields an apparent brightness of 1.3 × 10^3 W/m^2.

b. To get the Sun's apparent brightness at 10 billion light-years, we need to repeat the calculation from part (a), only with a different distance. Since 1 light-year is 9.46 × 10^{15} m, 10 billion light-years is 9.46 × 10^{25} m. Plugging in, we find an apparent brightness of 3.4 × 10^{-27} W/m^2.

c. The number of stars we would need at 10 billion light-years to make the same total apparent brightness as the Sun is the ratio of the Sun's apparent brightness to the brightness of a single Sun at 10 billion light-years:

$$\frac{1.3\times10^3\ \text{W/m}^2}{3.4\times10^{-27}\ \text{W/m}^2} = 3.8\times10^{29}$$

We would require 3.8 × 10^{29} stars to do the job. We estimated the number of stars in the observable universe in Mathematical Insight 1.3 to be around 10^{22}, so we would need many times more stars than we have in the observable universe to make the night sky as bright as day. This explains why the night sky is dark, and why even a finite number of stars would make the sky as bright as the Sun (and life as we know it impossible).

53. One remaining puzzle for cosmologists is, "Why did the Big Bang happen?" Answering this question at the very minimum will require the successful unification of a theory of gravity with a theory of quantum physics. Will string theory show the way?

54. What are the strengths and failures of the current Big Bang theory? We have spent much time discussing the successes, and here we can highlight some of its shortcomings ("failures" may be too strong a word). Certainly there is a gap at the beginning of the story (as suggested in Problem 53), in the very first fractions of a second where our knowledge of physics peters out. The current theory cannot tell us whether we're the only universe or one of a finite or an infinite set of universes. The current theory cannot tell us where we're headed—we don't know what dark energy is or how it behaves. But recall that scientists do not in a sense care whether the Big Bang is "right" or not—they care about testing the consequences of the theory. If it fails a crucial test, it is not a correct theory. A scientist must be willing to abandon a nice theory in the face of incontrovertible data that contradict the theory.

Part VII: Life on Earth and Beyond

Chapter 24. Life in the Universe

This final chapter discusses the fascinating topic of life in the universe, and includes discussion of life on Earth; the search for life in the solar system, SETI; and interstellar travel and its implications to UFOs and to Fermi's paradox. *Note*: This chapter is essentially a very condensed version of topics covered in the text *Life in the Universe*, by Bennett, Shostak, and Jakosky. That text can be used for a full-semester course on life in the universe.

As always, when you prepare to teach this chapter, be sure you are familiar with the relevant media resources (see the complete, section-by-section resource grid in Appendix 3 of this Instructor's Guide) and the online quizzes and other study resources available on the Mastering Astronomy Web site.

What's New in the Fourth Edition That Will Affect My Lecture Notes?

Those who have taught from previous editions of *The Cosmic Perspective* should be aware of the following organizational or pedagogical changes to this chapter (i.e., changes that will influence the way you teach) from the Third edition:

- This is essentially a new chapter on astrobiology, combining elements of the old Chapter 24 and material on life on Earth from the old Chapter 14.

Teaching Notes (By Section)

Section 24.1 Life on Earth

This section covers life on Earth so that students can understand what we are looking for when we consider the possibilities of life elsewhere.

- By necessity, this section introduces the fossil record, the geological time scale, and the theory of evolution. Sadly, these ideas have become flashpoints of controversy in our society, so be prepared to answer student objections. We hope that the Special Topic box about evolution and the schools will be of at least some help on this issue.
- You may particularly wish to emphasize that the scientific view presented here need not contradict personal religious beliefs. For example, the theory does not require or exclude the existence of a Creator; the scientific theory may be nothing more than the description of how the Creator created. Be sensitive to the possibility that student opinions on the matter may differ significantly from your own and from other students'.

- You may also wish to remind students of the distinction made in Chapter 3 between science and nonscience. Since religious beliefs, including creationism, generally do not try to claim scientific support, science can say little about their validity. In fact, it is quite possible for students to learn and appreciate the scientific theory of evolution without necessarily believing that it actually happened. While this outcome may not be ideal from a scientific viewpoint, it still means that students can be open-minded enough to consider the scientific evidence at the level that is possible in this course.

Section 24.2 Life in the Solar System

This section covers possibilities for finding life elsewhere in our solar system, with the greatest emphasis on Mars and Europa. Note that it also includes a discussion of the controversy over whether a Martian meteorite contains evidence of past life on Mars and possible implications of the recently announced methane detection in the Martian atmosphere.

Section 24.3 Life around Other Stars

This section considers prospects for life on planets around other stars. We also discuss the "rare Earth hypothesis" in this section, emphasizing that arguments can be made on both sides of this hypothesis, and therefore that we cannot reach any conclusions about it until more data are available.

Section 24.4 The Search for Extraterrestrial Intelligence

This section provides a very short discussion of SETI efforts.

- Note that we use a modified version of the Drake equation that is easier for students to understand. We still call it the "Drake equation," thanks to approval from Frank Drake himself! FYI: The three key reasons why we created this modified equation rather than using the original are:
  1. Units: We make a big point throughout the book that units should always work out properly in equations. However, Drake's key result when he analyzes the equation, which he states as $N = L$, violates this rule because N and L do not have the same units; N is a number and L is a time.
  2. While all of the terms in the original equation made sense in light of the current knowledge when Drake first wrote it in 1961, they do not all seem quite as appropriate today. In particular, the term R_*, for the rate of star formation, was assumed to be fairly constant when Drake wrote his equation. Today, we know that it can vary significantly over time and in different galaxies. For example, the rate of star formation is essentially zero in elliptical galaxies. The original equation therefore would tell us that there is zero chance of finding a civilization in an elliptical galaxy—a conclusion that clearly cannot be supported in light of current understanding.
  3. We believe that the overall equation is easier to understand with fewer terms, as long as nothing is lost when the terms are combined. We've therefore combined sets of terms: First, we combined his $f_{\text{planet}} \times n_{\text{e}}$ into our single term n_{HP}, for the number of habitable planets. We think that it is easier to think about this single number than to consider separately the fraction of star systems with planets and the average number of Earth-like planets in those systems. Second, we combined his $f_{\text{intell}} \times f_{\text{civ}}$ into the single term f_{civ}. Our rationale here is that there are many

different levels of "intelligence" we could conceivably use in defining the former term, but from a SETI point of view the intelligence matters only if it is able to communicate across interstellar distances. Thus, again, the single term keeps students focused on the key point.

- We recommend to our students that they watch the movie *Contact*, or read the book by Carl Sagan. The movie can generate very interesting class discussions.

Section 24.5 Interstellar Travel and Its Implications to Civilization

This section offers a brief overview of the difficulty of interstellar travel and several potential technologies for interstellar travel, then discusses Fermi's question "Where is everybody?" We find that students are particularly intrigued by this paradox and its astonishing implications. This discussion also makes a nice wrap-up to a course.

- One of the main reasons for covering interstellar travel is so that students can understand both why scientists are so skeptical of UFO claims and why Fermi's paradox has a rational basis.
- The box about UFOs (Are Aliens Already Here?) should be interesting to many students and can be discussed thoughtfully once students understand something about the challenge of interstellar travel.

Answers/Discussion Points for Think About It Questions

The Think About It questions are not numbered in the book, so we list them in the order in which they appear, keyed by section number.

Section 24.1

- (p. 710) Plants and animals have lived for about the last $\frac{1}{8}$ of the history of life on Earth. Human existence, if we take it to be 100,000 years (for Homo Sapiens), is about 1/2,000 of the 200+ million years since early dinosaurs and mammals.

Section 24.2

- (p. 719) A world can be habitable but not have life. Habitability speaks only to the *potential* for having life. A world could not have life without being habitable—by definition, life can survive only on habitable worlds.

Section 24.4

- (p. 725) For this example we have $N_{HP} = 1{,}000$; $f_{\text{life}} = 1/10$; $f_{\text{civ}} = 1/4$, and $f_{\text{now}} = 1/5$. In that case, we find: number of civilizations = $1{,}000 \times 1/10 \times 1/4 \times 1/5 = 5$. That is, with these numbers, the Drake equation tells us that there are five civilizations (with which we could potentially communicate) out there now.
- (p. 727) It should be easy for students to come up with examples of technologies that can be used both for progress and destruction. The rest of the question is subjective and should generate interesting discussion.
- (p. 728) This is a very subjective question that can generate a good discussion about SETI efforts. You might point students to the book or movie *Contact* as a way of thinking about the question.

Solutions to End-of-Chapter Problems (Chapter 24)

1. Recent developments in the study of life on Earth that make it seem more likely that we could find life elsewhere include:
   - We have found that life arose on Earth early in the Earth's history, suggesting that life might form quickly on other worlds.
   - Laboratory experiments have shown that chemical constituents thought to have been common on the young Earth readily combine to form organic molecules
   - We have found living organisms that can survive in conditions similar to those found on some other worlds in our solar system.
2. We study the history of life on Earth by using fossils, relics of organisms that lived long ago. The geological time scale is a set of distinct intervals that break up the Earth's history. Important events along the geological time scale include:
   - Formation of the solar system and the Earth (4.5 billion years ago)
   - End of the heavy bombardment (4 billion years ago)
   - Oldest rocks on Earth formed (4 billion years ago)
   - Carbon isotope evidence for life (3.85 billion years ago)
   - First fossil microbes (3.5 billion years ago)
   - Oldest eukaryotic fossils (2.1 billion years ago)
   - Oxygen accumulates in the atmosphere (2 billion years ago to half a billion years ago)
   - Cambrian explosion (530 million years ago)
   - Plans and fungi colonize the land (480 million years ago)
   - Animals colonize the land (410 million years ago)
   - Mammals and dinosaurs appear (200 million years ago)
   - Dinosaurs prominent (200 million years ago to 65 million years ago)
   - K-T event causes dinosaur extinction (65 million years ago)
   - Mammals prominent (65 million years ago to present)
3. Evidence pointing to an early origin of life on Earth includes 3.5-billion-year-old stromatolites, rocks with nearly identical structure to bacterial mats found today. Evidence for an even earlier origin for life comes from carbon isotope ratios that indicate that life may have appeared as far back as 3.85 billion years ago, although it is difficult to date these rocks.
4. The theory of evolution is the unifying theory through which scientists understand the history of life on Earth. Evolution means "change with time" and Darwin's theory explains how this occurs.
5. Natural selection refers to the unequal reproductive success in a population where nature "selects" the advantageous traits to be passed along to future generations. This explains evolution of life on Earth because as new traits appear through mutations, the advantageous ones will be selected to be passed on. Over time, this will result in new species. DNA is the genetic material of all life on Earth. Mutations in DNA alter the genetic code of the organism, changing some traits in a way that can be passed along through natural selection.
6. Life began as simple, one-celled organisms. Although plenty of genetic evolution must have occurred over time, more complex life, such as animals and plants, evolved much more recently. The evidence that points to a common ancestor for all life on Earth includes the fact that all life on Earth uses DNA to pass along genetic information, makes proteins from the same amino acids, and uses essentially the same genetic code.

Oxygen accumulated in Earth's atmosphere starting about 2 billion years ago and continued to accumulate until about a few hundred million years ago. This accumulation occurred because photosynthesis produced molecular oxygen as a by-product. At first, this oxygen reacted with rocks on Earth's surface. Eventually, the rocks were saturated and the oxygen remained in the atmosphere, building up. Larger animals diversified around 540 to 500 million years ago as part of the Cambrian explosion.

7. We think that life first arose on the seafloor or in hot springs, using energy released as heat from Earth's interior. Comparisons of DNA sequences from a variety of organisms have indicated that the organisms that are closest to the common ancestor on the tree of life are the ones that live in these environments today. This suggests that life may have begun there in the first place.
8. Laboratory experiments have been done with chemical ingredients similar to what we think were found on the early Earth and with sparks or other energy sources. The results show that these simple processes can produce the major molecules of life including amino acids and DNA bases. Laboratory experiments also show that mixing warm, dilute solutions of organic molecules with naturally occurring clays or sands allows the molecules to assemble themselves into more complex molecules such as RNA. Other experiments have shown that microscopic, enclosed membranes can also form under conditions similar to what we expect for the early Earth. These membranes can enclose RNA molecules, letting the RNA replicate faster and more accurately.
9. Life may have migrated to Earth from elsewhere, perhaps being carried on meteorites. If life arose on Mars or Venus, meteorites knocked off of those planets could have traveled to Earth and seeded life here.
10. Life on Earth has been found in the cold, dry valleys of the Antarctic, several kilometers underground, and in hot water. We have found organisms that can survive acidic, alkaline, and salty environments that would be lethal to humans. We have even found organisms that can survive in high doses of radiation.

    All life on Earth seems to require three things:

    - A source of nutrients from which to build living cells
    - Energy to fuel activities, either from sunlight, chemical reactions, or the heat of the Earth itself
    - Liquid water
11. Habitable worlds are worlds that contain the basic necessities for life, including liquid water. Apart from the Earth, the only places that seem potentially habitable are Mars and some of the large moons of the jovian planets. We are fairly certain that water once flowed on Mars, making it a good candidate. The icy moons Europa, Ganymede, and Callisto all appear to have subsurface oceans and they may have volcanic vents at the bottoms of the oceans. Titan also offers a possibility for life, although it would have to use liquid methane in place of liquid water, or be located deep underground where it might be warm enough for ice to melt into water.
12. The Martian meteorite, ALH84001, shows chemical indications of life such as microscopic chains of magnetite crystals and complex organic molecules. It also has what appear to be fossilized nanobacteria. However, there are nonbiological ways to explain this evidence and terrestrial bacteria have been found inside the meteorite, indicating that it might have been contaminated. Right now, most scientists doubt that the meteorite shows true evidence of Martian life.

13. A star's habitable zone is the region around a star in which a terrestrial planet of the right size could have a surface temperature suitable for liquid water and life. Based on our understanding of habitable zones, it is reasonable to expect that many and perhaps most stars are capable of having habitable planets, though we do not know whether such planets really are common.
14. The rare Earth hypothesis is the idea that Earth's hospitality is the result of rare planetary luck. The arguments in favor of this hypothesis are that there may be a fairly narrow ring at about our solar system's distance from the center of the galaxy where habitable planets might have enough heavy elements to form but not be sterilized by nearby supernovae too frequently. They also argue that without a Jupiter-like planet, impact rates may remain high longer, posing a danger to life on terrestrial planets. Finally, they argue that plate tectonics, which keeps Earth's climate stable, may be rare since Venus does not have evidence of such activity.

    Counterarguments to these points are that it does not seem that a large abundance of heavy elements is necessary to form planets and that atmospheres may protect organisms on the surfaces from the radiation of supernovae. It is also pointed out that jovian planets appear to be common in our studies of extrasolar planets, so the probability of a Jupiter-like planet in a system seems reasonably good. Finally, plate tectonics may not occur on Venus because of its runaway greenhouse. Since the runaway greenhouse is caused by Venus's proximity to the Sun, it is possible than any Earth-like planet in the habitable zone could have plate tectonics.
15. The Drake equation gives a simple way to calculate the number of civilizations capable of interstellar communication that are currently sharing the Milky Way with us, at least in principle. It calculates this quantity by multiplying the number of planets capable of bearing life by the probability of such a planet actually having life, times the fraction of life-bearing planets where civilizations capable of interstellar communication have at some point arisen, times the fraction of those civilizations that exist now. We can make a reasonably educated guess about the number of potential habitable planets in the galaxy based on our studies of planet formation. The probability of life arising is harder to estimate, although the fact that life originated soon after Earth was able to sustain it suggests that the probability is high. If we assume that the time it took for intelligence to develop on Earth is typical, then intelligent species may be quite common, making the probability of developing a species capable of interstellar communication pretty high. The value of the final factor, the probability of such a civilization being around now, depends on how long civilizations survive. This factor is extremely difficult to estimate.
16. SETI is the search for extraterrestrial intelligence. SETI researchers generally use large radio telescopes to search for alien radio signals, although some researchers are starting to search other parts of the spectrum. Current SETI efforts could detect deliberate, strong signals, but not signals as weak as those from our own current radio communications.
17. Interstellar travel is made difficult by several things. The first is the enormous energy required to push a spacecraft up to reasonable speeds for this sort of voyage. This problem is made worse by the fact that fuel is heavy, adding to the weight of the spacecraft and therefore the amount of energy needed to propel the craft. The second is the limitation of traveling slower than light (according to Einstein's theory of special relativity). Finally, there are social problems with leaving all of one's family behind on Earth, knowing that when one returns, they will all be dead and the world will be a very different place.

There are some promising technologies that have been proposed, however. The first is project Orion, which suggests using hydrogen bombs detonated just behind a spacecraft to propel the ship to even faster speeds. Also, if we could find a way to make antimatter in usable amounts, we could use it to fuel the ships without adding a lot of mass for fuel. Finally, ramjets have been suggested. Ramjets would scoop up interstellar hydrogen and use it to power fusion engines. By collecting fuel along the way, the ship would not need to carry the weight of fuel onboard.

18. Fermi's paradox says that if it is likely that a galactic civilization should already exist, why have they not yet visited? There are at least three possible solutions:
    - We are alone—This suggests that our rise to intelligence and civilization is a remarkable achievement and that humanity is that much more precious for being unique.
    - Civilizations are common, but no one has colonized the galaxy—This suggests that either other civilizations do not have the drives that we have, that colonizing the galaxy is even more difficult than we think, or that civilizations tend to destroy themselves before colonizing the galaxy.
    - There is civilization, but it has chosen not to contact us—Perhaps they are leaving us alone to let us develop and may someday contact us.
19. *The first human explorers on Mars discover that the surface is littered with the ruins of an ancient civilization, including remnants of tall buildings and temples.* This statement is fantasy. If any such civilization ever existed on the surface of Mars, we would know about it bynow.
20. *The first human explorers on Mars drill a hole into a Martian volcano to collect a sample of soil from several meters underground. Upon analysis of the soil, they discover that it holds living microbes resembling terrestrial bacteria but with a different biochemistry.* This is conceivable: If there is life on Mars, it could have a different biochemistry from life on Earth.
21. *In 2020, a spacecraft lands on Europa and melts its way through the ice into the Europan ocean. It finds numerous strange, living microbes, along with a few larger organisms that feed on the microbes.* This is conceivable, assuming we are correct about the existence of an ocean under the icy surface of Europa.
22. *It's the year 2075. A giant telescope on the Moon, consisting of hundreds of small telescopes linked together across a distance of 500 kilometers, has just captured a series of images of a planet around a distant star that clearly show seasonal changes in vegetation.* This could really happen if we really build such large sets of interlinked telescopes (interferometers).
23. *A century from now, after completing a careful study of planets around stars within 100 light-years of Earth, we've discovered that the most diverse life exists on a planet orbiting a young star that formed just 100 million years ago.* This could not happen according to our present understanding of the origin of planets and life, because during the first 100 million years of a star system's history, we expect the planet to be pelted by many large impacts. If life could arise so quickly at all, it would almost certainly be extinguished.
24. *In 2030, a brilliant teenager discovers a way to build a rocket that burns coal as its fuel and can travel at half the speed of light.* This statement is fantasy, because chemical burning cannot possibly release enough energy to achieve such speeds.

25. *In the year 2750, we receive a signal from a civilization around a nearby star telling us that the* Voyager 2 *spacecraft recently crash-landed on their planet.* This statement is fantasy, because *Voyager* will take tens of thousands of years to reach the distance of even the nearest stars.
26. *Crew members of the matter-antimatter spacecraft* Star Apollo, *which left Earth in the year 2165, return to Earth in the year 2450, looking only a few years older than when they left.* This is possible on a spacecraft that travels at a speed very close to the speed of light, thanks to effects of time dilation as explained by Einstein's theory of relativity.
27. *Aliens from a distant star system invade Earth with intent to destroy us and occupy our planet, but we successfully fight them off when their technology proves no match for ours.* This is fantasy. The ability to travel through interstellar space requires technology far beyond that which we possess today. Clearly, a war between us and such advanced aliens would be a terrible mismatch—and not in our favor.
28. *A single, great galactic civilization exists. It originated on a single planet long ago but is now made up of beings from many different planets, each of which was assimilated into the galactic culture in turn.* Based on what we know today, this certainly seems like it could be possible—and how amazing it would be if it turns out to be true!
29. b; 30. a; 31. c; 32. a; 33. b;
34. c; 35. a; 36. c. 37. b; 38. c.
39. You would choose to breed the cows that produced the most milk. This artificial selection is similar to natural selection because the animals with the desired trait are most likely to reproduce. It differs from natural selection in that it is humans choosing the trait, rather than a natural response to the environment.
40. This is an essay question. The key point is that statistics of one cannot really be considered reliable, but we have no alternative as long as we know life only on our single world.
41. This is a subjective question, so the key is how well students defend their choice.
42. The largest habitable zone should be around the hottest star, which is Sirius with spectral type A1; the second largest would be around the second hottest star, Procyon (spectral type F5). However, both Sirius and Procyon are binary systems. If we rule them out and look only at single stars, the hottest is Tau Ceti, with spectral type G8—making this the most likely to be orbited by habitable planets.
43. This is a subjective question, so the key is how well students defend their answer.
44. This is a subjective question, so the key is how well students defend their choice.
45. It is not realistic because it presumes that all the civilizations are at about the same technological level. Historically, this has not even been true of different civilizations occupying our own world at the same time. Given the different ages of star systems, it is completely implausible for multiple civilizations that have arisen on different worlds.
46. Answers will vary depending on the movie chosen.
47. We will start this problem by determining the area per civilization in the galaxy and then find the distance between civilizations from that. The galaxy is 100,000 light-years in diameter or 50,000 light-years in radius. If we assume that it is a disk, the total area is πr^2, 8 billion square light-years. If there are 10,000 civilizations, we divide the area by the number of civilizations to find the area per civilization.

This comes out to be 800 thousand square light-years per civilization. If we assume that each civilization has a square piece of the galactic disk, the average distance between civilizations is the square root of the area per civilization. Plugging in the numbers, this comes out to be about 900 light-years between civilizations.

48. We will start this problem by determining the area per civilization in the galaxy and then find the distance between civilizations from that. The galaxy is 100,000 light-years in diameter or 50,000 light-years in radius. If we assume that it is a disk, the total area is πr^2, 8 billion square light-years. If there are 100 civilizations, we divide the area by the number of civilizations to find the area per civilization. This comes out to be 80 million square light-years per civilization. If we assume that each civilization has a square piece of the galactic disk, the average distance between civilizations is the square root of the area per civilization. Plugging in the numbers, this comes out to be about 9,000 light-years between civilizations.

49. The number of stars we expect to have to search to find a signal is the total number of stars in the galaxy divided by the number of civilizations in the galaxy. We are told that there are about 500 billion stars in the galaxy and 10,000 civilizations. With these values, we expect to survey 50 million stars before we find one with a civilization. If instead there are only 100 civilizations in the galaxy, we expect to survey 5 billion stars before we find a civilization.

50. To find out how large a radio disk we would need to detect this signal, we need to recall that the signal's intensity decreases as follows:

$$B = \frac{L}{4\ \pi d^2}$$

where B is the signal intensity per area at the telescope, L is the absolute intensity of the signal, and d is the distance. We also must remember that the radio dish gathers a signal proportional to its area so that the total signal detected for a dish with radius r will be:

$$\begin{aligned} S &= \pi r^2 B \\ &= \frac{L\pi r^2}{4\ \pi d^2} \\ &= \frac{L}{4}\left(\frac{r}{d}\right)^2 \end{aligned}$$

We can now equate the minimum signal that Arecibo can receive to the minimum signal our hypothetical dish can receive:

$$\frac{L}{4}\left(\frac{r_A}{d_A}\right)^2 = \frac{L}{4}\left(\frac{r_h}{d_h}\right)^2$$

Here the subscript A refers to Arecibo and h refers to the hypothetical dish. The quantity L is the same for both sides of the equation since we assume that both dishes are detecting the same signal. We can therefore solve for r_h the radius if the hypothetical dish:

$$r_h = r_A\left(\frac{d_h}{d_A}\right)$$

We know that the Arecibo dish is 150 meters in radius and can detect the particular signal up to 100 light-years away. We seek a disk that can detect that same signal up to 70,000 light-years away. Plugging in the numbers:

$$r_h = (150\text{ m})\left(\frac{70{,}000\text{ ly}}{100\text{ ly}}\right)$$
$$= 1 \times 10^5\text{ m}$$

The dish would need to be 1×10^5 m in radius or 2×10^5 m in diameter. That is equivalent to a dish 200 kilometers in diameter, a very large telescope.

51. a. Using the formula for kinetic energy, $\frac{1}{2}mv^2$, we can find the energy needed to accelerate the ship. First, the desired speed is 10% the speed of light, or 3×10^7 m/s. The mass of the ship is 100 million kilograms. We can plug these values in to get the energy required: 4.5×10^{22} joules.

    b. This is nearly the same as the total energy use for the entire world in 1 year, 5×10^{22} joules.

    c. If energy costs 5 cents per million joules, then the total cost of the energy need to accelerate the ship would be 2.3×10^{15} dollars.

52. From Problem 51 we know that we require 4.5×10^{22} joules to accelerate the spaceship. Using Einstein's equation, $E = mc^2$, we can calculate how much matter and antimatter combined are needed to supply this energy by solving for the mass: $m = E/c^2$. Putting in the energy from Problem 51, we find that we will need 5×10^5 kg of matter and antimatter combined to provide this energy. We need half of this mass to be matter and half of it to be antimatter, so we need 2.5×10^5 kg of antimatter.

Appendix 1
Using *Voyage SkyGazer, College Edition*

Voyager SkyGazer, College Edition, combines exceptional planetarium software with informative tutorials. Here are just a few of the exciting things your students can do with this powerful tool:

- Use the planetarium features to learn the constellations.
- Use the built-in animations to explore astronomical phenomena such as meteor showers, eclipses, and more.
- Identify nebulae and galaxies that can be observed through a small telescope from your campus.
- Study and map how the sky varies with time of day, time of year, and latitude.
- Observe the Earth/Moon system from another planet within our solar system.

Appendix 3 lists particular resources within *SkyGazer* that you may wish to integrate with various topics in the book. To use these features, note that:

- To access Basic Files, start from the main menu, choose *File, Open Settings* and then select the desired file.
- To access Demo Files, start from the main menu, choose *File, Open Settings* and then select the desired file.
- To access Explore files, start from the main menu and choose *Explore*.

In addition, remember that you can find numerous assignable activities for *SkyGazer* in the *Astronomy Media Workbook* supplement (ISBN 0-8053-9206-8).

Appendix 2
Using the *Cosmos* Series

If you've watched the *Cosmos* series, you won't be surprised to know that the series and Carl Sagan were major influences on the authors of this textbook. (In fact, the lead author changed his graduate study plans from biophysics to astrophysics as a result of watching the series.) Thus, while the series does not correspond directly to the textbook, it makes an outstanding resource for reinforcing key ideas. There are at least two basic ways in which you can use the *Cosmos* series with this textbook:

1. If you can make the series available for students to watch on reserve at your department library, you may wish to assign the *Cosmos* episodes for individual viewing on students' own time. For example, you might assign one episode per week or one episode every two weeks. Although this means that the topics in the episodes might not correspond directly to what you are covering in the book at the same time, it makes it easy for students to plan their viewing time and still serves as a great way to reinforce ideas that students cover in the text at some other time.
   - For a one-semester course focusing on the solar system, we suggest assigning Episodes 1–6.
   - For a one-semester course focusing on stars, galaxies, and cosmology, we suggest assigning Episodes 7–10 and 12 (with 11 and 13 optional).
   - For a one-semester "everything" course, we suggest assigning Episodes 1, 3–6, and 9–10 (with 2, 7, and 8 optional).
2. You may wish to assign shorter segments (or show them in class) at times when their subject matter matches what you are covering in the text. The Section-by-Section Resource Grid (Appendix 3) lists appropriate segments for each section in the textbook.

For reference, the following table lists all the *Cosmos* episodes, each broken down into 12 "scene" titles that correspond to the scene selections on the DVD version of the series.

The *Cosmos* Series: Episode/Scene Titles

Episode 1 The Shores of the Cosmic Ocean	**Episode 2 One Voice in the Cosmic Fugue**	**Episode 3 The Harmony of Worlds**
1-1 Druyan Intro	2-1 Opening	3-1 Opening
1-2 Opening	2-2 Spaceship Cosmic Matter	3-2 Astronomers vs. Astrologers
1-3 The Cosmos	2-3 Haike Crab	3-3 Astrology
1-4 Spaceship Universe	2-4 Artificial Selection	3-4 Laws of Nature
1-5 Spaceship Galaxy	2-5 Natural Selection	3-5 Constellations
1-6 Spaceship Stars	2-6 Watchmaker	3-6 Astronomers
1-7 Spaceship Solar System	2-7 Cosmic Calendar	3-7 Ptolemy/Copernicus
1-8 Planet Earth	2-8 Evolution	3-8 Kepler
1-9 Alexandrian Library	2-9 Kew Gardens—DNA	3-9 Kepler and Tycho Brahe
1-10 Ages of Science	2-10 Miller–Urey Experiment	3-10 Kepler's Law
1-11 Cosmic Calendar	2-11 Alien Life	3-11 The Somnium
1-12 End Credits	2-12 Update/End Credits	3-12 End Credits

Episode 4 Heaven and Hell	**Episode 5 Blues for a Red Planet**	**Episode 6 Travelers' Tales**
4-1 Opening	5-1 Opening	6-1 Opening
4-2 Heaven and Hell	5-2 Martians	6-2 Voyager, JPL
4-3 Tunguska Event	5-3 Lowell	6-3 Traveller's Routes
4-4 Comets	5-4 Edgar Rice Burroughs	6-4 Dutch Renaissance
4-5 Collisions with Earth	5-5 Goddard	6-5 Huygens
4-6 Planetary Evolution	5-6 Inhabited Planets	6-6 Huygens—Conclusion
4-7 Venus	5-7 Mars	6-7 Traveller's Tales
4-8 Descent to Venus	5-8 Viking Lander	6-8 Jovian System
4-9 Change	5-9 Life on Mars?	6-9 Europa and Io
4-10 Death of Worlds	5-10 Mars Rover	6-10 Voyager Ship's Log
4-11 Conclusion	5-11 Terraforming Mars	6-11 Saturn and Titan
4-12 Update/End Credits	5-12 Update/End Credits	6-12 Update/End Credits

Episode 7 The Backbone of Night	**Episode 8 Travels in Space and Time**	**Episode 9 The Lives of Stars**
7-1 Opening	8-1 Opening	9-1 Opening
7-2 What Are the Stars?	8-2 Constellations	9-2 Apple Pie
7-3 Brooklyn Schoolroom	8-3 Time and Space	9-3 The Very Large
7-4 Mythology of Stars	8-4 Relativity	9-4 Atoms
7-5 Ancient Greek Scientists	8-5 Leonardo da Vinci	9-5 Chemical Elements
7-6 Science Blossoms	8-6 Interstellar Travel	9-6 Nuclear Forces
7-7 Democritus	8-7 Time Travel	9-7 The Stars and Our Sun
7-8 Pythagorus	8-8 Solar Systems	9-8 Death of Stars
7-9 Plato and Others	8-9 Cosmic Time Frame	9-9 Star Stuff
7-10 Distance to Stars	8-10 Dinosaurs	9-10 Gravity in Wonderland
7-11 Evidence of Other Planets	8-11 Immensity of Space	9-11 Children of the Stars
7-12 End Credits	8-12 Update/End Credits	9-12 Update/End Credits

— continued on next page —

The *Cosmos* Series: Episode/Scene Titles (continued)

Episode 10 **The Edge of Forever**	
10-1	Opening
10-2	Big Bang
10-3	Galaxies
10-4	Astronomical Anomalies
10-5	Doppler Effect
10-6	Humeson
10-7	Dimensions
10-8	The Universe
10-9	India
10-10	Oscillating Universe
10-11	VLA
10-12	Update/End Credits

Episode 11 **The Persistence of Memory**	
11-1	Opening
11-2	Intelligence
11-3	Whales
11-4	Genes and DNA
11-5	The Brain
11-6	The City
11-7	Libraries
11-8	Books
11-9	Computers
11-10	Other Brains
11-11	Voyager
11-12	End Credits

Episode 12 **Encyclopedia Galactica**	
12-1	Opening
12-2	Close Encounters
12-3	Refutations
12-4	UFOs
12-5	Champollion's Egypt
12-6	Hieroglyphics
12-7	Rosetta Stone
12-8	SETI
12-9	Arecibo
12-10	Drake Equation and Contact
12-11	Encyclopedia Galactica
12-12	Update/End Credits

Episode 13 **Who Speaks for Earth?**	
13-1	Opening
13-2	Tlingit and Aztec Indians
13-3	Who Speaks for Earth?
13-4	Nuclear War and Balance of Terror
13-5	Alexandrian Library
13-6	Hypatia
13-7	Big Bang and the Stuff of Life
13-8	Evolution of Life
13-9	Star Stuff
13-10	What Humans Have Done
13-11	We Speak for Earth
13-12	Update/End Credits

Appendix 3:
Complete Section-by-Section Resource Grid

Four of the key resources available with *The Cosmic Perspective* are the interactive tutorials at the Mastering Astronomy Web site, the topical movies also available at the Mastering Astronomy site, the built-in features of *SkyGazer, College Edition*, and the *Cosmos* video series. To help you in deciding when it is appropriate to make use of these resources in your teaching, the following grid matches these resources to sections of the textbook. Note that, if you choose simply to assign entire episodes of the *Cosmos* series, you should refer to the guidelines in Appendix 2.

The Cosmic Perspective 4e CHAPTER/ SECTION	MasteringAstronomy Online TUTORIAL	Cosmic Lecture Launcher CD-ROM APPLETS	Mastering-Astronomy Online MOVIE	*Voyager: SkyGazer* CD-ROM, v. 3.6	Carl Sagan's *COSMOS* SEGMENT VHS or DVD
1. Our Place in the Universe					
1.1 Our Modern View of the Universe			*From the Big Bang to Galaxies*	**File: Basics:** Chicago 10000AD Dragging the Sky **File: Explore Menu:** Solar Neighborhood Paths of the Planets	1-2 Opening 1-3 The Cosmos 1-4 Spaceship Universe 1-5 Spaceship Galaxy 1-6 Spaceship Stars 1-7 Spaceship Solar System 1-8 Planet Earth 1-11 Cosmic Calendar
1.2 The Scale of the Universe	**Scale of the Universe** Lesson 1 Distances scales: the solar system Lesson 2 Distances scales: stars and galaxies Lesson 3 Powers of 10 *Special: A Tour of the Solar System*	relative_dist_earth_moon sun_relative_moon_orbit size_of_mara_orbit size_of_jupiter_orbit size_of_pluto_orbit accurate_model_of_solar_sys relative_dist_nearest_star relation_dist_speed_time size_of_the_milky_way dist_to_andromeda_galaxy galaxy_clusters_and_struct Zooming_26_orders_of_mag			
1.3 Spaceship Earth					
1.4 The Human Adventure of Astronomy					
2. Discovering the Universe for Yourself					
2.1 Patterns in the Night Sky				**File: Basics:** Wide Field Milky Way Eclipse 1991–1992 Views Winter sky **File: Demo:** Russian Midnight Sun Earth Orbiting the Moon Mars in Retrograde **File: Explore Menu:** Shadows on Earth Phases of the Planets	

The Cosmic Perspective 4e CHAPTER/ SECTION	MasteringAstronomy Online TUTORIAL	Cosmic Lecture Launcher CD-ROM APPLETS	Mastering-Astronomy Online MOVIE	*Voyager: SkyGazer* CD-ROM, v. 3.6	Carl Sagan's *COSMOS* SEGMENT VHS or DVD
2.2 The Reason for Seasons	**Seasons** Lesson 1 Factors affecting seasonal changes Lesson 2 The solstices and equinoxes Lesson 3 The Sun's position in the sky	temperature_vs_distance directness_of_light flashlight_beams flux_of_light_vs_latitude why_does_flux_ sunlight_vary cause_of_seasons dist_to_aun_and_earth_ tilt sun_over_equator_at_ equinox arotic_circle_sunrise_set sun_altitude_vs_lat_ season			
2.3 The Moon, Our Constant Companion	**Phases of the Moon** Lesson 1 The causes of lunar phases Lesson 2 Time of day and horizons Lesson 3 When the moon rises and sets	innar_ss_and_moon_ orbit cause_of_lunar_phase moon_orbit_from_earth how_simulate_lunar_ phases phases_of_the_moon the_horizon time_and_location_of_ sun moon_rise_and_set_vs_ phase			
	Eclipses Lesson 1 Why and when do eclipses occur? Lesson 2 Types of solar eclipses Lesson 3 Lunar eclipses	earth_and_lunar_orbits cause_of_eclipses_anim cause_of_eclipses_lool eclipses_twice_a_month ecliptic_plane_anim moon_orbit_till_vs_ ecliptic moon_orbit_with_nodes eclipse_seasons precession_of_moon_ orbit angular_size_of_moon annuler_vs_total_solar partial_vs_total_solar evolution_of_partial_ solar evolution_of_total_lunar			
2.4 The Ancient Mystery of the Planets					

The Cosmic Perspective 4e CHAPTER/ SECTION	MasteringAstronomy Online TUTORIAL	Cosmic Lecture Launcher CD-ROM APPLETS	Mastering-Astronomy Online MOVIE	*Voyager: SkyGazer* CD-ROM, v. 3.6	Carl Sagan's *COSMOS* SEGMENT VHS or DVD
3. The Science of Astronomy					
3.1 The Ancient Roots of Science				**File: Basics:** Ptolemy on Venus Phase of Mercury Pluto's Orbit **File: Demo:** Hale-Bopp Path Hyakutake Nears Earth Venus-Earth-Moon File: Explore Menu: Solar System Paths of the Planets	1-9 Alexandrian Library 1-10 Ages of Science 3-2 Astronomers vs. Astrologers 3-3 Astrology 3-7 Ptolemy/ Copernicus 3-8 Kepler 3-9 Kepler and Tycho Braha 3-10 Kepler's Laws 3-11 The Somnium 13-5 Alexandrian Library 13-6 Hypatia
3.2 Ancient Greek Science					
3.3 The Compemican Revolution	**Orbits and Kepler's Laws** Lesson 2 Kepler's firat laws Lesson 3 Kepler's second law Lesson 4 Keplar's third law	drawing_ellipse_with_string what_is_a_circle Orbital Radius & Orbital Position eccentrcty_and_semimjr_axis kepler_2_velocity_vs_orbit_r kepler_2_area_and_time_int orbit_vs_orbit_period_vs_r			
3.4 The Nature of Science					
3.5 Astrology					
S1. Celestial Timekeeping and Navigation					
S1.1 Astronomical Time Periods	**Seasons** Lesson 2 The solstices and equinoxes	sun_over_equator_and_tropic sun_over_equator_at_equinox	*The Celestial Sphere* *Time and Seasons*	**File: Basics:** Analemma Rubber Horizon Three Cities **File: Demo:** Venus Transit of 1769 Celestial Poles Russian Midnight Sun	
S1.2 Cetestial Coordinates and Motion in the Sky	**Seasons** Lesson 3 The Sun's position in the sky	arctic_circle_sunrise_set sun_altitude_vs_lat_season			
S1.3 Principles of Celestial Navigation					

The Cosmic Perspective 4e CHAPTER/ SECTION	MasteringAstronomy Online TUTORIAL	Cosmic Lecture Launcher CD-ROM APPLETS	Mastering-Astronomy Online MOVIE	*Voyager: SkyGazer* CD-ROM, v. 3.6	Carl Sagan's *COSMOS* SEGMENT VHS or DVD
4. Making Sense of the Universe: Understanding Motion, Energy, and Gravity					
4.1 Describing Motion: Examples from Daily Life			*Orbits in the Solar System*	**File: Basics:** Planet Paths Planet Orrery Follow a Planet **File: Demo:** Earth and Venus Hyakutake at Perihelion Pluto's Orbit **File: Explore Menu:** Solar System Paths of the Planets	3-4 Laws of Nature
4.2 Newton's Laws of Motion	**Motion and Gravity** Lesson 1: Newton's Laws of Motion	pushing_cart_2ndLaw gravity_acceleration			
4.3 Conservation Laws in Astronomy	**Energy** Lesson 1: Kinetic Energy and Gravitational Potential	energy_pushed_cart impact_velocity_egg mass_vs_initial_velocity cannonball_drop cannonball_orbit_fire hand_light_temperature mass_to_energy_conversion			
4.4 The Universal Law of Gravitation	**Motion and Gravity** Lesson 2: The Force of Gravity	planet_mass_radius_weight inverse_square_gravity			
	Orbits and Kepler's Laws				
	Lesson 1 Gravity and orbits	orbit_trajectory_cannonball cannonball_mass_vs_orbit			
	Lesson 2 Kepler's first law	acceleration_due_to_gravity feather_and_hammer_on_moon drawing_ellipse_with_string			
	Lesson 3 Kepler's second law	what_is_a_circle orbit radius & orbital position			
	Lesson 4 Kepler's third law	eccentrcly_and_semimjr_axis kepler_2_velocty_vs_orbit_r kepler_2_area_and_time_int orbit_vs_init_velocty_and_r kepler_3_orbit_period_vs_r			
4.5 Orbits, Tides, and the Acceleration of Gravity					

The Cosmic Perspective 4e CHAPTER/ SECTION	MasteringAstronomy Online TUTORIAL	Cosmic Lecture Launcher CD-ROM APPLETS	Mastering-Astronomy Online MOVIE	*Voyager: SkyGazer* CD-ROM, v. 3.6	Carl Sagan's *COSMOS* SEGMENT VHS or DVD
5. Light and Matter: Reading Messages from the Cosmos					
5.1 Light in Everyday Life	**Light and Spectroscopy**				9-2 Apple Pie 9-3 The Very Large 9-4 Atoms 9-5 Chemical Elements
	Lesson 1 Radiation, light and waves	surface_waves_in_pond anatomy_of_a_wave visible_light_waves			
5.2 Properties of Light					
5.3 Properties of Matter					
5.4 Learning from Light	**Light and Spectroscopy**	electromag_spectrum			
	Lesson 2 Spectroscopy	intro_to_spectroscopy			
	Lesson 3 Atomic spectra—emission and absorption lines	de_excitation_and_emission production_of_emission_line energy_level_diagrams composition_mystery_gas photo_excitation_of_atom			
	Lesson 4 Thermal radiation	production_of_absorp_line spectrum_of_low-dens_cloud thermal_radiation wiens_law			
5.5 The Doppler Shift	**The Doppler Effect**				
	Lesson 1 Understanding the Doppler Shift	examples_of_motion hearing_the_doppler_effect cause_of_doppler_effect doppler_shilt_vs_velocity			
	Lesson 2 Using emission and absorption lines to measure the Doppler Shift	position_observer_vs_source doppler_effect_for_light doppler_shilt_emission_line determine_velocity_gas determine_velocity_cold_gas			

The Cosmic Perspective 4e CHAPTER/ SECTION	MasteringAstronomy Online TUTORIAL		Cosmic Lecture Launcher CD-ROM APPLETS	Mastering-Astronomy Online MOVIE	*Voyager: SkyGazer* CD-ROM, v. 3.6	Carl Sagan's *COSMOS* SEGMENT VHS or DVD
6. Telescopes: Portals of Discovery						
6.1 Eyes and Cameras: Everyday Light Sensors						10-11 VLA
6.2 Telescopes: Giant Eyes	**Telescopes** Lesson 1–2:	Optics and Light-gathering Power, Angular Resolution	mirror_position_focus diffraction_rings wave_effect_resol.			
6.3 Telescopes and the Atmosphere						
6.4 Telescopes and Technology						
7. Our Planetary System						
7.1 Studying the Solar System	**Scale of the Universe** Lesson 1	Distances of scale: our solar system	relative_dist_earth_ moon sun_relative_moon_orbit size_of_mars_orbit size_of_jupiter_orbit size_of_pluto_orbit accurate_model_of_sola r_sys	*Orbits in the Solar System* *History of the Solar System*	**File: Basics** Saturn's Phases Tracking Venus Planet Panel **File: Demo:** Earth and Venus Trailing Saturn Triple Conjunctio n of 7 BC **File: Explore Menu:** Solar System Paths of the Planets	6-2 Voyager, JPL
	Formation of the Solar System Lesson 1	Comparative planetology	comparative_ planetology orbit_and_rotation_ planets			

The Cosmic Perspective 4e **CHAPTER/ SECTION**	**MasteringAstronomy Online TUTORIAL**	**Cosmic Lecture Launcher CD-ROM APPLETS**	**Mastering-Astronomy Online MOVIE**	***Voyager: SkyGazer* CD-ROM, v. 3.6**	**Carl Sagan's *COSMOS* SEGMENT VHS or DVD**
7.2 Patterns in the Solar System	*Special: A Tour of the Solar System*				
	Orbits and Kepler's Laws				
	Lesson 2 Kepler's first law	drawing_ellipse_with_string what_is_a_circle orbital radius & orbital position			
	Lesson 3 Kepler's second law	eccentrety_and_semimjr_axis kepler_2_velocity_vs_orbit_r			
	Lesson 4 Kepler's third law	kepler_2_area_and_time_int orbit_vs_init_velocity_and_r kepler_3_orbit_period_vs_r			
7.3 Spacecraft Exploration of the Solar System					
8. Formation of the Solar System					
8.1 The Search for Origins			*Orbits in the Solar System* *History of the Solar System*	**File: Basics:** Saturn's Phases Tracking Venus Planet Panel **File: Demo:** Earth and Venus Trailing Saturn Triple Conjunction of 7 BC **File: Explore Menu:** Solar System Paths of the Planets	
8.2 The Birth of the Solar System	**Formation of the Solar System**				
	Lesson 1 Comparative planetology	comparative_planetology orbit_and_rotation_planets			
	Lesson 2 Formation of the protoplanetary disk	collapse_of_solar_nebula formation_protoplanet_disk why_does_disk_flatten formation_circular_orbits temp_distribution_of_disk condensate_regions_in_disk			

The Cosmic Perspective 4e CHAPTER/ SECTION	MasteringAstronomy Online TUTORIAL	Cosmic Lecture Launcher CD-ROM APPLETS	Mastering-Astronomy Online MOVIE	*Voyager: SkyGazer* CD-ROM, v. 3.6	Carl Sagan's *COSMOS* SEGMENT VHS or DVD
8.3 The Formation of Planets					
8.4 The Aftermath of Planetary Formation					
8.5 The Age of the Solar System					
9. Planetary Geology: Earth and the Other Terrestrial Worlds					
9.1 Connecting Planetary Interiors and Surfaces	**Formation of the Solar System** Lesson 1 Comperative Planetology	Planetinfo1 Planetinfo2	*History of the Solar System*	**File: Demo:** Earth and Venus Venus-Earth-Moon **File: Explore Menu:** Solar System Paths of the Planets	4-7 Venus 4-8 Descent to Venus 4-9 Change 4-10 Death of Worlds 4-11 Conclusion 5-2 Martiana 5-3 Lowell 5-4 Edgar Rice Burroughs 5-5 Goddard 5-6 Inhabited Planets 5-7 Mars
9.2 Shaping Planetary Surfaces	**Shaping Planetary Surfaces** Lesson 1 The four geological processes Lesson 2 What do geological processes depend on? Lesson 3 Planet surface evolution	production_of_a_crater volcanic_eruption_and_lava tectonics_convect_of_mantle plate_tectonics_on_earth water_erosion history_of_cratering tectonics_and_heat_transfer history_volcanism_tectonics history_of_erosion evolution_of_planet_surface			
9.3 Geology of the Moon and Memory					
9.4 Geology of Mars					
9.5 Geology of Venus					
9.6 The Unique Geology of Earth					

The Cosmic Perspective 4e CHAPTER/ SECTION	MasteringAstronomy Online TUTORIAL	Cosmic Lecture Launcher CD-ROM APPLETS	Mastering-Astronomy Online MOVIE	*Voyager: SkyGazer* CD-ROM, v. 3.6	Carl Sagan's *COSMOS* SEGMENT VHS or DVD
10. Planetary Atmospheres: Earth and the Other Terrestrial Worlds					
10.1 Atmospheric Basics			*History of the Solar System*	**File: Basics** Planet Paths Planet Orrery Follow a Planet **File: Demo:** Earth and Venus Hyakutake at Periheilon Pluto's Orbit **File Explore Menu:** Solar System Paths of the Planets	
10.2 Weather and Climate	**Surface Temperature of Terrestrial Planets** Lesson 1 Energy balance	thermal_equilibrium temp_vs_size_and_day_length temp_vs_day_length			
10.3 Atmospheres of the Moon and Mercury					
10.4 The Atmospheric History of Mars					
10.5 The Atmospheric History of Venus					
10.6 Earth's Unique Atmosphere					
11. Jovian Planet Systems					
11.1 A Different Kind of Planet	**Formation of the Solar System** Lesson 1 Comparative Planetology	Planetinfo 1 Planetinfo 2		**File: Basics:** Tracking Jupiter and Io Salum **File: Demo:** Backside of Jupiter Locked on Dione Three Moons on Jupiter	6-7 Traveller's Tales 6-8 Jovian System 6-9 Europa and Io 6-10 Voyager Ship's Log 6-11 Saturn and Titan
11.2 A Wealth of Worlds: Satellites of Ice and Rock					
11.3 Jovian Planet Rings					

The Cosmic Perspective 4e CHAPTER/ SECTION	MasteringAstronomy Online TUTORIAL		Cosmic Lecture Launcher CD-ROM APPLETS	Mastering-Astronomy Online MOVIE	*Voyager: SkyGazer* CD-ROM, v. 3.6	Carl Sagan's *COSMOS* SEGMENT VHS or DVD
12. Remnants of Rock and Ice: Asteroids, Comets, and the Kuiper Belt						
12.1 Asteroids and Meteorites				*History of the Solar System*	**File: Basics:** Orbit of Hale-Bopp Pluto's Orbit **File: Demo Folder** Hale-Bopp Path Hyakutake at Perihelion Hyakutake nears Earth Pluto's Orbit	4-3 Tunguska Event
12.2 Comets				*Orbits in the Solar System*		4-4 Comets 4-5 Collisions with Earth
12.3 Pluto: Lone Dog, or Part of a Pack?	**Formation of the Solar System**					
	Lesson 3	Formation of Planets	summary_condensates_in_desk accretion_and_planets nebular_capture_and_jovians the_solar_wind			
12.4 Cosmic Collisions: Small Bodies versus the Planets						
13. Other Planetary Systems: The New Science of Distant Worlds						
13.1 Detecting Extrasolar Planets	**Dectecting Extrasolar Planets**			*History of the Solar System*		7-10 Distance of Stars 7-11 Evidence of Other Planets
	Lesson 1	Taking a picture of a planet	luminosity_of_planet angular_sep_vs_distance angular_sep_jupiter_sun			
	Lesson 2	Stars' wobbles and properties of planets	stellar_motion_and_planets oscillation_of_absorp_line determine_star_velocty_vs_t star_orbit_vs_planet_mass determine_planet_mass_orbit			
	Lesson 3	Planetary transits	Planetary_transits			
13.2 The Nature of Extrasolar Planets	**Detecting Extrasolar Planets**					

The Cosmic Perspective 4e CHAPTER/ SECTION	MasteringAstronomy Online TUTORIAL	Cosmic Lecture Launcher CD-ROM APPLETS	Mastering-Astronomy Online MOVIE	*Voyager: SkyGazer* CD-ROM, v. 3.6	Carl Sagan's *COSMOS* SEGMENT VHS or DVD
	Lesson 1 Taking a picture of a planet	luminosity_of_planet angular_sep_vs_distance angular_sep_jupiter_sun			
	Lesson 2 Stars' wobblee and properties of planets	stellar_motion_and_planets oscillation_of_absorp_line determine_star_velocty_vs_t star_orbit_vs_planet_mass determine_planet_mass_orbit			
	Lesson 3 Planetary transits	planetary_transits			
13.3 The Formation of Other Solar Systems					
13.4 Finding More New Worlds					
S2. Space and Time					
S2.1 Esinstain's Revolution					8-2 Constellations 8-3 Time and Space 8-5 Leonardo da Vinci 8-6 Interstellar Travel 8-7 Time Travel
S2.2 Relative Motion					
S2.3 The Reality of Space and Time					
S2.4 Toward a New Common Sense					
S3. Spacetime and Gravity					
S3.1 Einstein's Second Revolution					
S3.2 understanding Spacetime					
S3.3 A New View of Gravity					
S3.4 Testing General Relativity					
S3.5 Hyperspace, Wormholes, and Warp Drive					

The Cosmic Perspective 4e CHAPTER/ SECTION	MasteringAstronomy Online TUTORIAL	Cosmic Lecture Launcher CD-ROM APPLETS	Mastering-Astronomy Online MOVIE	*Voyager: SkyGazer* CD-ROM, v. 3.6	Carl Sagan's *COSMOS* SEGMENT VHS or DVD
S4. Building Blocks of the Universe					
S4.1 The Quantum Revolution					
S4.2 Fundamental Particles and Forces					
S4.3 Uncertainty and Exclusion in the Quantum Realm					
S4.4 Key Quantum Effects in Astronomy					
14. Our Star					
14.1 A Closer Look at the Sun	**The Sun** Lesson 1 Structure and Gravitational Equilibrium				9-6 Nuclear Forces 9-7 The Stars and Our Sun
14.2 The Cosmic Crucible	**The Sun** Lesson 2 Fusion of Hydrogen into Helium Lesson 3 Why Does Fusion Only Occur at High Temperatures?				
14.3 The Sun-Earth Connection					
15. Surveying the Stars					
15.1 Properties of Stars	**Measuring Cosmic Distances** Lesson 2 Stellar parallax	intro_to_parallex parallax_of_nearby_star Parallax_angle_vs_ distance Measuring_parallax_ angle		**File: Basics:** Large Stars More Stars Star Color and Size **File: Demo Folder** Circling the Hyades Flying around Pleiades The Tail of Scorpius	

The Cosmic Perspective 4e CHAPTER/ SECTION	MasteringAstronomy Online TUTORIAL	Cosmic Lecture Launcher CD-ROM APPLETS	Mastering-Astronomy Online MOVIE	*Voyager: SkyGazer* CD-ROM, v. 3.6	Carl Sagan's *COSMOS* SEGMENT VHS or DVD
15.2 Patterns among Stars	**The Hertzsprung-Russell Diagram**				
	Lesson 1 The Hertzsprung-Russell (H-R) Diagram	generate_hr_diagr temp_and_luminosity determine_stellar_sizes stellar_mass_and_hr_diagr hydrostatic_equilibrium			
	Lesson 2 Determining stellar radii				
	Lesson 3 The main sequence				
15.3 Star Clusters	**Stellar Evolution**				
	Lesson 1 Main-sequence lifetimes	main_seq_lifetime_and_mass			
	Lesson 4 Cluster Dating	star_cluster_evolving hr_diagr_and_age_of_cluster			
16. Star Birth					
16.1 Stellar Nurseries			*Lives of Stars*		
16.2 Stages of Star Birth					
16.3 Masses of Newborn Stars					
17. Star Stuff					
17.1 Lives in the Balance			*Live of Stars* *Double Stars*	**File: Demo Folder** Crab from Finland	9-8 Death of Stars 9-9 Star Stuff
17.2 Life as a Low-Mass Star	**Stellar Evolution**				
	Lesson 2 Evolution of a low-mass star	death_sequence_of_sun stages_low-mass_death_aeq			
17.3 Life as a High-Mass Star	**Stellar Evolution**				
	Lesson 3 Late stages of a high-mass star	death_seq_of_high-mass_star			
17.4 The Roles of Mass and Mass Exchange					
18. The Bizarre Stellar Graveyard					
18.1 White Dwarfs	**Stellar Evolution**				9-10 Gravity in Wonderland 9-11 Children of the Stars
	Lesson 1 Main-sequence lifetimes	main-seq_lifetime_and_mass			

The Cosmic Perspective 4e CHAPTER/ SECTION		MasteringAstronomy Online TUTORIAL		Cosmic Lecture Launcher CD-ROM APPLETS	Mastering-Astronomy Online MOVIE	*Voyager: SkyGazer* CD-ROM, v. 3.6	Carl Sagan's *COSMOS* SEGMENT VHS or DVD
18.2	Neutron Stars	**Stellar Evolution**					
		Lesson 3	Late stages of a high-mass star	death_seq_of_high-mass_star			
18.3	Black Holes: Gravity's Ultimate Victory	**Black Holes**					
		Lesson 1	What are black holes?	escape_velocity_earth orbital_trajectory_and_r escape_velocity_and_r orbital_r_vs_planet_r g_vs_dist_black_hole determine_event_horizon schwarzechild_r			
		Lesson 2	The search for black holes	formation_xray_bin rotation_galactic_center orbital_velocity_mass_and_r evidence_of_black_hole			
18.4	The Mystery of Gamma-Ray Bursts						
19. Our Galaxy							
19.1	The Milky Way Revealed	**Detecting Dark Matter in a Spiral Galaxy**			*The Milky Way Galaxy*	**File: Basics:** Milky Way Wide Field Milky way Winter Milky Way Lagoon Nebulae **File: Explore Menu:** Solar Neighborhood	
		Lesson 1	Introduction to rotation	motion_merrygoround rotation_merrygoround rotation_of_solar_system adjust_mass_of_sun rotation_of_spiral_galaxy			
		Lesson 2	Determining the mass distribution	edge_and_face_spiral_gal mass_doppler_shifts_for_gal orbital_velocity_mass_and_r mass_vs_dist_solar_system mass_vs_dist_galaxy			
19.2	Galactic Recycling						

The Cosmic Perspective 4e CHAPTER/ SECTION	MasteringAstronomy Online TUTORIAL	Cosmic Lecture Launcher CD-ROM APPLETS	Mastering-Astronomy Online MOVIE	*Voyager: SkyGazer* CD-ROM, v. 3.6	Carl Sagan's *COSMOS* SEGMENT VHS or DVD
19.3 The History of the Milky Way	**Black Holes**				
	Lesson 1 What are black holes?	excape_velocity_earth orbital_trajectory_and_r escape_velocity_and_r orbital_r_vs_planet_r g_vs_dist_black_hole determine_event_horizon			
	Lesson 2 The search for black holes	schwarzschild_r formation_xray_bin rotation_galatic_center orbital_velocity_mass_and_r evidence_of_black_hole			
19.4 The Mysterious Galactic Center					
20. Galaxies and the Foundation of Modern Cosmology					
20.1 Islands of Stars			*From the Big Bang to Galaxies*	**File: Basics:** Galaxies in Come	10-2 Big Bang 10-3 Galaxies 10-4 Astronomical Anomalles 10-5 Doppler Effect 10-6 Hurneson 10-7 Dimensions 10-8 The Universe
20.2 Measuring Galactic Distances	**Measuring Cosmic Distances**				
	Lesson 1 Radar	radar_pulses Intro_to_parallax			
	Lesson 2 Stellar parallax	parallax_of_nearby_star parallax_angle_vs_distance			
	Lesson 3 Standard candles: main sequence stars and Cepheid variables	measuring_parallax_angle flux_of_star_vs_distance bright_stars_near_or_lum			
	Lesson 4 Standard candles: white dwarf supernovas and spiral galaxies	main_seq_as_standard_candle cepheid_as_standard_candle suprnova_as standard_candle fully_fisher_relationship galaxy_as_standard_candle summary_of_distance_methods			

The Cosmic Perspective 4e CHAPTER/ SECTION	MasteringAstronomy Online TUTORIAL	Cosmic Lecture Launcher CD-ROM APPLETS	Mastering-Astronomy Online MOVIE	*Voyager: SkyGazer* CD-ROM, v. 3.6	Carl Sagan's *COSMOS* SEGMENT VHS or DVD
20.3 Hubble's Law	**Hubble's Law**				
	Lesson 1 Hubble's Law	discover_hubble_law measure_hubble_constant			
	Hubble's Law Lesson 2 The expansion of the Universe	cause_of_hubble_law expansion_and_hubble_law relation_dist_and_velocity			
	Lesson 3 The age of the universe	peculiar_velocities estimate_age_of_universe age_and_hubble_constant			
21. Galaxy Evolution					
21.1 Looking Back Through Time			*From the Big Bang to Galaxies*	**File: Basics:** Galaxies in Coma	
21.2 The Lives of Galaxies					
21.3 Quasars and Other Active Galactic Nuclei	**Black Holes**				
	Lesson 1 What are black holes?	escape_velocity_earth orbital_trajectory_and_r escape_velocity_and_r orbital_r_vs_planet_r g_vs_dist_black_hole determine_event_horizon schwarzschild_r			
	Lesson 2 The search for black holes	formation_xray_bin rotation_galactic_center orbital_velocity_mass_and_r evidence_of_black_hole			
22. Dark Matter, Dark Energy, and the Fate of the Universe					
22.1 Unseen Influences in the Cosmos			*From the Big Bang to Galaxies*		

The Cosmic Perspective 4e CHAPTER/ SECTION	MasteringAstronomy Online TUTORIAL		Cosmic Lecture Launcher CD-ROM APPLETS	Mastering-Astronomy Online MOVIE	*Voyager: SkyGazer* CD-ROM, v. 3.6	Carl Sagan's *COSMOS* SEGMENT VHS or DVD
22.2 Evidence for Dark Matter	**Detecting Dark Matter in a Spiral Galaxy**					
	Lesson 1	Introduction to rotation curves	motion_merrygoround rotation_merrygoround rotation_of_solar_ system adjust_mass_of_sun rotation_of_spiral_ galaxy			
	Lesson 2	Determining the mass distribution	edge_and_face_spiral_ gal mass_doppler_shifts_ for_gal			
	Lesson 3	Where is the dark matter?	orbital_velocity_mass_ and_r mass_vs_dist_solar_ system mass_vs_dist_galaxy stellar_mass_vs_dist_ galaxy evidence_of_dark_ matter determine_distrib_dark_ mat			
22.3 Structure Formation						
22.4 The Fate of the Universe	**Fate of the Universe**					
	Lesson 1	The role of gravity	fate_of_launched_ cannonball escape_velocity_vs_ mass determine_velocity_ cannonbll			
	Lesson 2	The role of dark energy	universe_and_mass_ density universe_and_dark_ energy universe_history_and_ fate			
	Lesson 3	Fate and history of the universe				
23. The Beginning of Time						
23.1 The Big Bang	**Hubble's Law**				*From the Big Bang to Galaxies*	13-7 Big Bang and the Stuff of Life 13-8 Evolution of Life 13-9 Star Stuff 13-10 What Humans Have Done 13-11 We Speak for Earth

The Cosmic Perspective 4e CHAPTER/ SECTION	MasteringAstronomy Online TUTORIAL		Cosmic Lecture Launcher CD-ROM APPLETS	Mastering-Astronomy Online MOVIE	*Voyager: SkyGazer* CD-ROM, v. 3.6	Carl Sagan's *COSMOS* SEGMENT VHS or DVD
	Lesson 1	Hubble's Law	discover_hubble_law measure_hubble_constant			
	Lesson 2	The expansion of the universe	cause_of_hubble_law expansion_and_hubble_law			
	Lesson 3	The age of the universe	relation_dist_and_velocity peculiar_velocities estimate_age_of_universe age_and_hubble_constant			
23.2 Evidence for the Big Bang						
23.3 The Big Bang and Inflation						
23.4 Observing the Big Bang for Yourself						
24. Life in the Universe						
24.4 Life on Earth				*Search for Extraterrestrial life*		2-7 Cosmic Calendar 2-8 Evolution 2-9 Kew Gardens—DNA 2-10 Miller–Urey Experiment 2-11 Alien Life 5-8 Viking Lander 5-9 Life on Mars? 12-2 Close Encounters 12-3 Refutations 12-4 UFOs 12-8 SETI 12-9 Arecibo 12-10 Drake Equation and Contact 12-11 Encyclopedia Galactica
24.2 Life in the Solar System						
24.3 Life around Other Stars	**Detecting Extrasolar Planets**					
	Lesson 1	Taking a picture of a planet	luminosity_of_planet angular_sep_vs_distance angular_sep_jupiter_sun stellar_motion_and_planets			
	Lesson 2	Stars' wobbles and properties of planets	oscillation_of_absorp_line determine_star_velocity_vs_t star_orbit_vs_planet_mass			
	Lesson 3	Planetary transits	determine_planet_mass_orbit planetary-transits			
24.4 The Search for Extraterrestrial Intelligence						
24.5 Interstellar Travel and Its Implications to Civilization						

Appendix 4
Sample Syllabus

Many problems that tend to arise in classes can be alleviated if you are very clear about your expectations of students. One way to make your expectations clear is with a detailed syllabus. On the pages that follow, we offer a sample syllabus, adapted from one of lead author Jeffrey Bennett's courses. If you've never taught before, you might wish to use this as a starting point for creating your own syllabus. Otherwise, it might simply provide you with a few ideas of things to add or change in your current syllabus format. Most of the sample syllabus should be self-explanatory, but we offer a few notes about particular elements and modifications you might want to consider:

- Office hours/open review sessions: Notice the implementation of the ideas discussed under Personalizing the Impersonal Classroom (p. 47), bullet 4—referring to office hours as "review sessions" and holding some of them in a less intimidating location than an office.
- Course Requirements and Grading: Obviously, this is just one model of how grades might be assigned. The important part is that the requirements and grading policy are spelled out clearly, so that students know exactly what you expect of them. Here are a few notes on specific parts of our requirements:
  - Regarding the online quizzes: We have chosen to require the basic quiz for each chapter as a way of making sure students come to class prepared. We would also like students to complete the conceptual quizzes, but we don't require them. Instead, we let students know that many of the questions on their midterm and final exams will be taken verbatim from the conceptual quizzes. This gives them ample incentive to complete them as part of their studying.
  - We also include "class participation" as a way of encouraging students to attend class and to come prepared. Obviously, this is easier to implement in smaller classes. However, you can still have a class participation component to grades in larger classes if you use interactive lecturing techniques (see p. 45). For example, collecting worksheets or using electronic transmitters (see book by Doug Duncan, *Clickers in the Astronomy Classroom*) for short in-class activities will give you both an attendance record and some indication of whether students came to class prepared. *Note:* It's a good idea to make the class participation grade as objective as possible, since subjective grades are more likely to generate complaints and arguments. One strategy we've used is to assume that everyone starts with a perfect score of 10 points for their class participation, which is subject to reduction for absences or being clearly unprepared. For example, we usually allow two absences or unprepared days without penalty, but each absence thereafter is a 1-point deduction, and being unprepared (e.g., being unable to even begin on a class activity) is a 1/2-point deduction. For students with legitimate excuses, you may want to have a make-up policy for absences; we generally ask students to let us know *in advance* if they are going to miss class. An alternate strategy works if you use clickers (transmitters) for in-class questions: give, say, 2 points for participating at all in a question, and 3 points for getting the right answer.

- Observing Sessions: Unless it's truly impossible, we hope that you will find a way to give your students some type of evening observing opportunity. If your campus has telescopes available, perhaps you can have a few nights reserved for your class. If not, perhaps you can do some naked-eye observing, teaching the students some prominent constellations (which they usually love to learn). Also try to take advantage of any "special" observing opportunities such as meteor showers or eclipses. We've sometimes had students meet in the early morning, well outside of town, to observe meteor showers or bright planets. Such events can be a lot of fun for both you and the students.
- We did not include any major project in our requirements, but some teachers like to have a project component to final grades. You can easily make a project from some of our end-of-chapter Web Projects or from observing projects. Another project that we have sometimes used and that students seem to enjoy is a Book Review. For this project, students select a nonfiction book that is relevant to the course (e.g., topic areas might include the history of astronomy, recent discoveries in astronomy, or books about the space program or space policy) and then write a three-to-five-page critical book review. (Many students have never written a critical book review, so we suggest they look at book reviews in the Sunday *New York Times*. Also, you may wish to have students get your "OK" on their book selections before they begin, so that you can make sure they've chosen a real science book rather than something quacky.)

- Regarding the Common Courtesy Guidelines: It would be nice if we could assume that all students would treat each other and you with proper respect—but we all know that this does not occur automatically these days. We therefore include these explicit guidelines and we've found far fewer problems in class since we started including them. Perhaps the root of most classroom behavior problems is simply that students have gotten away with so much in high school that they have no idea what constitutes "normal" classroom behavior until you spell it out for them.
- The section entitled "Can I Get the Grade I Really Want?" should get students' attention. It is all part of our ongoing emphasis on the fact that the key to student success is hard work.
- The schedule is designed to fit on one page so that students can pin it to a wall and keep track easily. It is meant to serve as a template that you can adapt if you wish, since it is already sized properly to fit on one page. If you have a M/W/F class rather than a T/Th class, you can instead use the following cells as your basic template:

Mondays		**Wednesdays**		**Fridays**	
Mar 4	Reading: Chapter 8 Online quiz: Chapter 8 basic	Mar 6	Viewing (optional): *Cosmos* ep. 4	Mar 8	Homework 4 due

- *Note:* On the first day of class, you may also wish to hand out the Assignment 1 that we describe on page 41; this can be a good way to get a sense of where your students are coming from, as well as to get students excited about what they'll be learning in the class.
- *Note:* If you choose to make use of the *Cosmos* videos, as we do on this schedule, you can do so in a variety of ways. For example, you can have the videos on reserve at the library for students to watch on their own time. You might also arrange evening showings at a time that works for most of your students (and

others can watch them on their own). If you really want the students to come to the viewings, provide cookies or other refreshments that will encourage mingling after the video.

- *Note:* If you really want to be sure that students have read and understood your guidelines, add a page to the syllabus that asks them to sign and certify that they have read it and agree to abide by it.

Introductory Astronomy 1: The Solar System Syllabus

Dr. Jeffrey Bennett

Tu, Th 12:30–1:45 p.m., Duane Physics Building, Room 1B30

Office: Stadium room 119. Phone: 303-440-9313
E-mail: jbennett@casa.colorado.edu; **personal Web page**: www.jeffreybennett.com
Office hours/open review sessions:

- Tu, Th: 2–3 p.m., at my office.
- W, F: 12:30–1:30 p.m.—Look for me in the main dining area of the Student Union; I'll try to be at a table near the northwest corner.
- If these hours do not work for you, e-mail me to make an appointment for a time that will be convenient.

General Information

Astronomy 1 is one of two general courses in introductory astronomy. In this class we concentrate on the development of human understanding of the universe and survey current understanding of our planetary system. The other semester (Astronomy 2) explores our understanding of the structure and evolution of stars and galaxies, and current scientific theories concerning the history of the Universe.
No scientific or mathematical background is assumed, beyond the entrance requirements to the University. Astronomy is a *science*, however, so you will be expected to develop your critical thinking skills in order to understand and apply the scientific method. In terms of mathematics, we will use only arithmetic and a bit of simple algebra.
Although I have taught this course many times previously, there is always room for improvement. Please feel free to make comments, criticisms, or suggestions at any time. I will make any adjustments that are necessary to ensure that you find the course both challenging and rewarding.

Required Textbooks/Media

The textbook for this course is *The Cosmic Perspective, Fourth Edition*, by Bennett, Donahue, Schneider, and Voit. You will also need a personal access kit for The Mastering Astronomy Web site and the *SkyGazer* CD, both of which should have come with your book if you purchased a new copy. (*Note:* If you purchased a used copy of the book, you can buy access to the Web site online at www.masteringastronomy.com.)

Course Requirements and Grading

Your final grade will be based on the following work:

- Six homework assignments. *Late homework will be accepted only if you have made prior arrangements and there is a very good reason for the lateness.*
- Scores from online quizzes. *You may take a quiz as many times as you wish BEFORE its due date, and you will be credited with your highest score. If you take a quiz late, you will be credited with the first score you get, minus a 10% late penalty.*
- Class Participation: During classes, we will engage in discussions and occasional activities, some of which may involve completing worksheets. Participation in these activities will form part of your final grade.
- Observing Sessions: We will have several nights where the campus observatory is reserved for our class. You are required to attend at least one of these observing sessions and complete the observing worksheet that will be given to you when you arrive.
- Exams: We will have two in-class midterms and one final exam. There are no make-ups. If you miss an exam due to an extenuating circumstance and can provide documentation, you may discuss with your instructor the possibility of writing a term paper in place of the missed exam.

Calculating Your Final Grade

Your final course grade will be weighted as follows:

Homework	25%
Quizzes	10%
Class Participation	10%
Observing Sessions	5%
Midterm 1	10%
Midterm 2	15%
Final Exam	25%
Total	100%

A final score of 99–100% will be an A+; 92–98 is an A; 90–91 is an A–; the pattern continues for each lower grade.

Common Courtesy Guidelines

For the benefit of your fellow students and your instructors, you are expected to practice common courtesy with regard to all course interactions. For example:

- Show up for class on time.
- Turn off your cell phones before class begins!
- Do not leave class early, and do not rustle papers in preparation to leave before class is dismissed.
- Be attentive in class; stay awake, don't read newspapers, etc.
- If you must be late or leave early on any particular day, please inform your instructor or TA in advance.
- Play well with others. Be kind and respectful to your fellow students and your teachers.

You can expect your grade to be lowered if you do not practice common courtesy.

Can I Get the Grade I Really Want?

Yes—but it will depend on your effort. It does not matter whether you have even learned anything about astronomy before or whether you are "good" in science. What does matter is your willingness to work hard. Astronomy is a demanding course, in which we will move quickly and each new topic will build on concepts covered previously. If you fall behind at any time, you will find it extremely difficult to get caught back up. If you want to get a good grade in this class, be sure to pay special attention to the following:

- Carefully read the section in the Preface of your textbook called "How to Succeed in Your Astronomy Course." It describes how much time you should expect to spend studying outside class and lists a number of useful suggestions about how to study efficiently.
- When you turn in assignments of any kind, make sure they are done clearly and carefully. Refer to the separate handout on "Presenting Homework and Writing Assignments."
- Don't procrastinate. The homework assignments will take you several hours, so if you leave them to the last minute, you'll be in trouble—and it will be too late for you to ask for help. Both quizzes and homework need to be completed on time if you want to avoid late penalties.
- Don't miss class, and make sure you come to class prepared, having completed the assignments due by that date.
- Don't be a stranger to your instructor—come see me in office hours, even if you don't have any specific questions.
- If you find yourself confused or falling behind for any reason at any time, let me know immediately! No matter what is causing your difficulty, I am quite willing to work with you to find a way for you to succeed—but I can't help if I don't know there's a problem.

A Closing Promise

All the hard work described above might sound a bit intimidating, but I can make you this promise: Few topics have inspired humans throughout the ages as much as the mysteries of the heavens. Thisclass offers you the opportunity to explore these mysteries in depth, learning both about our tremendous modern understanding of the universe and about the mysteries that remain. If you work hard and learn the material well, this class will be one of the most rewarding classes of your college career.

Schedule

The indicated assignments should be completed *before* class on the listed date.
Listen in class and check your e-mail for updates to the schedule or syllabus.

Tuesdays		Thursdays	
Aug 24	First day of class	Aug 26	Reading: Chapter 1 Viewing (optional): *Cosmos* ep. 1
Aug 31	Online Quiz: Ch. 1 Basic Viewing (optional): *Cosmos* ep. 2	Sep 2	Reading: Chapter 2 Online Quiz: Ch. 2 Basic
Sep 7	Reading: Chapter 3 Viewing (required): *Cosmos* ep. 3 Online Quiz: Ch. 3 Basic	Sep 9	HOMEWORK 1 DUE
Sep 14	Reading: Chapter S1 Online quiz: Ch. S1 Basic	Sep 16	Reading: Chapter 4 Online Quiz: Ch. 4 basic
Sep 21	Reading: Chapter 5 Online Quiz: Ch. 5 basic	Sep 23	HOMEWORK 2 DUE
Sep 28	Reading: Chapters 6 Online Quiz: Ch. 6 basic	Sep 30	
Oct 5	FIRST MIDTERM (IN CLASS)	Oct 7	Reading: Chapter 7 Online Quiz: Ch. 7 basic
Oct 12	Reading: Chapter 8 Online Quiz: Ch. 8 Basic	Oct 14	HOMEWORK 3 DUE Viewing (optional): *Cosmos* ep. 4
Oct 19	Reading: Chapter 9 Online Quiz: Ch. 9 Basic	Oct 21	
Oct 26	Reading: Chapter 10 Online Quiz: Ch. 10 Basic	Oct 28	HOMEWORK 4 DUE
Nov 2	Reading: Chapter 11 Online Quiz: Ch. 11 Basic	Nov 4	SECOND MIDTERM (IN CLASS)
Nov 9	Viewing (optional): *Cosmos* ep. 5	Nov 11	HOMEWORK 5 DUE
Nov 16	Reading: Chapter 12 Online Quiz: Ch. 12 Basic	Nov 18	Viewing (optional): *Cosmos* ep. 6
Nov 23	Reading: Chapter 13 Online Quiz: Ch. 13 Basic	Nov 25	Thanksgiving Holiday—No class!
Nov 30		Dec 2	HOMEWORK 6 DUE
Dec 7	Reading: Chapter 24 Online Quiz: Ch. 24 Basic		
Final Exam: Monday, Dec. 14, 3:30 p.m. – 6:30 p.m.			

Observatory nights (weather dependent): Aug. 30, Sep. 22, Oct. 6, Nov. 4, Dec. 1.

Appendix 5
Handout on Homework Presentation

If you assign written work to your students, you'll find that it is far easier to grade if it is completed in a form that is easy for you to read. For our own classes, we have developed a one-page handout that describes clearly what we expect of our students when they turn in written work. The handout appears *on the next page*. You may feel free to photocopy it and hand it out to all your students in any of your courses. Note that these guidelines apply both to printed and e-mailed assignments (including assignments completed within the Mastering Astronomy system).

— Handout appears on next page —

Presenting Homework and Writing Assignments

All work that you turn in should be of *collegiate quality:* neat and easy to read, well organized, and demonstrating mastery of the subject matter. Future employers and teachers will expect this quality of work. Moreover, although submitting homework of collegiate quality requires "extra" effort, it serves two important purposes directly related to learning:

1. The effort you expend in clearly explaining your work solidifies your learning. In particular, research has shown that writing and speaking trigger different areas of your brain. By writing something down—even when you think you already understand it—your learning is reinforced by involving other areas of your brain.
2. By making your work clear and self-contained (that is, making it a document that you can read without referring to the questions in the text), you will have a much more useful study guide when you review for a quiz or exam.

The following guidelines will help ensure that your assignments meet the standards of collegiate quality:

- Always use proper grammar, proper sentence and paragraph structure, and proper spelling.
- All answers and other writing should be fully self-contained. A good test is to imagine that a friend is reading your work and to ask yourself whether the friend would understand exactly what you are trying to say. It is also helpful to read your work out loud to yourself, making sure that it sounds clear and coherent.
- In problems that require calculation:
  - Be sure to *show your work* clearly. By doing so, both you and your instructor can follow the process you used to obtain an answer. Also, please use standard mathematical symbols, rather than "calculator-ese." For example, show multiplication with the x symbol (not with an asterisk), and write 10^5, not 10^5 or 10E5.
  - *Word problems should have word answers.* That is, after you have completed any necessary calculations, any problem stated in words should be answered with one or more *complete sentences* that describe the point of the problem and the meaning of your solution.
  - Express your word answers in a way that would be *meaningful* to most people. For example, most people would find it more meaningful if you express a result of 720 hours as 1 month. Similarly, if a precise calculation yields an answer of 9,745,600 years, it may be more meaningful in words as "nearly 10 million years."
- Finally, pay attention to details that will make your assignments *look* good. For example:
  - If you are turning in your work electronically (e.g., by e-mail), be sure that you still follow standard rules of writing. For example, avoid typing your work in all caps or using the shorthand that you may use when sending instant messages to friends.
  - If you are turning in your work on printed paper, try to make it look as professional as possible. For example, use standard-size white paper with clean edges (that is, do not tear paper out of notebooks, because it will have ragged edges), and staple all pages together rather than using paper clips or folded corners (because clips and corners tend to get caught with other

students' papers). Ideally, turn in your work as typed pages. If you must handwrite it, please print neatly—I will not grade papers that are difficult to read.

- Include illustrations whenever they help explain your answer, and make sure your illustrations are neat and clear. For example, if you graph by hand, use a ruler to make straight lines. If you use software to make illustrations, be careful not to make them overly cluttered with unnecessary features.